Colonial Botany

Colonial Botany

Science, Commerce, and Politics in the Early Modern World

EDITED BY LONDA SCHIEBINGER
AND CLAUDIA SWAN

PENN

University of Pennsylvania Press

Philadelphia

Copyright © 2005 University of Pennsylvania Press
All rights reserved
Printed in the United States of America on acid-free paper

10 9 8 7 6 5 4 3 2 1

First paperback edition 2007

Published by
University of Pennsylvania Press
Philadelphia, Pennsylvania 19104-4112

Library of Congress Cataloging-in-Publication Data

Colonial botany : science, commerce, and politics in the early modern world / edited by
    Londa Schiebinger and Claudia Swan.
        p.   cm.
    Includes bibliographical references and index.
    ISBN-13: 978-0-8122-2009-4 (pbk : alk. paper)
    ISBN-10: 0-8122-2009-9 (pbk : alk. paper)
    1. Botany, Economic—Europe—History.   2. Botany, Economic—Economic
aspects—Europe—History.   3. Botany, Economic—Political aspects—Europe—History.
4. Plant introduction—Europe—History.   I. Schiebinger, Londa L.   II. Swan, Claudia.
SB108.E85C64   2004
581.6'094—dc22   2004052646

Contents

Introduction

Londa Schiebinger and Claudia Swan

In 1735 the Académie Royale des Sciences in Paris commissioned an expedition to the equatorial regions of South America to measure the length of a degree of meridian near the equator in order to determine the earth's size and shape. The French were among the earliest foreign scientists to penetrate the interior of Spanish Peru: for centuries Spain had guarded the secrets of its American natural resources. Two emissaries of the Spanish governor accompanied the expedition to make independent observations and to keep an eye on the foreigners. Among the members of the expedition was the French explorer and mathematician Charles-Marie de La Condamine (1701–74). While La Condamine entered Spanish territory on official business—to find these measurements and to chart the course of the Amazon—he is famous for having taken this opportunity to spirit away seedlings of the precious Peruvian bark trees (*Cinchona officinalis*), from which quinine is derived, and of trees yielding valuable *caoutchouc* (rubber). Among other things, he wished to test the fabled Amazonian botanical poison curare, with which natives of the region poisoned their arrows, and to settle the question of whether Amazons (the warlike women said to inhabit the wilds of the river named after them) actually existed. Joseph de Jussieu (1704–79), a member of the dynasty of French naturalists, accompanied La Condamine, who called him his "botanical eyes."[1] Their dream was a familiar one: to procure (legally or not) valuable foreign botanicals—often precious remedies; in this case, cinchona—for production in some part of the French Empire. La Condamine wrote of his prospecting for the valuable Peruvian bark:

On June 3rd I spent the whole day on one of these mountains [near Loja in present-day Ecuador]. Though assisted by two Americans of the region whom I took with me as guides, I was able to collect no more than eight or nine young plants of *Quinquina* [cinchona] in a proper state for transportation. These I had planted in earth taken from the spot in a case of suitable size and had them carried on the shoulders of a man whom I kept constantly in my sight, and then

by canoe. I hoped to leave some of the plants at Cayenne [in Guiana] for cultivation and to transport the others to the King's garden in France.[2]

Despite his care, the plants did not prosper (La Condamine was un-
aware that cinchona grows only at high altitudes). Nonetheless, while
gathering geographical information he also collected seeds of potentially valuable plants—he mentioned ipecacuanha, simarouba, sarsapa-
rilla, guaiacum, cacaoa, and vanilla—and kept his eyes open for treasures yet unknown to Europeans.

La Condamine's interests, motivations, aims, and failures are representative of the volatile nexus of botanical science, commerce, and state
politics that is the focus of this volume. Throughout the early modern
period, from the earliest voyages of discovery, naturalists sought profitable plants for king and country, personal and corporate profit. By sur-
reptitiously acquiring seedlings of the valuable Peruvian bark, La Condamine sought to undercut the Spanish monopoly on this antimalarial,
which was valuable to Europeans in their efforts to colonize tropical
areas. Like so many voyaging naturalists, La Condamine depended on
but did not quite trust his "native guides," whom he kept under constant surveillance. Moreover, like other naturalists, La Condamine over-
estimated the extent to which plants could be appropriated and reaccli-
matized; his efforts to transplant the delicate cinchona plants were in
vain. In a similarly ill-fated assertion of the hegemony of European botanical practices and the unity of global botany, the great Carl Linnaeus
(1707–78) sought to grow tea in the frigid wilds of Sweden.[3]

Colonial botany—the study, naming, cultivation, and marketing of
plants in colonial contexts—was born of and supported European voyages, conquests, global trade, and scientific exploration. The expanding
science of plants depended on access to ever farther-flung regions of the
globe; at the same time, colonial profits depended largely on natural
historical exploration and the precise identification and effective cultivation of profitable plants. Costly spices and valuable medicinal plants—
nutmeg, tobacco, sugar, Peruvian bark, peppers, cloves, cinnamon,
tea—ranked prominently among the motivations for voyages of discovery. Christopher Columbus (1451–1506) and Vasco da Gama (1469–
1524) aimed to secure sea routes to the rich spices, silks, and dyes of the
Moluccas, China, and India that would enable their countries to conduct trade without the intermediary of Middle Eastern and Venetian
merchants.[4] Plants also figured in generating funds for European colonial expansion. On his second voyage, in 1494, Columbus brought to
the West Indies sugar-cane cuttings, eventually one of the world's most
lucrative cash crops.[5] Colonial endeavors moved plants and knowledge
of plants promiscuously around the world. At the height of its powers

the Verenigde Oostindische Compagnie (Dutch East India Company or VOC)—colloquially known as the first multinational company— imported as many as six million pounds of black pepper to the Netherlands annually. Botany was "big science" in the early modern world; it was also big business, enabled by and critical to Europe's burgeoning trade and colonialism.[6]

This volume presents case studies that, together, chart the shifting relationship between botany, commerce, and state politics in the early modern period. The essays gathered here, written by scholars as international as their subjects, study botanical endeavors in Europe and its colonies as well as in Siberia. Chronologically they cover three centuries (roughly 1550–1800) of varying colonial and botanical theories and practices. Colonial practices, scientific organizations, and commercial connections differed not only over time but also from place to place: the absolute monarchies of Spain and France operated differently in this regard from the United Provinces of the Netherlands, the constitutional monarchy in England, or other states. It is our thesis that early modern botany both facilitated and profited from colonialism and long-distance trade, and that the development of botany and Europe's commercial and territorial expansion are closely associated developments. In ways that the remainder of this introduction will suggest, the essays presented here adumbrate an emerging cultural history of plants and botanical practices in Europe and its far-flung colonies and possessions.

Colonial Politics of Botany

One of the primary aims of this volume is to chart a new map of European botany along colonial coordinates, and in this sense the book offers a lively challenge to the historiography of early modern botany. A resilient and long-standing narrative in the history of botany has characterized its rise as coincident with and dependent on the development of taxonomy, standardized nomenclature, and "pure" systems of classification.[7] Indeed, the late seventeenth and eighteenth centuries witnessed key developments in the systematization of many fields. But to isolate the science of botany is to overlook the dynamic relationships among plants, peoples, states, and economies in this period. Recent studies by Richard Drayton, Paula Findlen, Richard Grove, Steven Harris, Lisbet Koerner, Roy MacLeod, James McClellan, David Miller, François Regourd, Pamela Smith, Emma Spary, and others have revealed how early modern science and especially natural history, of which botany was a subfield, remained strategically important in global struggles among emerging nation-states for land and resources.[8]

While it is our general thesis that early modern botany intimately supported and profited from European expansion, Europe's emerging nation-states engaged in different types of colonial arrangements and in different relationships with traders and naturalists. According to historian Philip Curtin's cogent characterization of these interactions, Spain established a "territorial empire" in the New World. Spanish conquistadors took New Spain (Mexico) and Peru by military conquest and attempted to reproduce the sociopolitical structures of the Iberian Peninsula in these American domains. Spain's American holdings were not conceived as "colonies" but organized as viceroyalties or kingdoms within a greater federated monarchy. These Spanish American viceroyalties contrasted with Portuguese and Dutch "trading-post empires."[9] The Dutch, for example, managed their trade through a network of fortified seaports in Goa, Southern Africa (the Cape of Good Hope), Batavia, and in North and South America. Their powerful trading companies—the VOC and the Westindische Compagnie (Dutch West India Company or WIC)—functioned until the last quarter of the seventeenth century as "war instruments" designed to monopolize the spice trade in the East and to undermine Iberian power in the West.[10] These companies held the right to govern, administer justice, conclude treaties, and maintain an army as well as a fleet. Curtin has restricted "settlement empires," or what he calls "true colonization," to North America. Large numbers of Europeans emigrated and settled in North America. Native populations were run out so that these colonies consisted primarily of relocated English and French. Curtin characterizes European settlement in the Caribbean as the "plantation complex," where Europeans conquered and then replaced vanishing native peoples with settlers, not primarily from Europe as in New England, but from Africa.[11]

While scientific investigation was often bound up with colonialist projects, these intimate connections were pronouncedly varied.[12] The Dutch VOC, for example, was a conglomerate born of small investment initiatives; its ties to the government of the Dutch republic were forged of individual connections. Scholars have argued that the VOC was an "unwilling Maecenas" vis-à-vis science because it only grudgingly allowed for the pursuit of natural historical knowledge by its employees.[13] Instead, enterprising individuals such as Jacobus Bontius (1592–1631), Hendrik van Reede tot Drakenstein (1636–91), and George Everhard Rumphius (1627–1702) collated and transferred the knowledge of eastern plants to European audiences. The absence of state sponsorship as such in the East merits comparison, however, with the relationship between Dutch governance of Brazil in the mid-seventeenth century and science there.[14]

French science, by contrast, was largely initiated and financed by the king and his ministers.[15] McClellan and Regourd have characterized the highly bureaucratic and centralized organization of French science from the time of Louis XIV onward as a "scientifico-colonial machine." Administered by the Ministère de la Marine et des Colonies and centered in the Jardin du Roi, the government and its naturalists coordinated bioprospecting around the globe to advance the colonial, national, and dynastic interests of the Bourbon monarchy. The efficacy of this "machine" to mobilize material and intellectual resources more than fulfilled royal colonial ambitions: by the 1780s Saint Domingue, the jewel in the French Crown's sugar islands, was the single richest and most productive colony in the world.[16] French absolutism set the tone for the association between science and government throughout much of the eighteenth century in France and in Europe generally.[17]

Some of the essays in this volume contribute further to the study of early modern colonial governance and botanical practices. Chandra Mukerji discusses how plant collection, cultivation, gardening, and engineering contributed to building the French territorial state. The administration of the Jardin du Roi, in the heart of Paris, worked hand in hand with the navy, outfitting ships with medicinal plants required for long voyages and, in turn, receiving specimens from around the globe. The purpose of the king's garden, founded in 1635, was to bring useful and glorious plants under the control of the state, represent the power of the monarch as part of a lawful system of nature, improve French forestry, increase French silk production, and generally yield profits. As Mukerji points out, French botanical efforts were intended to consolidate power within France and not necessarily overseas. In other European countries, however, the relationships among colonizing efforts, state governance, and botanical practices were not so direct. In his essay Staffan Müller-Wille addresses the paradox of Linnaeus's system of binomial nomenclature (useful in forging botany into a global science) as it emerged in Sweden, a country without significant colonies. More specifically, he situates Linnaean systematics within the context of Swedish "cameralism," a political economy that sought to create a (in Lisbet Koerner's words) "miniaturized mercantile empire" within Swedish borders, thus staunching the flow of bullion out of the country. Botany was to serve this cause by identifying and acclimatizing to Swedish soil plants—such as tea, cinnamon, and rice—that could substitute for expensive imports. Müller-Wille labels Linnaeus's efforts to create for Sweden a science-based, self-sufficient state economy "colonialism turned inward."

The relationship between individual states and botanical reconnaissance shifted over time. As the historian Richard Drayton has shown, English natural history was promoted through individual initiative along Dutch lines until the latter part of the eighteenth century, when the English adopted the successful French model. It is significant that much botanizing was initiated and directed by the British War Office. Along with military intelligence, the secretary of war asked colonial governors for "botanical dispatches" detailing the management of natural resources in their territories.[18]

Michael T. Bravo's chapter on Moravian naturalists highlights the fact that colonial governance did not originate only in states, trading companies, and metropolitan botanical gardens. Bravo is careful to emphasize that missionary societies cultivated distinctive relationships with their imperial patrons: Catholic Franciscans in California, for example, became cogs in the machinery of the Spanish Empire, while Protestant Moravians in Greenland received little more than permission from the Danish Crown to settle and purchase land in its territories. Focusing on missionary naturalists also undercuts the view of colonial botany as tied exclusively to large-scale economic activities: most mission gardens were laid to provide only those foods and medicines necessary for survival. The Moravians in the Danish West Indies were an exception. Hardship drew them into the same plantation complex of sugar cane and slave trading as their nonreligious counterparts. Unable to achieve an internally self-sufficient community, these servants of God produced sugarcane for cash—and with the labor of African slaves.

Variations on the relations between colonial governance and botanical practices abound in the early modern period. There are as many sorts of colonial botany as colonies, in the sense that different state structures or companies deployed or produced differing modes of scientific practice. Colonial botany developed along with a web of trade routes and was informed by patterns of commerce and naval prowess that kept them open.

Local Knowledge(s), Global Science

Vast quantities of botanical specimens and data were brought back to Europe throughout the early modern period from voyages sponsored by colonialist powers. In many cases data was assimilated to dominant European paradigms for understanding and organizing the natural world, but in some cases important new paradigms were developed. In her essay Daniela Bleichmar studies the fascinating example of early modern Spain's foremost authority on the plants of the New World, the Sevillean physician Nicolás Monardes (c. 1512–88). His text on New World *materia medica*, items that Monardes sold as well as prescribed, was a classic of its time. And yet Monardes never crossed the Atlantic. Bleichmar examines the representation of botanical encounters between Old and New World agents and the extent to which they were staged to "cleanse" New World finds of local (often religious) implication, to render the plants and other substances "global goods." Her essay, like others in this volume, raises questions about local knowledge, its accessibility by Europeans, and its transport to the Old World.

Such transport depended on acquiring knowledge of far-flung species in the first place. Acquiring knowledge of this sort necessitated complex encounters among cultures and often resulted in the overwriting of indigenous or local knowledge and practices. Harold J. Cook's essay on Dutch botany in Java adduces the model of European botany riding a wave of "objectification" by which specimens were wiped clean of cultural complexities in order to be pasted neatly into folios of European herbaria, shipped effortlessly to European botanical gardens, and included efficiently in European classificatory systems. Cook invokes the metaphor of the palimpsest—wax tablets that even when erased show traces of earlier texts—to characterize European collecting practices in the field. Although naturalists such as Jacob Bontius, whom Cook profiles, in many cases valued the knowledge of their local informants, these informants' words were written on top of erasures so that behind and structuring them were European intellectual and commercial frameworks. According to Cook, new knowledge was made "objective and exchangeable" on European terms.

Müller-Wille, a Linnaean scholar, argues that the political and economic infrastructure of colonialism provided the material base for creating a universal botany and that Linnaean classification enabled all the stuff of nature to be represented in the same "code": Linnaean taxonomy and nomenclature, though "non-colonial in origin, later became one of the prime instruments of colonial exploration," in Müller-Wille's words. Müller-Wille shows how the achievement of European botany depended on systems of taxonomy that reduced plants to specimens, numbers, and names so that a specimen, once identified, represented any plant of its type anywhere in the world where it is found. This, and this alone, enabled a system of global botanical exchange.

While Bleichmar, Cook, and others discuss how Europeans gave short shrift to the metaphysics associated with herbal regimes in many cultures, Antonio Lafuente and Nuria Valverde note that the Linnaean system "worked" only by disregarding the material circumstances in particular locales—climate, soils, altitude, and so forth—important for the cultivation of plants. They argue that the Swede's system was not just "good science," allowing for comparison of species across cultures, but also became both a "technoscope" and a "teletechnique" that underpinned the success of imperial undertakings. They point out that Spanish colonial administrators required the use of Linnaean taxonomy in the colonies largely for the purposes that Müller-Wille identifies—namely, because it allowed for clear and certain plant identification be-

tween the metropole and its colonies. Lafuente and Valverde present
the case of Creoles in eighteenth-century New Spain who rejected the
Linnaean system of classification and embraced taxonomies that better
served their own purpose of effective cultivation of plants in various terrains and under various climatic conditions.

In his essay Jorge Cañizares-Esguerra shows how Creole interests in
horticulture and profit in their own lands led to scientific innovations
that have wrongly been credited to Europeans. It was, he demonstrates,
Francisco José de Caldas (1768–1816) and his colleagues in New Spain
who contributed the foundations for the understanding Alexander von
Humboldt (1769–1859) had of the role of climate, altitude, and soil
composition in the biogeography of plants. Cañizares also reveals that
the Creoles of New Granada imagined their kingdom—not the Spanish
metropole—as "the warehouse of the universe" that could become
wealthy by supplying the world with natural products. They saw their
lands as a microcosm able to supply the world with every food, medicine,
and luxury through global commerce.

While several of the authors of this volume dispute the benefits and
deficits of the Linnaean system, Emma Spary complicates the story by
showing that plant identification depended on a process of cultural negotiations as much as on principles of scientific taxonomy. In a recon-
struction of the nutmeg skirmishes on the Isle de France (Mauritius) in
the 1750s, she shows that the explosive rivalry among colonial botanists
for metropolitan crown patronage, rather than disinterested comparison of species across continents, was central to the scientific identification of the commercially valuable nutmeg. Botanical "truths" (clear
identification of species), she posits, were grounded as much in a botanist's standing in colonial hierarchies as in a plant's morphology. Spary
emphasizes that species are not born but made in a process of negotiation between botanists, their patrons, and the expediencies of the marketplace. Cultural negotiations, whether in the case of nutmeg or that
of the Michoacan root, which Monardes described and sold in Seville,
deeply conditioned the production of colonial botanical knowledge.

Columbian Exchange Revised: Making and Remaking Nature

Early modern European botany and botanists participated not only in
knowing nature but in reconfiguring it as well. Environments and floral
terrains—both inside and outside Europe—changed radically as markets shifted and governments came and went. Europeans' efforts to develop scientific taxonomies to capture the order of nature coincided, at
times paradoxically, with large-scale alteration of nature by European
global botanical, economic, and military operations. Since ancient times

the spread of useful cultigens has been a constant feature of human history. From the fifteenth century on, however, European colonial expansion touched off an unprecedented widespread movement of flora globally that, as Marie-Noëlle Bourguet reminds us in her essay, deeply restructured the world agricultural map. Alfred Crosby captured this frenzy of European activity in his analysis of the "Columbian exchange."[19] Maize, the common bean, potatoes, tomatoes, squash, and sweet potatoes all moved by ship from the Americas to Europe and Africa. Wheat, rye, oats, and Old World vegetables were shipped from the Old World to the New. The seventeenth-century English slaver Richard Ligon, for example, carried from England to Barbados the seeds of "Rosemary, Time, Winter Savory, Sweet Margerom, Pot Marjerom, Parsley, Penniroyall, Camomile, Sage, Tansie, Lavender, Cotten, Garlicks, Onyons, Colworts, Cabbage, Turnips, Redishes, Marigolds, Lettice, Taragon, Southernwood"; all the seeds, he noted, grew and "prospered well."[20] The historian Jack Kloppenburg has remarked that this exchange of foodstuffs was of such magnitude that it enabled European industrialization after 1750.[21]

The story of European Columbian exchange is well known. As Judith Carney stresses in her *Black Rice* and in her essay in this volume, Europeans were not the only force moving useful cultigens from continent to continent.[22] Africans, in their forced migration on European vessels from their homelands into the Americas, carried not only seeds and plants but also the technologies for their cultivation. Slaves' dooryard gardens and provision fields—"botanical gardens of the dispossessed," she calls them—provided the staging ground for this African "diaspora" of cultigens. Carney also argues that these gardens should be, but rarely are, conceptualized as part of the colonial network of botanical gardens. Already in the seventeenth century slavers noted that Africans often brought seeds with them. These included Guinea corn, oil bush (used as a remedy for ulcers), prickly yam vine, and gourd vine. As Carney points out, slaves also cultivated African domesticates in their new homelands, including okra (used along with gully root as an abortifacient), black-eyed peas, and watermelons.[23] She designates African slaves "active botanists," who along with Amerindians and Europeans shaped the agricultural systems of the Americas.

While European colonization of territory and knowledge alike cannot be denied, the vexed and vexing relationships between knowledge systems developed by indigenous peoples of Europe's commercial and territorial colonies and European efforts to gather, understand, and transport that knowledge remain tremendously fruitful areas of research. Taking a cue from Richard Grove's work, which emphasizes the influence of extra-European natural systems on European naturalists, authors in this volume pay close attention to the agents of knowledge, asking whose knowledge is represented in accounts compiled by colonial botanists and how knowledge systems from other parts of the world influenced European academic botany and colonial practices.[24]

Who Were Colonial Botanists?

When we refer to botanists in the early modern period, we tend to have in mind academically educated Europeans; the terms *botaniste* and *botany* embody European concepts developed in the late seventeenth century to denote specialists learned in the taxonomy and nomenclature of plants.[25] Prior to that time the term *botanicus* was used to qualify interest in the plant world common among medical professionals, naturalists, and amateurs alike. The newer, more specialized usage of the term *botanist* accompanied a fundamental shift in Europe in the early modern period, when the academic study of plants split away from a practice-based tradition.[26] Increasing professionalization of engagement with the natural world went hand in hand with greater standardization of methods and means within medicine and pharmacology. Distinctions sharpened in this period between text-based, learned, academic practices and oral, indigenous traditions.

European botanists voyaging out to the colonies were a varied lot. Some were traveling missionaries; many were academically trained as physicians or apothecaries. A few paid their own passage; most were sent by trading companies, kings, or scientific academies. Historical accounts often cite the voyagers' national origins and professions. Such distinctions are important, but religions also forced important divides.[27]

Other important differences arose in the colonies between naturalists who merely passed through and those who settled in those areas. Many voyagers and naturalists, such as Hans Sloane (1660–1753), went to the colonies as young men, attempting to amass fortunes and return as quickly as possible to Europe. Others, such as Jacobus Bontius or Jean-Baptiste-Christophe Fusée Aublet (1720–78), put down deep roots, intending to stay for extended periods of time. As noted above, Spanish Creoles born and educated in the Americas held attitudes toward nature and its representations that differed significantly from those of Iberians. Bleichmar points out that some of the greatest prospectors—Monardes, for example—never left Europe but, as did those individuals who would come in the eighteenth century to be known as *botanistes de cabinet*, collected, studied, classified, experimented with, and popularized the bounty of nature from university posts and botanical gardens at home.

Stories of colonial science often recount the deeds—heroic or ignoble—of European men. In this volume women too emerge as active agents of natural knowledge; but because they seldom wrote and published books, we have access to their knowledge (like that of many indigenous peoples of colonized regions) only through the texts of European men. Women hardly served as European "colonial botanists"; they rarely numbered among academically trained naturalists who voyaged in the service of god, crown, or country. Even the celebrated Maria Sibylla Merian (1647–1717), who provided much information about New World flora in the early eighteenth century, was primarily interested in insects. Indigenous women, however, played a central role as informants to Europeans in the East Indies, as per Cook's account of Bontius's herbalizing in and around Batavia, and in the West Indies, as per Londa Schiebinger's analysis in this book, of European bioprospecting among the Tainos, Arawaks, Caribs, and African slaves. The study of colonial botany, and of science more generally, has tended to privilege texts and hence the knowledge of the Europeans producing them. Careful perusal, however, often reveals women's learning and practices represented in academic compilations produced by learned men.

This volume also complicates received notions of "encounters" between European voyagers and the indigenous peoples of foreign and faraway lands, who were frequently considered "simple" and even "barbarous." In this book Kapil Raj argues that historians often construe Europeans as the producers of knowledge and indigenous peoples as mere suppliers of material artifacts from which that knowledge is born, and in so doing diminish the accomplishments of non-European peoples. He argues that ancient and highly developed knowledge systems existed and circulated in South Asia long before Europeans touched those shores. According to Raj, the French naturalist Nicolas L'Empereur (c. 1660–1742), whose unpublished manuscript he recently found in the archives of the Muséum National d'Histoire Naturelle in Paris, simply translated and reconfigured Orivan, Telugu, Tamil, and other ancient knowledge about plants and their uses for a European market. Raj shows that collecting in the East Indies, where Europeans formed but a small and new commercial group among many long-established trading communities, was far different from collecting in the West Indies, where Europeans quickly overpowered many native peoples. Europeans' "encounters" in the East followed already formalized relations and highly stylized civilities that predated their incursion into these ancient trading networks.

It is important to note that "encounters" took place on European soil as well as in foreign territory. From the sixteenth century on, academic naturalists sought to record and codify "indigenous" traditions among European lay herbalists, many of them women, in a process similar to that undertaken in the colonies. In her classic text on European herbals, the historian Agnes Arber underscores how countrywomen in Europe served as repositories of knowledge concerning medicinal herbs.[28] Such encounters crossed gender and professional lines; in most cases it was academic botanists and medical professionals who sought out herbalists and their practical knowledge. Anton Schneeberger, a Polish botanist, declared in 1557 that he "was not ashamed to be the pupil of an old peasant woman."[29] Müller-Wille, in his essay in this book, notes that Linnaeus also valued the knowledge of the "vulgars," as he called them— common people without formal botanical training. Schiebinger discusses the process by which naturalists, often with government sponsorship, tested and then purchased "marvellous secrets"—sometimes at exorbitant prices—from lay healers of both sexes.

Some essays in this volume draw on instances in which indigenous knowledge—whether European or foreign—was highly valued by European botanists. Bontius, for example, considered the knowledge of the Javanese superior to that of Greek and Roman authorities. He objected that Europeans called the locals in Batavia "barbarians," finding that their knowledge of herbs "leaves our own far behind."[30] Pierre-Louis Moreau de Maupertuis (1698–1759), president of the Berlin-Brandenburg Akademie der Wissenschaften, also praised the medical knowledge of non-European peoples, stating that Europeans owed their knowledge of effective tropical medicines to them. European racist tendencies were tempered by the recognition that inhabitants in the colonies—whether Bengali slaves, Malayan women in Indonesia, devout fakirs in India, or the peoples of the Americas—often possessed knowledge worth recruiting.[31] In the early years of voyaging, Europeans' tenuous foothold in tropical areas compelled them to seek out the local peoples' knowledge on which their very survival often depended. Male voyagers—the majority in the early modern period—often sought care from local women who, familiar with the plants of the regions, were able to prepare suitable foods for them and administer appropriate medicines when they fell ill.

Europeans' respect, however, for traditional knowledges gradually waned over the course of the eighteenth and nineteenth centuries. In his essay on botany in the new Republic of the United States, Andrew J. Lewis notes the beginnings of this trend. By the early years of the nineteenth century, academic botanists, who had previously sought out Native American knowledge concerning plants and their uses, began disparaging Native American knowledge as "superstition." While many U.S. farmers and artisans continued to value traditional medical practices and folk knowledge, urban academic botanists became increasingly interested in developing the science of plants in ways that made plant identification more exact and supplies of natural products to industry more dependable.

Material Culture of Colonial Botany

Fast-running ships, collecting boxes, books, gardens, and greenhouses, both portable and fixed, were among the materials botanists mobilized to collect, record, and move nature from place to place. Marie-Noëlle Bourguet adds to this inventory of instruments and artifacts thermometers and barometers, useful for keeping plants alive during transfers from one climate to another. Thermometers—gas, mercury, or alcohol—also provided a map of daily temperatures collected from around the world and supplied the metropolis with valuable information for setting out new trade routes and colonial settlements. Setting plantations at the right altitude and in the right soil and climatic conditions allowed cash crops such as coffee, for example, not merely to grow but also to flourish and produce a maximized yield. Acclimatization was a key concern, whether plants were to be grown in Europe or to be produced in colonial territories.

By the end of the eighteenth century, Europe possessed some sixteen hundred botanical gardens connecting scientific enterprises, plant acclimatization, plant transfers, and experimentation around the world.[32] Bourguet calls naturalists "gardeners of the earth" as they reshaped global flora by moving plants across seas and climates. The skills of botanists and horticulturalists were crucial to establishing revictualing gardens along trade routes to restock trading company and naval vessels with familiar European fare. In turn, botanical gardens served as the laboratories of colonial botany. Botanists transferred plants from garden to garden around the world, building inventories and stocks of natural goods and thus facilitating the study, cultivation, and experimentation with profitable plants from all parts of the globe. Thomas Dancer (c. 1750–1811), the "island botanist" in Jamaica, wrote in 1804, "The necessity of a botanical garden for promoting the knowledge of plants in general, and for the introduction and cultivation of exotics that are rare, curious, and useful, whether in medicine or the arts, is, in the present age, so universally apparent, that there is hardly in any part of the civilized world wanting some such establishment."[33] The value of botanical gardens has been proved, Dancer stated, by the study and introduction of several varieties of sugarcane that improved yield and profits.

Commerce provided not only the rationale, capital, and routes for botanical exploration but also techniques for recording the wealth of the vegetable kingdom. In her essay on the German scholar and physician Daniel Gottlieb Messerschmidt (1685–1735), who traveled to Siberia on behalf of the Russian czar Peter the Great, Anke te Heesen analyzes the association in the early modern period between mercantile bookkeeping methods and botanists' techniques for inventorying natural goods. Mercantile patterns of thought and behavior structured the experiences and records of collectors such as Messerschmidt who deployed, as te Heesen shows, cognitive and material techniques for recording the natural world as they encountered it. The double-entry bookkeeping technique he adapted to his own ends allowed Messerschmidt to maintain astoundingly precise records of his experiences in the field. Over the course of his seven-year expedition he assembled manuscripts, catalogs, and notebooks, cross-referenced to his daily encounters and to a burgeoning collection of natural historical specimens he would take back to Saint Petersburg with him.

Revisiting the history of the objects and practices of colonial botany requires us to think about the ways in which things from far-flung places were amassed, transported, collected, bought and sold, processed, and otherwise put to use. In most cases this concerns the storage and cultivation of specimens in gardens and other collections; colonial botany was also practiced through pictorial representation, indexing and classifying practices, and display. By way of addressing how botanical specimens were acquired and circulated and how they came to be known and understood, Claudia Swan examines early seventeenth-century Dutch collections of exotic specimens as an instance of botany's material culture. Her essay asks how collecting and knowledge production were associated and treats the social bonds crucial to the production of botanical knowledge in particular. Naturalists and medical professionals alike performed botanical work in the context of and process of accumulating collections of *naturalia*, many of which can be traced to the trade voyages that filled Dutch coffers.

Pictorial and written accounts of new natural histories were generally integral to efforts to assimilate them—or, at least, to understand them. Visual representation was a long-standing means of accounting for new experiences of the plant world. In many cases voyagers were accompanied by artists or made images themselves to record specimens as they encountered them. The fifteenth-century book that has been called "the first printed illustrated account of the results of a journey undertaken with scientific purposes in mind," the *Gart der Gesundheit* (1485, Mainz), was written by an anonymous gentleman who traveled to the Middle East to study plants in the company of a painter.[34] Similarly, Dutch-trained artists accompanied Johan Maurits, Dutch governor in Brazil, in the 1630s and 1640s—along with naturalists, they worked to produce what Maurits would call a "portrait" of the country's peoples, animals, birds, fish, and fruits. As Rebecca Parker Brienen has recently argued, the efforts to capture and record Brazilian flora and fauna "strengthened Johan Maurits's position as a colonial and cultural leader."[35]

In this volume authors Raj and Julie Berger Hochstrasser offer the most sustained analyses of the visual culture of early modern botany. Hochstrasser unpacks the common (to the point of commonly over-looked) inclusion of pictorial references to colonial botany in still life paintings of the Dutch Golden Age. Seventeenth-century paintings served up colonial luxuries—lemons from the Mediterranean, salt from Brazil, silver (in tableware) from Mexico and Peru, grain (in bread) from the Baltic, and peppercorns from Malabar. Densely strewn table-tops extol Dutch trading prowess by showing its fruits. As Hochstrasser shows, however, what is represented in these pictures is as significant as what is omitted: a cruel story of oppression undergirds the availability of pepper, the commodity on which she trains her gaze. The celebratory tone of many of these *pronk*—or sumptuous—still life paintings masks the means by which these goods were procured. Raj's essay brings to light an unpublished manuscript whose extraordinary illuminations bear witness to a completely different tradition of representation of co-lonial botany. Whereas many Europeans produced images of colonial bounty themselves or had them made by European-trained artists, other traditions and conventions were also in play. The fact that so few records of encounters between alternative modes of scientific representation exist or survive makes the L'Empereur manuscript all the more signifi-cant. As L'Empereur's work reveals, illustration of colonial botanics also consisted in hybrids of European pictorial idioms and indigenous Asian artistic styles.[36]

In his foundational *Nature and Empire,* Roy MacLeod calls for a history of colonial science that "tests similarities and differences between and among imperial systems" and, more important, that begins to sketch "the process of multiple engagements—between Europeans at home and abroad, between European and indigenous [colonial] peoples, and between Western and non-Western science."[37] In concert with Mac-Leod's appeal, this volume sets out the great variety in colonial gover-nance in different parts of the world, as well as the diversity of commer-cial and scientific establishments planted around the globe. Authors in this volume have also been sensitive to exactly how knowledge was pro-duced—in both material and intellectual terms. This anthology offers the groundwork for a reappraisal of the relations between indigenous and academic knowledge. What emerges is a picture of the active contri-butions of indigenous people, whether in the West Indies, India, or Java. The rebellious character and active contributions of many Europeans who became native to colonial areas emerge as well. The Spanish Creoles, living in New Spain, in particular, challenged the value of European systems of classification for developing medicinal and agrarian plants in their country. In the early modern world, European botany was one system of plant knowledge among many.

As much as this collection of essays examines how knowledge traveled and was translated between cultures, it also highlights how nature—in the form of seeds, cuttings, and cash crops—traveled between continents and peoples. Europeans' desire to develop commercial crops brought cuttings of nutmeg from Dutch colonial areas in the East Indies (the islands of Amboina and Banda) to French colonies in the Indian Ocean (Mauritius), where they were cultivated in large monocultures. The same is true of sugarcane, which transformed the landscape of the Caribbean, and of rice, which remade coastal areas of the Carolinas. The story of colonial botany is as much a story of transplanting nature as it is one of transforming knowledge. While this volume cannot comprehensively cover all times, places, and topics important to early modern colonial botany, we offer it as the beginnings of what we hope will come to constitute a larger cultural history of plants and botanical practices in Europe and its colonies.

I.
Colonial Governance and Botanical Practices

Chapter 1

Dominion, Demonstration, and Domination

Religious Doctrine, Territorial Politics, and French Plant Collection

Chandra Mukerji

During the seventeenth and eighteenth centuries, gardens played key roles in French political culture as acquiring and controlling territory became central to state government. While fortresses were constructed around the perimeter of France and forests were surveyed for reform, formal gardens were laid out over vast areas around royal residences as symbols of territorial domination and exemplars of orderly land management. Their designers demonstrated and experimented with the French capacity to control land and its resources, and to use them for advantage.[1] In this context, the collection and display of rare and exotic plants took on strategic significance. Colonial botany became embroiled in state politics as finding exotic species, learning their uses, collecting their various names, sending them to France, displaying them in botanical gardens, and comparing them systematically constituted practices of territorial governance as well as of natural history.

Imported plants played a small but important role in the military articulation of the state's boundaries. Trees were used in fortress construction and shipbuilding; they were tools for defending state borders along coasts and in the interior. Forest reform was, in turn, fueled by the need for timber, and importations of new species of trees and the development of techniques for transferring large specimens into French gardens were visible exercises in military management as well as horticulture. The botanist Pierre Belon (1517?–64) brought plane trees into France and worked tirelessly (if ineffectively) to adapt them to the French climate. Others later succeeded, and plane trees became ubiquitous elements of French gardens, canal banks, and country roads. From the forests of North America, botanist-explorers from René-Robert Cavelier, sieur de La Salle (1643–87), to Michel Sarrazin (1659–1734) sent walnuts, mulberries, oaks, and other trees from Canada to the Jardin du Roi for the reforestation of France.[2]

At a more symbolic level, the collection and study of imported plants had much greater political significance. French gardens displayed *intelligence* in the use of territorial resources, as well as a legitimating capacity to manage nature and "improve" it through knowledgeable human action, which included collecting, naming, and finding the useful properties of new species. Botanical gardens at Montpellier, Paris, and some port cities were laid out strategically to contain a wide range of plants and to embellish and improve French life.[3]

Formal gardens, particularly at Versailles, took advantage of and publicly demonstrated French abilities to raise lush forests, acquire exotics, and use horticulture effectively. Versailles housed a huge collection of tender citrus and palms kept warm in the Orangerie in the winter and trundled out to scent the halls of the chateau in winter or to decorate the Parterre du Midi during warm weather. These botanical assets, tender plants of the Mediterranean world, exemplified French interests in and control over southern France. The flower parterres near the house were simultaneously filled with bulbs and other exotics that brought color and other sensual (and Edenic) qualities to the king's domain—among them, plants from the Baltic and Mediterranean regions as well as far-flung trading stations. The *potager,* or kitchen garden, was equally a collector's paradise, containing hundreds of varieties of pears, apples, and melons as well as vegetables forced behind glass to mature out of season. The royal collections of rare plants and tender species spoke of the ambition and intelligence of French naturalists and horticulturalists, and of the political aspirations of the monarchy to develop power over the natural world and deploy it to glorify France.[4]

Land management as a legitimating principle of power took on a unique role in French politics. Walks through the gardens at Versailles were used under Louis XIV (1638–1715) to demonstrate French power, using this grand park as a microcosm of territorial France for diplomatic purposes.[5] Knowledgeable and orderly use of state land—both in royal gardens and in the countryside beyond—became a measure and moral foundation of political territoriality.

French political attention to territorial improvement intensified on account of the importance of land control during the Reformation and Counter-Reformation in France. During the Wars of Religion the strategic parity of Protestants and Catholics made the conflicts over faith particularly hard to resolve. Fighting was repetitive and intense, and it destroyed human life as well as the landscape. One solution to the problem was to allocate places to faiths. In this context, some writers began to urge study and improvement of the countryside (God's works) as an alternative spiritual practice to contemplation of the Word—the source of doctrinal conflicts.[6] In the words of Bernard Palissy (1509–90):

I came to consider the marvellous deeds which the Sovereign has commanded
Nature to perform; and among other things I contemplated the branches of the
vines, of peas, gourds, which seemed as though they had some sense of their
weak nature; for being unable to sustain themselves, they stretched certain little
arms like threads into the air, and finding some small branch or twig, came to
unite and attach themselves, never again to part thence, that they might sustain
the parts of their weak nature . . . when I had seen and contemplated such a
thing I could find nothing better than to employ oneself in the art of agriculture, and to glorify God, and to recognize Him in His marvels.[7]

Studying nature and using land effectively was a spiritual path with
social virtues in a country devastated by war. During the reign of Henri
IV (1553–1610), Olivier de Serres (1539–1619) made this *mesnagement*
(rational land management) philosophy explicitly political. He argued
that the legitimacy of a regime and the moral and political fiber of a
good ruler were (or ought to be) reflected in the orderliness and abundance of the countryside.[8] De Serres convinced Henri IV to put the rebuilding of French lands (in part through strategic plant collection and
cultivation) at the heart of his administration.[9]

By the mid-seventeenth century, when the French monarchy was once
again overtly Catholic, *mesnagement* ideas about the restoration of Eden
seemed too Protestant, and stewardship was abandoned as an explicit
principle of power in France. Still, demonstrations of *intelligence* in using
the natural world remained marks of virtuous administration; stewardship was reduced to reasoned use of political territory.[10]

The garden theorist Jacques Boyceau de la Barauderie (1560–1633)
was crucial in engineering the shift from the moral language of *mesnagement* politics to a learned language of territorial governance. This ad-
mired and charming courtier was from one of the famous Huguenot
gardening families living by the Tuileries in Paris. He was also an educated man who had ambitions for his profession, wanting to make elite
garden design a recognized art form and, as such, closer to architecture
than agriculture. Treating rational land use à la *mesnagement* tradition as
an intellectual pursuit, he argued that young gardeners of talent should
acquire formal education in mathematics, engineering, and classical design. Proper schooling would allow them to build French gardens with
a measured orderliness that would highlight the continuities between
French and classical culture and display the lawful orderliness of nature.
Boyceau argued that parterres or formal garden beds should be laid out
in complex and symmetrical ways to embody the abundance and orderliness of nature as it was known to science. This design strategy was implic-
itly a program of "restoration," intended to bring gardens closer to per-
fection using human intelligence, but Boyceau's writings presented the
work as intellectual, a fruit of knowledge rather than faith. With this shift

in register, he melded antiquarian and scientific forms of "rationality" with *mesnagement* techniques.[11]

The gardens built for Louis XIV by Boyceau's protégé André Le Nôtre (1613–1700) were experiments in and massive displays of this "intelligent" territorial management.[12] The publicist André Félibien (1619–95) praised them for improving on nature with art, just as Boyceau had prescribed. Even as Louis XIV turned against the Huguenots, revoking the Edict of Nantes and making faith once again the dominant moral mandate for Christian governance, the material techniques and political logic of *mesnagement* land use, stripped of religious connotations by Boyceau, remained central to French governance. Stewardship was folded into territorial administration, and botanical study and horticulture remained important aspects of government.[13]

French rule over its colonies was shaped by these moral currents, which animated French territorial politics—the association of legitimacy with visible and effective land management. The great statesman Jean-Baptiste Colbert's (1619–83) early hope was that colonies would quickly flourish and gain the same political status, social customs, and material control as other regions of France. However, colonial lands required a level of mastery and orderly management of the landscape that was hard to realize—particularly outside of France. Stewardship worked well enough in Saint Domingue (Haiti), where disease and the Spanish had decimated the indigenous population (see Londa Schiebinger's essay in this volume), so there were no traditional land management strategies to displace. The labor force was made up of slaves who had been wrested from their own lands to populate the island. The sugar plantations in Saint Domingue acquired both the "natural orderliness" associated with virtuous management and the "abundance" needed to subsidize the requisite infrastructure.[14]

In contrast, territoriality worked less well in Canada, Louisiana, and Madagascar, in spite of the botanical assets of these colonies, precisely because there were still indigenous groups in these regions intent on preserving their own ways of life. Illness, warfare, and the more marginal success of economic ventures made these sites poor examples of "dominion" or good administration. In these colonies, reforming the populations through faith became more pressing than land reform, and plants with useful or sensual properties were packed off to France, where they could be properly placed both intellectually and practically inside French culture. During the seventeenth and eighteenth centuries, plants from the colonies were "rescued" in this way from botanical backwaters and folded into territorial politics. At the Jardin du Roi both the glories of exotic plants and the marvels of human intellect were demonstrated to Parisians, and specimens from abroad were ceremonially translated into assets of France.[15]

The organization of French botanical research changed with the erosion of dreams of re-creating Eden in the colonies. Botanists sent to these regions in the mid-seventeenth century (if they were lucky) became collectors for and "correspondents" of the Académie Royale des Sciences.[16] They might or might not set up their own gardens (without state support), but they were certainly not asked to be stewards of their own colonies. Their collections were transferred to botanical gardens in French cities. Plants were physically assimilated to French soil, where they could be readied for participation in French intellectual and political life.[17] This movement of plants might be understood as part of an abstract centralizing tendency in French state-based government, although it was closer to an expression of French territoriality with its religious roots and their attachment to botanical and horticultural knowledge.

Mesnagement Principles and Botanical Gardening

Beginning around the turn of the seventeenth century, French *mesnagement* principles came to shape the history of botany and not just public administration in France. Henri IV authorized the construction of the first major botanical garden in France for the medical school of the university at Montpellier. This garden was meant to be a place to acclimatize new species, study the pharmacological qualities of plants, and provide France with a center for exercising the intellectual and practical skills required for good medical practice throughout and virtuous stewardship of the kingdom. It was not only designed to make useful medicinal plants more commonly available in France but was also charged with promoting rural life through the strategic cultivation of new imports with refined horticultural techniques.[18] This use of plant collection for political advantage fit perfectly with the *mesnagement* policies prescribed by Olivier de Serres.

The garden at Montpellier was modeled on existing Italian botanical gardens built near medical schools (as at Pisa and Padua) and has often been described as a replica of them. The program of education was comparably academic and similarly a product of the commercial culture around the Mediterranean that brought imported species of plants and books about medicine into circulation. Pedagogy in medical botany at Montpellier centered on plant demonstrations too. These instructional walks through the garden taught students how to identify plants and were connected to lectures on botany, which grappled with the problems of botanical nomenclature that likewise interested antiquarian natural philosophers in both Italy and France.[19] The use of live plants in these early medical gardens was designed to improve on the empirically limiting practice of using only dried herbs for pharmacology. Dead plants were hard to study for botanical purposes, were of limited help in plant collecting, and often had to be obtained at high cost through drug sellers who had secret sources. The use of live plants was forcefully advocated by the Greek physician and pharmacologist Dioscorides (first century C.E.), whose *De Materia Medica* was first published in 1499 in Venice and appeared in a French edition by Jean Ruel (1474–1537) in 1516. Dioscorides argued that in order to understand plants one had to grow them and observe their development rather than cut them and collect them as dead specimens. He even advocated methods of field study, looking at how different flora grew in their natural habitat. His ideas, in turn, convinced the Roman physician Galen (b. 129 C.E.) of the importance of botanical field studies and careful observation of the physical properties of plants. As Galen's works entered into the sixteenth-century medical curriculum, they supported a turn throughout Europe toward plant collection and cultivation, and the continued development of botanical gardening at Montpellier.[20]

Plant demonstrations at Montpellier and later in Paris served to teach medicine but also articulated the intellectual and ceremonial underpinnings of territorial politics. The botanical garden was a built environment (like a political territory)—an intentionally designed combination of artifice and natural features with marked boundaries, internal infrastructure, and natural assets. It was a place to impose order on the natural world and to show the effects of human "improvements." The botanical garden at Montpellier brought together naming practices, specimens, and the organization of the garden beds too. Different parts of the garden were designated by their microclimates—humid, dry, warm, cool—and plants were sorted among them just as *mesnagement* writers recommended in cases of planning estates and administering states. As a collection for study, as well, the garden stood between the countryside with its array of wild plants, on the one hand, and a pharmacist's medicine jars with their dried herbs, on the other. It brought the order of knowledge to the disorder of plants—the fundamental act of stewardship. Plant demonstrations taught students how to learn from these collections and how to use their intellects to enhance life in France.[21]

The Montpellier tradition of botanical gardening and medical education was brought to Paris at the end of the seventeenth century, where it found a home at the newly conceived Jardin du Roi. This botanical garden was developed under the authority of the king's household (officially, his physician) and had state as well as academic functions. It was primarily a medical facility but was explicitly not associated with the University of Paris; nor was it an expanded version of the small but respectable botanical garden (also called the Jardin du Roi) developed in Paris by the Robins, father and son.[22] The project of developing the new Jardin du Roi in Paris, mainly using staff from Montpellier, not surprisingly offended University of Paris scholars who were suspicious of the "intellectual merit" of the botany taught in the south of France. The analytic cosmopolitanism and pedagogical empiricism that made Montpellier so exciting intellectually also made it a threat to the orthodox medical faculty in Paris with its relative lack of interest in botanical collecting. The rivalry between Paris and Montpellier reflected differences in intellectual life between northern and southern France: the humanism and antiquarianism that animated the Mediterranean basin were worlds away from the more scholastic intellectual culture of the North. By promoting empirical study of botany, Montpellier resisted the scholasticism that characterized medical education in Paris. Montpellier with its *mesnagement* roots also supported a religious liberalism that was anathema to Paris.[23] Precisely because of these differences, the Jardin du Roi with its personnel from Montpellier was awkwardly but efficiently alienated from the University of Paris and more easily wedded to the king's household and state politics.

Colbert was the political champion of this institution—as he was of academies and territorial governance in general. He wanted the Parisian garden to eclipse its counterparts at Montpellier and in Italy while providing better medicines for the metropolis, promoting botany in the Académie Royale des Sciences, bringing useful plants into France, and enhancing the efficiency and stature of his navy. The new Jardin du Roi brought the plant trade (and especially trade in medicines) more directly under control of the state, using the navy to assemble exotics from all over the world.[24]

Pharmacopoeia and Botanical Gardening

The territorial ambitions of this regime had complex connections to pharmacopoeia and botanical collection. Knowledge of medicine clearly held the general promise of providing a better life for inhabitants of the kingdom. Proper understanding of plants could counteract illness and help match diseases with the cures that botanists argued had been provided at the Creation. Because of the spiritual and political significance of health and illness, *mesnagement* writers had prescribed the development of herb gardens devoted to pharmacopoeia, and this had placed botanical gardening squarely in the tradition of territorial politics. Colbert, at the same time, wanted his navy to carry botanists and doctors to the colonies; he furnished the ships with medicines and physicians for the crews, as well as collectors to acquire useful medicinal herbs and beautiful exotics for the Jardin du Roi.[25]

Specific concern about the ill health of Parisians was a primary (and symbolically consequential) argument for building a great botanical garden in Paris. France's major city had a reputation for being disease-prone. The Romans had characterized the city as wet, cold, and bad for the lungs, and nothing in subsequent centuries had changed this characterization. Henri IV had been active in improving the city, building a water supply for the royal residences so that they would not tax the city's water system and setting out mulberry trees in the royal gardens to promote the silk industry. Still, in the seventeenth century Paris was easy to characterize as a sickly center of France, hardly what Colbert wanted for the "New Rome." He was worried about Louis XIV's neglect of Paris too—particularly his disinterest in the Louvre but also the monarch's general aversion to the city. Letting Paris degenerate would raise questions of governmental legitimacy and efficacy. So Colbert risked the king's wrath when he pushed for rebuilding the Louvre, making the Tuileries garden a center of taste and fashion in Paris and establishing in the city the academies, botanical garden, and observatory intended to stimulate and display French intellectual life. The Jardin du Roi, like the observatory, was meant to make visible (on the world stage) the cultural and intellectual depth of France's greatest city, while the garden gave Parisian doctors better medicines for addressing endemic maladies.[26]

The Jardin du Roi was also, crucially, the king's medicine cabinet, run by his physician, and an institution that addressed the politically pressing question of how to maintain the well-being of the kingdom through the health of the king. The king's body and French lands were symbolically aligned in the propaganda and political rituals of the period. The famous portrait of Louis XIV (oil on canvas, 1701, Musée du Louvre, Paris) by Hyacinthe Rigaud (1659–1743), which was venerated at Versailles during the monarch's periodic absences, significantly represented his body as young and healthy with the dancer's legs—covered only in white stockings—he had coveted from his youth.[27] Keeping the king healthy was not a simple project. It took an active search for new ingredients for his morning tonic, new cures for his constipation, and novel medicines to reduce the chronic pain he suffered in his jaw.[28] The Jardin du Roi was precisely the kind of institution that could help with this complex, daunting, and very political work.

The Jardin du Roi was also a vital study center, an important institution for public display of French intelligence in botany and horticulture. It was affiliated with the botanical gardens in French port cities that helped organize the plant trade and raised useful species for both the navy and the Jardin du Roi. These regional gardens gained power in large part because of their utility for the navy but also because they were part of the territorial administration set up by Colbert. A system of local intendants had been dispatched by Colbert to the major regions of France to counter the political control of local elites and the influence of the army. Not surprisingly, the intendants of port cities became particularly powerful advocates of naval facilities in their jurisdictions and the gardens and medical corps that grew with the ports. Major cities on the Atlantic coast supplied the ships' surgeons and medicinal plants needed for long voyages and in turn received plants from the colonies.[29]

The central botanical garden serving the navy was at Nantes, the major port used for provisioning vessels destined to cross the Atlantic. Its moderate climate made it particularly suitable for raising plants from various provenances and natural habitats, so the garden could indeed supply many of the simples that were needed aboard ship. In addition, specimens from the colonies that arrived in poor condition in Nantes succeeded better when they were able to heal in this garden. Once they reached France they still needed to go via the Loire River first to Orléans and then to the Jardin du Roi in Paris. Sometimes this journey—particularly in bad weather—could kill off rare and delicate plants. Decisions about the fates of newly acquired specimens were orchestrated through correspondence between the two cities, weighing the relative danger of travel, bad weather, and salt air. Rochefort was also an important entrepôt for new plants, as were other Atlantic cities. The ports of entry for colonial plant specimens were those used more generally for colonial trade.[30] Botanical gardens were thus arrayed along the coast of France, helping to mark the boundaries of the kingdom with institutions designed for the territorial improvement and potential extension of French lands.

The proliferation of botanical gardens in France resulted in a dearth of these sites in the colonies, where they were sorely needed. Maintaining and restoring good health was a terrible problem in colonial areas where European diseases mixed with local ones, and poverty and violence periodically made things worse. Hospitals were built in Canada, Saint Domingue, Martinique, Guadeloupe, and Guiana, among other places, and were staffed with trained doctors, many of them from Montpellier. These men had some experience in botanical collection and were frequently eager to learn about local cures for diseases, writing home about how skillful indigenous people were in using native plants for cures. They were chagrined when there was no official encouragement to set up medical gardens in the colonies. But botanical gardening was a closed system designed for the improvement of territorial France.[31] In spite of the fact that many botanical publications cited the religious tenet that God provided the cures for diseases in precisely those places where the illnesses arose, doctors in Canada and Louisiana patiently sent specimens and writings to the Jardin du Roi and waited for supplies from France.[32]

The story of Michel Sarrazin in Canada provides a view into how this system of colonial medicine, military power, and academic botany worked (and failed). Sarrazin arrived in Canada in 1685 as a ship's surgeon interested in conducting botanical studies there. He was of a noble family from Burgundy and appropriately served the military with his medical skills. In fact, he became the surgeon general of the colony and was one of the few people at the time to practice both surgery and medicine. Arriving in this colony at a time when war with the Iroquois was particularly bloody, his skills as a surgeon were tested. Finding the supplies of medicines limited, he also was pushed to consider how to serve the colonists better. His *Mémoire des Médicaments nécessaires pour les troupes du Roy en Canada à envoyer en 1693* enumerated the kinds of medicines needed in the colonies and, in principle if not in practice, meant to come from France.[33]

Sarrazin returned to France after his first trip to Canada and became *correspondant* to the great systematic botanist and plant collector Joseph Pitton de Tournefort (1656–1708).[34] Because of Sarrazin's knowledge of Canada he was urged to return there; there was need in the colony for both his medical skills and his botanical studies. When he did return in 1697 one of a series of epidemics had started, giving Sarrazin little time for research. He battled typhus, smallpox, and a disease thought to be yellow fever.[35] This experience was frustrating because he had inadequate medical supplies and no time for study, but it also drew his attention to the need for medicines in the colonies and the difficulty in obtaining them in adequate supply. It was clearly possible to build botanical gardens in the colonies, he reasoned, but the lack of medical personnel made it impossible for those who could study the medicinal properties of plants to find time to build gardens adequate to the task.[36]

Sarrazin's skills in collecting indigenous species in North America were most evident from the large shipment of botanical specimens he sent from Canada in 1704. He clearly also experimented with useful plants in the vicinity of Quebec since he wrote about them in 1725.[37] Given that his collections were exemplary and his studies and skills well respected, there was no obvious reason for the Académie Royale des Sciences or the French state to resist his entreaties to build a botanical garden. However, while the French considered Quebec an interesting source of plants, they hardly saw it fit for demonstrating and developing French territorial power. The political stage was European, and botanical knowledge was centered there; therefore, France, not Canada, was where botanical gardening could be made a visible and useful tool of power.

Botanical Demonstration at the Jardin du Roi

Botanical demonstrations at the Jardin du Roi were ceremonial occasions on which the public witnessed formal exercises in the identification, collection, and uses of plants. These events became popular over the course of the seventeenth and eighteenth centuries as territorial politics gained force. They were the foci for the *botanophilia* that developed in Paris as exotics were increasingly imported into the city and housed in the Jardin du Roi and the collections of elites. Plants cultivated in the king's garden were known by their various names, sorted by uses, raised by habitat, labeled by provenance, understood by developmental patterns, classified by characteristics, associated with families, and appreciated for their effects on the human body.[38] Mastery of nature was both a natural philosophical project and emblematic of effective government—over plants as well as people. In the king's garden botanical and horticulture capacities became measures of monarchical authority as well as learning.

The use of public ceremonies in Paris to signify and ratify the power of the king was nothing new. Even territorial claims had traditionally been exercised through ceremonial practices. Since the reign of Catherine de Medici (1519–89, r. 1547–74) garden events such as tournaments, parties, ballets, and the like had proliferated in Paris. In the Luxembourg gardens and the Tuileries nobles enacted stories about gods and heroes, simultaneously representing and naturalizing their superiority.[39]

Seventeenth- and eighteenth-century botanical demonstrations deployed rare plants rather than classical literature, but they still enacted stories of natural superiority. The Jardin du Roi presented visitors with an extraordinary collection of rare and surprising specimens; the plants in the garden might have been identified by their names and their provenances, but this information also carried implicit stories of distant travels, difficult voyages, and erudite learning. Plant demonstrations were spectacles of natural knowledge that foregrounded exotics with amazing qualities—scents, colors, sizes, or uses.[40] If plant demonstrations were also displays of social difference, those at the Jardin du Roi spoke quietly but insistently of greatness—of France, its reach, its intelligence, and its administration.[41]

The symbolic power of botanical demonstrations is well illustrated by Charles Plumier's (1646–1704) *Description des Plantes d'Amérique* (1695).[42] Plumier declares that he became a botanist after seeing plant demonstrations in Rome. He writes with passion of the remote places where collectors traveled and how they encountered botanical wonders in doing their work. He also describes individual plants using the vocabulary of growth patterns and provenance, showing how the language of botany could stabilize the identities of plants while also stimulating dreams in readers about the world. Plumier makes the wonders of the Creation seem at once palpable and knowable. According to him, botanical collection and demonstration made such ineffables more accessible.

The most parochial political forces affecting botanical research during the seventeenth century were startlingly evident in the great natural history project overseen by the celebrated architect and physician Claude Perrault (1613–88).[43] Perrault organized colleagues from the Académie Royale des Sciences to use techniques of dissection common in animal studies on botanical specimens in the Jardin du Roi. A political caveat was imposed on the work, in the sense that the research had to focus on French native plants. This restriction was officially accepted by the scientists but had little effect as no definition of what constituted French plants was provided. Species that had been imported into France already, the scientists thought, were already assimilated to French agriculture and gardening and were now French plants. Given the use of trade as a resource for French territorial development and the organization of the colonial plant trade to serve the Jardin du Roi, this position made sense.[44]

Botanical demonstration served not only a political function at the Jardin du Roi but also the growth of systematic botany. The most notable Jardin du Roi demonstrator was Joseph Pitton de Tournefort, a powerful figure in French botany whose system of classification was so esteemed in France that it found support well after the introduction of the classification system devised by Linnaeus. Tournefort also helped to further systematic botany in France. He grew up in the intellectually sophisticated city of Aix-en-Provence and learned much of his botany at Montpellier, where he worked as a plant demonstrator for the professor of botany Pierre Magnol (1638–1715). In this capacity Tournefort faced the increasingly difficult challenge of plant classification that accompanied the growth of collections and the naming of new plants. He wrote, "We must therefore apply a precise method to the naming of plants, for fear that the number of names does not eventually equal the number of plants. This is exactly what would happen if everyone could name each plant as he wished. The result would be not only great confusion but an astonishing burden for the memory, weighted down by an infinite number of denominations."[45] He chose to try to classify plants by the structure of their flowers, spoke of them as identifying natural families, and used this way of seeing plants to generate a single system of classification.[46]

This work was so interesting to Guy-Crescent Fagon (1638–1718), the king's physician, that he brought Tournefort to the royal garden in Paris in 1683. Demonstrations at the garden became even more popular as Tournefort made flowers the center of observation and taught his systematics.[47] He provided a unified vocabulary for plants that defined groupings and organized how to observe plants and their familial relationships. The simplification of names and groups provided means for seeing the plant kingdom as more lawful and orderly. This fit an understanding of the earth as the object of the Creation and an understanding of the plant world as a model kingdom.

The ultimate meeting of the plant kingdom and the French state occurred when a botanical garden organized around Bernard de Jussieu's (1699–1777) principles of classification was built at Versailles. This new garden was set out by the Trianon, the part of Versailles symbolically associated with paradise, or the garden of perpetual spring. In this corner of the royal park the garden of Eden met the orderly land of botanists, a place brought closer to paradise by human knowledge. Stewardship merged with scholarship in the most important of French formal gardens as territorial politics was ceremonially united with botanical study.[48]

The Return of Stewardship to French Colonial Botany

This brief history of botanical gardening and the plant trade may give the impression that French interest in dominion and stewardship—the moral dimensions of land, people, and plant control—was permanently replaced in the seventeenth century with a system of territorial domination served by botanical study. But surprisingly, the language of dominion and stewardship started to reappear in the mid-eighteenth century, suggesting that this tradition had been suppressed rather than erased in the previous century. The head of the Jardin du Roi, Georges Louis Leclerc, comte de Buffon (1707–88), expressed these vivid and ambitious feelings:

Brute Nature is hideous and dying; I and I alone, can render her pleasant and living. Let us drain these marshes, bring to life these stagnant waters, by making them flow, forming them into streams and canals, utilizing this active and devouring element, hitherto hidden from us, and which we have learnt to use by ourselves. Let us set fire to this useless growth, these old, half-decayed forests, then cut away what the fire has not consumed. Soon, in place of the reed and water lily, from which toads make their poison, we shall see the ranunculus and the trefoil, sweet and healthful herbs. . . . A new Nature will be shaped by our hands.[49]

Buffon articulated a (deeply gendered) model of good stewardship and legitimacy as a fundamental critique of the (failed) partnership of botany and territorial government in France.

Questions about legitimacy and authority in the colonies were also raised in books on colonial botany, many of them written by the naval officers who were seeking new opportunities for colonial development and asking how it should be done. Precisely the men who were explicitly charged with exercising power in the name of the state were the ones posing questions about moral relations to the natural and social worlds. They used familiar Christian rhetoric about the Creation, human intelligence, and dominion to think through the moral problems of colonial governance.

Jean-Baptiste Thibault de Chanvalon's (c. 1725–86) 1761 *Voyage à la Martinique* exemplified the new moral rhetoric. Chanvalon, who dedicated his text to Etienne-François, duc de Choiseul (1719–85), minister of war and the Navy, argued that the simple extraction of local resources for the well-being of the monarchy and motherland was not what rendered colonies assets to the state. He claimed that a merchant might view a colony as a mine, as a source of materials to exploit, but a military man would recognize that the welfare of the nation depended on careful management of resources—*mesnagement*, so to speak—in the colonies. It was the capacity to cultivate the local assets in their own habitat that could turn colonies such as Martinique into sources of commerce and power for France.[50]

Chanvalon went on to argue that French colonies had been less profitable than they should have, not because of a lack of horticultural skill or botanical knowledge, but because of a perceived distance between people in the colonies and those in the French state. Those in the colonies should have been treated like those of the metropole—as brothers under the same hand of the sovereign. The distance of colonists from the king had rendered them distant from prosperity and prevented Martinique from being properly managed to take advantage of its soil, climate, and natural assets and to increase its agricultural and industrial capacity.[51]

Chanvalon contended that colonialism had been a moral assault on people and places. He wrote in heart-wrenching terms about the diseases that plagued the Negroes in Martinique, diseases he claimed were the results of the way they had been transported to America. They had been captured, fed unfamiliar foods, held on vessels like criminals, and taken across the ocean with no hope of seeing their families again. This made them prone to stomach ailments that killed them slowly. Chanvalon's moral accounting was his explanation of economic failure and his guide for improvement. To Chanvalon, a kind of chauvinism and cruelty had grown along with French territorial politics. The country had to have its moral foundations restored to make France a more effective colonial nation.[52]

The naturalist Pierre Sonnerat's (1745–1814) *Voyage à la Nouvelle Guinée* of 1776 echoed many of these themes. The book is a report on his trip to the Philippines and Papua New Guinea, where he had been sent to study what treasures the French could take from this part of the world. The author declared, however, that his interest in the voyage was different—namely, to observe the spectacularly novel plants, animals, and land that could only be found so far away. He described the poor treatment of the people of the Philippines, forced to live in ignorance and to give up their own culture to please the Spanish missionaries. They had naturally sought revenge against their oppressors because people universally seek liberty. While championing their freedom, Sonnerat also argued that the indigenous inhabitants of the Philippines did not understand the treasures of their lands that were sought by Europeans and so outsiders such as the French were needed to make better use of these areas that were now "abandoned," he said, to "stupid savages."[53] The odd mix of predatory opportunism and stewardship revolved around claims to universalism and particularism: what was universal to men and what was particular to places. With their superior knowledge of the particulars of places (natural history), the French could exercise a kind of stewardship that locals would not be able to achieve by themselves.

Conclusion

The return of the language of stewardship to the literature on botany in the late eighteenth and early nineteenth centuries was a reminder that proper land management was still a deeply held French measure of morality and justice. The secular language of territorial politics, which shaped public policy under Colbert and Louis XIV, might have masked this tradition, but stewardship remained important. The submersion of the spiritual stakes behind *mesnagement* politics only appeared to make territoriality into a secularized and rational system of administrative action. The legitimacy of good government still depended on an orderly landscape and healthy people. The secular language of land administration may have made state territoriality a more robust political philosophy during periods of religious strife, but once political change and territorial expansion seemed possible, the moral language of French politics returned to the botany literature that could serve it. This surprising turn indicated how deeply questions of stewardship had been felt before this period and how earnestly such concerns had been maintained, connecting the intellectual orderliness of botany to the political order of French government.

Chapter 2

Walnuts at Hudson Bay, Coral Reefs in Gotland

The Colonialism of Linnaean Botany

Staffan Müller-Wille

> *A system emerged out of the technology of healing in a particular part of Europe. . . . This system has spread over the world, India included, without meeting a rival. Its origin was, as it were, a sublimation in which "Man" was displaced from the focus of thought that the "plant" might be placed there.*
>
> —I. H. Burkill, "Chapters on the History of Botany in India," 1953

Carl Linnaeus (1707–78) figures prominently among eighteenth-century naturalists who influenced botany in its nineteenth-century association with colonialism. Professor of botany and medicine from 1741 to 1778 at the University of Uppsala (Sweden), Linnaeus devised a universal system of botanical nomenclature epitomized by the introduction of binomial names for plant species—that is, Latinized designations composed of a generic name and a specific epithet, intended to remain stable and universally applicable once introduced. Historians consider Linnaeus's binomial nomenclature important to colonial botany because it allowed for the integration of the results of prior accounts of non-European flora and provided the model for subsequent inventories of colonial flora. By the first half of the nineteenth century, moreover, Linnaean nomenclature had become institutionalized to the degree that no botanist could claim to produce scientific knowledge who did not follow its rules. Most historians have therefore taken for granted the role of Linnaean botanical nomenclature as an instrument of colonialism.[1]

This chapter explores an enigma of Linnaeus's relation to scientific colonialism. The cultural and political context in which Linnaeus developed his nomenclatural reform can hardly be characterized as "colonialist." In the wake of two short-lived colonial adventures (a well-known one in Delaware from 1638 to 1655 and a less well-known one at Cape Coast in Ghana, West Africa, from 1649 to 1663) and the economically devastating results of the "great power" politics of Charles XII (1682–1718), Sweden had given up any aspirations to hegemony in its "era of freedom" (1718–72), during which Linnaeus lived and worked. Imperial ambitions had definitively exhausted Sweden's meager economic resources.[2]

In Sweden a peculiar form of political-economic ideology developed in lieu of colonial aspirations. As Lisbet Koerner has shown, it was this ideology, Swedish "cameralism," that inspired Linnaeus's achievements. Like many of his compatriots, he held the view that Sweden's prosperity depended on strategies of import substitution. Imports should be supplanted either by domestic substitutes or by the importation and subsequent acclimatization of foreign natural products. As Koerner succinctly puts it, "The idea was that science would create a miniaturized mercantile empire within the borders of the European state."[3] Linnaeus initially developed and applied binomial nomenclature in the context of "patriotic" projects that he undertook in explicit accordance with cameralism.[4] Even if Sweden were to excel as a science-based, self-sufficient state economy, the political-economic strategies adopted to achieve this goal would not result in a colonial empire but rather in the coexistence of internally rationalized and homogeneous, yet externally autarkic and independent political units. (How different this strategy was from mercantile strategies in other European countries, such as France and the Dutch Low Countries, can be gathered from the chapters in this volume by Chandra Mukerji and Harold J. Cook.)

How, historiographically speaking, can we then explain that Linnaean nomenclature became a powerful instrument of colonialist plant exploration if the context in which it originated was not colonial but one of national development? Given the cameralist context, would it not have been more rational to devise a system of nomenclature that was not universal but entirely adapted to the cultural and natural idiosyncrasies of Sweden? These questions gain more weight if one considers that Linnaeus's attempts at import substitution were largely unsuccessful (as Koerner amply demonstrates in her book and as Marie-Noëlle Bourguet recalls in her chapter in this volume) and that there was widespread initial resistance to Linnaean nomenclature among colonial elites (see chapters by Londa Schiebinger, Andrew Lewis, and Antonio Lafuente and Nuria Valverde in this volume). So how do we explain that Linnaean taxonomy and nomenclature became a universal tool of colonialism if it was not only not colonialist in origin but also failed initially in terms of practical application and acceptance in various colonialist contexts?

In what follows I answer this question in three steps. First, I offer an analysis of the rationale behind Linnaeus's patriotic projects that will reveal that the identification of domestic and exotic plants providing the basis for Linnaeus's strategy of import substitution was in turn based on a practice of worldwide plant collecting. Second, I investigate the activities of one of Linnaeus's students, Pehr Kalm (1716–79), who traveled in North America between 1747 and 1750 with the aim of collecting exotic plants for cultivation in Sweden. Third, surveying the results of his travel, as documented in Kalm's publications and in the *Species plantarum* (1753), the book in which Linnaeus first consistently applied binomial nomenclature, I propose that two social systems were operative in botanical exploration: one that provided the concrete infrastructure—resources, allowances, patronage, and so on—for exploratory activities; and one that was instituted by global exchange of specimens among botanists. While the latter system depended materially on the former, it was not determined by it. This, I argue, helps to account for the curious fact that Linnaeus's nomenclature, noncolonial as it was in origin, later became one of the prime instruments of colonial exploration.

The Great in the Little World

In 1741, on assuming the title of professor of botany and medicine at the University of Uppsala, Linnaeus gave a speech on "the necessity of traveling in one's own country." Rather than the grand tour to capitals of Europe, a well-established component of gentlemanly education in the eighteenth century, Linnaeus recommended study and travel within Sweden before turning, if at all, to foreign countries: "Whoever, before he sets out to visit regions warmed by other suns, has laid the first foundations of his studies in his native country, will be most likely to bring back materials of far greater price, than we usually see amongst the greatest part of our travellers, who seldom return home laden with any thing, but fine sounding, and empty words collected out of the European languages."[5] Linnaeus's critical view of foreign travel as an adequate means of education is surprising in view of the fact that he sent many of his disciples around the globe. The beginning of the passage just quoted does, however, allude to a cameralist justification for such travels. Where the wealth of a nation depended on developing domestic knowledge and technology while avoiding the import of luxuries, it was critical to send out trained disciples to bring back foreign goods for acculturation and reproduction back home. But what kind of knowledge could disciples acquire by traveling in their home country to prepare them for this task?

A peculiar rhetorical feature of Linnaeus's oration offers an answer to this question. After having criticized the grand tour, he goes on to describe the riches of his native country. These descriptions often feature a peculiar reversal: whenever possible Linnaeus identifies the resources of the kingdom of Sweden with exotic products. Consider, for example, the following passages:

We admire the abundance of coral on the Indian shores, yet the port of Capellus in Gotland alone equals, nay exceeds those riches of the East.

What traveler that is not totally ignorant of botany, does not go from Paris to Fontainebleau to see those very rare orchids . . . ? Who imagined these flowers grew in our countrey, and in such plenty in Oeland, that they are to be met with in every field?

What are those famous exotic remedies brought from either Indies and purchased at so great a price . . . , but remedies approved by long use amongst the vulgar? And are not innumerable remedies used among our own country people of the same nature?[6]

These passages all articulate a peculiar "taxonomic gaze." Linnaeus does not cite Swedish flora and fauna in regional terms but as representatives of the world at large. Relations of generic identity are established across the regional divisions of the world. What structures the world is not the order of habitats but rather the nested hierarchy of taxonomic classes: each representative of a certain class, whatever particular region it comes from, may represent all the other members of that same class. In short, the "taxonomic gaze" treats its objects as interchangeable.

Linnaeus's students were thoroughly trained in this gaze, and in the practice of putting it on record. Each academic year Linnaeus led botanical excursions of up to three hundred participants around Uppsala. They were organized in a quasi-military manner: equipment and dress were prescribed; instructions were given to hunt for specific objects; the party was split into troops to forage for plants; and students gathered at regular intervals to have Linnaeus "demonstrate" the findings. His demonstrations consisted of spelling out the name of a collected item, supplemented by reference to taxonomic literature, usually a page or serial number from one of Linnaeus's publications. Brief mention of particular features and a more detailed account of the uses to which the plants could be put, including and emphasizing the uses to which they were put in foreign countries, followed. Students were asked to keep numbered protocols, organized in the manner in which Linnaeus gave his information: name, literary references, particular features, and uses. The following example is taken from a protocol dated 1747:

3. *Acer*, Maple. 303 [that is species no. 303 in Linnaeus, *Flora suecica*, 1746]. Abroad ours [that is, the species growing in Sweden] is quite rare; where it is

found [abroad] it is of lesser quality. Has 8 stamina and 1 pistil. Is used for spoons, bowls and other things. Its juice is sweet, and sugar is cooked from it in America, in Canada.[7]

The taxonomic gaze characterized above is clearly in evidence here: the plant encountered is identified with a plant from North America. This generic identification suggests that the domestic species might be put to the same use as the foreign one or that the foreign species might be reproduced in Sweden. As a matter of fact, Swedish maple does not yield maple syrup. Yet, inasmuch as both species belong to the same genus, there is reason to believe that one could substitute for the other in terms of economic usefulness.

While the taxonomic gaze was applied in the field, it was neither inspired nor confined to this domain. The American maple, for example, is simply not found in Sweden. The local environment provides the material scope of the taxonomic gaze, but it is not its material source, as taxonomic identities simply do not correlate with the geographic distribution of plants. The material source for the taxonomic gaze, Linnaeus made amply clear in the introductory part of his oration, was provided by specific academic institutions—namely, botanical gardens.

> Physic gardens are . . . cultivated [at academies]; where the plants of various kinds are collected from all parts of the globe, that we may by this means behold, as it were, the great in the little world. . . .
> [. . . They] tend to this end, that though we be confined to one spot, one corner of the earth, we may examine the great and various stores of knowledge, and therein behold the immense domains of nature, and get acquainted with such things, as otherwise must be sought for, and oftentimes in vain, over the whole globe.[8]

The botanical garden, as described by Linnaeus, may be interpreted in Latourian terms as both a "center for calculation" that allows for the accumulation and shuffling around of inscriptions, and a center for training disciplined "agents" sent out to enhance collection in "cycles of accumulation." It is well known that Bruno Latour's sociology of science took its inspiration from John Law's study of long-distance trade.[9] To be sure, long-distance trade, as Claudia Swan points out in her essay in this volume, does not equal colonialism. Yet it facilitated and profited from the establishment of colonial empires. And it is in regard to long-distance trade that one can indeed identify a connection between Linnaeus and the colonial endeavors of his time. As professor of botany and medicine he had access to the protocols that captains and doctors traveling for the Swedish East India Company kept about their travels and transactions, and he received seeds and specimens from them. Linnaeus also engaged in a kind of long-distance trade through a correspondence network that reached across Europe and as far as the Americas (Virginia and Surinam).[10] Finally, as was already mentioned, Linnaeus sent about twenty students on long-distance travels to collect plants for him. Among them were Daniel Solander (1733–82), who accompanied Joseph Banks on James Cook's first circumnavigation of the globe (1768–71); Anders Sparrmann (1748–1820), who was with the Forsters on Cook's second circumnavigation (1772–75); and Carl Peter Thunberg (1743–1828), who traveled to Japan as a surgeon for the Dutch East India Company (1770–79). In the following section I review the itinerary of one of these traveling disciples, in order to analyze in more detail how he acted as an "agent" in a "cycle of accumulation." The journey I examine is that of Pehr Kalm to North America, which preceded the publication of Linnaeus's *Species plantarum*.

Turning into One Hundred People

The proposal for Kalm's research trip came in 1744 from Baron Sten Carl Bielke (1709–53), associate judge at the High Court in Åbo (now Turku, Finland), who had taken Kalm under his guardianship in 1740 at the suggestion of Johan Browallius (1707–55), then professor of physics at the University of Åbo and a close friend of Linnaeus. Bielke addressed his proposal to the Royal Academy of Sciences in Stockholm (alongside Linnaeus he had been one of its founders in 1739) after his return from an extended journey to Russia. Kalm, having completed studies in physics and natural history with Anders Celsius (1701–44) and Linnaeus at Uppsala University, had accompanied him.[11]

Bielke had initially planned to send Kalm to Siberia for the purpose of collecting seeds for cultivation trials on his estates. Linnaeus (and Celsius) welcomed Bielke's idea but were less enthusiastic about his plans for Kalm's destination: they contemplated Iceland and Greenland as well as China and the Cape of Good Hope. When Bielke made his proposal to the academy, Linnaeus intervened with a letter and proposed North America, arguing that as many useful plants could be expected from there as from Siberia.

The reason for this shift is unclear, but Linnaeus had probably been interested in North American plants since his years in Holland (1735–38) and from his correspondence with naturalists such as Peter Collinson (1694–1768), Jacob Dillen (1687–1747), and Philip Miller (1691–1771) in England as well as John Bartram (1699–1777) and Cadwallader Colden (1688–1776) in New England. In January 1746 Linnaeus submitted a memorandum to the academy in which he laid out what he expected from Kalm's North American visit:

Those plants which occur naturally at the same latitude enjoy life and grow without trouble wherever they are cultivated at the same latitude, as long as the soil is the same, just as plants from the East Indies grow easily in the West Indies at the same latitude.

That this is so is abundantly demonstrated in the Northern parts of Europe, Asia, and America, where a lot of similar plants occur, such as fir, spruce, juniper, and yes, even the blackberry from Norrland grows at Hudson Bay in America and in Samojedna above Siberia in Asia. On the other hand, each of them has their particulars, as Asia its curious cedar, from which they have timber and nuts and North America its many kinds of oak for food and timber.[12]

Linnaeus's argument was patently inconsistent in that he maintained both homogeneity and heterogeneity for regions at the same latitude. Yet it seems to have convinced the academy, for Kalm was soon sent to North America. Financial support was procured from the three Swedish universities and the manufacturing commission of the Swedish Reichstag. (Kalm's future was further secured by his appointment to the newly created chair of economics at the University of Åbo.) The principle of "latitudinal homogeneity" that Linnaeus formulated in support of sending Kalm to North America was also to shape his itinerary: Kalm was asked to travel first to Philadelphia, where a Swedish "cultural colony" had persisted since 1655. From there he was to proceed to the Hudson Bay, as it is on the same latitude as the more northerly and underdeveloped parts of Sweden and Finland. Linnaeus expected oaks, walnuts, chestnuts, and medicinal plants such as the famous *radix ninsi*, sassafras, and the rattlesnake root to grow there; and Kalm was supposed to gather and send back as many seeds as possible from such plants, to take detailed notes on their habitats, and to investigate how colonists and indigenous people used them.

Kalm set out for America in October 1747. Everything had been prepared to guarantee the success of his journey: he had been given detailed travel directives and equipped with money, instruments, books, and a servant by the name of Lars Jungström. Naturalists in England and Pennsylvania had been informed of his arrival, and he carried letters of recommendation from Linnaeus to John Bartram and Cadwallader Colden, at the time a member of the Governor's Council in New York. Kalm was even assured contacts at the highest diplomatic level. In order that he might be assisted by French authorities in Canada, Count Clas Ekeblad (1708–71), Swedish ambassador in Paris from 1742 to 1744 and one of Linnaeus's correspondents, contacted the governor-general of Canada, the marquis de la Galissonière (1693–1756). Backed by an array of recommendations and passports, Kalm could move freely across a region that, at the time, was plagued by resurgent military conflicts between the English and the French colonies.

Copy of the map drawn by Lewis Evans for Kalm in 1750. Kalm's journey from Raccoon to Cap aux Oyes in Canada is shown on this copy by a thick, broken line and his journey from Albany to Niagara by a dotted line.

Figure 2.1. Kalm's route through North America, showing the two linear north and westbound legs of his journey. From Martti Kerkkonen, *Peter Kalm's North American Journey: Its Ideological Background and Results* (Helsinki: Finnish Historical Society 1959).

Despite these preparations, Kalm would never reach his final destination, Hudson Bay (see Figure 2.1). He arrived in Philadelphia as late as 15 September 1748, a year after he had set out. To reach regions at the same latitude as Sweden in the remaining year of his two-year enterprise he would have had to set out for the North immediately. John Bartram and other local naturalists informed him that the vegetation was already dormant for the winter in this region. Kalm observed that the plants Linnaeus had asked him to look for were abundant in Pennsylvania, and he decided to wait with his travel until February of the following year and then go to New England only. These new plans, of which Kalm informed Linnaeus in a letter that was read to the academy, dissatisfied his supporters to the highest degree. In a letter to the academy's secretary dated 30 January 1749 Linnaeus expressed his annoyance:

From Kalm's letter I see that he aims to make New England his northern-most destination, and from there he intends, after having traveled westward in Canada, to return home. New England lies between 40 and 48 degrees latitude, which is equal to France, Spain, and Italy. According to his promises and our wish he obliged himself to travel in America to the same latitude as Sweden—between 55 and 60 degrees—where one infallibly should find such things as could serve our hard climate. . . . [I]f Prof. Kalm should not go further, it would have been better if he had never traveled there at all.[13]

Linnaeus must have sent Kalm a similar letter (which does not, however, survive) as Kalm responded to Linnaeus on 28 May 1749: "Never write again, good patron, such hard letters to me and do not embarrass me in front of others by letting them open them to forward them to others" (the letter had been passed on via Collinson and Bartram).[14] The reaction testifies to the maturing self-confidence of the disciple who, after more than half a year in North America, had learned about its climatic peculiarities. He knew from his own experience, but more so from his informants, that a trip to the Hudson Bay was simply impossible and that New England's climate was comparable to that of Scandinavia, despite the difference in latitude. Yet the pressure put on him seems to have been sufficient to change the course of his travel.

In his first year Kalm attempted to journey as far north as possible, on an almost straight, northbound course. From New York he followed the Hudson River past Albany and continued along Lake Champlain to Montreal and Quebec City. During this trip he made daily meteorological observations, gathered plant specimens and seeds, and took detailed notes on their natural habitat, cycles of vegetation, and the uses made of them by local inhabitants. These observations were compiled in a journal at regular intervals. One of the plants that Linnaeus had shown interest in was the walnut tree. Here are Kalm's observations:

Black Walnut Tree. At most farms black walnut trees were planted, which otherwise grow wild here and there in the woods; this tree was considered to be the most valuable, as it is sought after by carpenters, and the things made thereof look quite well. . . . How Indians make some kind of well-tasting milk from their nuts, see further down on p. 330. It is considered to be the one tree of all those growing at this place [that is, at Racoon, New Jersey] which is the most devastating and which, as it were, kills all kinds of trees and herbs that are found near it, such as apple trees, cherry trees, linden, etc.[15]

The linear course of such observations, unavoidable while traveling and clearly documented in the journal, resulted in a severe problem, however, which Kalm explained to Linnaeus on 30 August 1749, after having reached the northernmost point of his travel at Cap-aux-Oyes, seventy-five miles north of Quebec City (that is, at about 47° 20″ latitude): "[F]rom all I have seen, I cannot promise seed, because then I would have to divide myself into hundreds of people, as I have found an herb in bloom at one place only, and afterwards never had the chance to come back again."[16] The linearity of Kalm's trip resulted in his being able to observe plants in single developmental stages only. This severely hindered his ability to collect seeds, the prized objects for which he had been sent. The solution to this problem was, as Kalm put it, to divide himself "into hundreds of people," which, in a sense, he did. First, he sent his servant Jungström back at certain points during their travels to collect seeds from places they had visited earlier. He also had help from North American naturalists: Bartram in Pennsylvania, Colden in New York, and John Clayton (1694–1773) in Virginia all provided him with seeds. The most substantial assistance probably came from the vast network that the governor of Quebec had built up in his province. The marquis de la Galissonière had a strong interest in natural history and had requested the province's *medicus regius*, Jean François Gaultier (1708–56), to issue a directive to civil servants and military personnel in Quebec to collect and conduct research. As Kalm described it in his journal: "[De la Galissonière] had kindled such a fire in the hearts of all those under him, in high officers, fortress commandants, priests, Jesuits, soldiers, tradesmen and even in the nobler sex, that they were to be seen running about the forests gathering plants just as though they had been trained by Doctor Linnaeus himself."[17] The motives for engaging in these activities may have been as diverse as the sorts of people involved, spanning from direct order to amateurish enthusiasm. The crucial point is that Kalm explicitly compared his emerging network of collectors with Linnaeus's own in and around Uppsala. De la Galissonière had generously put his local network at Kalm's disposal, asking Gaultier to assist him and provide him with native guides. Even monasteries and nunneries, longtime sites of pharmaceutical knowledge, opened their doors for the Lutheran Kalm. He received the same good assistance during the second year of his travel, which took him westward to Niagara Falls.

Only with the help of such networks, spanning the multifarious sectors and levels of North American societies, including indigenous ones, could Kalm transcend the limitations of the linear course of his travels. Building up and maintaining such networks did not so much depend on behaving as a disciplined agent as on developing the requisite self-consciousness and flexibility to overcome the kinds of conflicts and difficulties characteristic of the "bio-contact zone" (as explored in Schie-binger's chapter in this volume). The information, on the other hand, that Kalm received from his networks of informants cannot be character-ized as mere corroboration or elaboration of the knowledge he started off with in Europe. Rather, the resulting knowledge formed a mosaic of bits and pieces extracted from the most diverse local frameworks. These frameworks were not recorded by Kalm. While he did take notice of Na-tive American food taboos, he did not care to chronicle the taxonomies that we now know are inherent to them. Detached from its cultural ori-gins, the information assembled in Kalm's journal acquired the abstract status of mere data collection with no inherent order. However, by col-lecting seeds Kalm accumulated the means to reproduce the objects to which these isolated pieces of knowledge referred, means that could later be used to produce and reproduce evidence in the form of speci-mens.

Specimens, Numbers, Names

Ultimately, it was specimens that convinced Linnaeus of Kalm's success. In February 1750 he was still unsatisfied: "It is strange that Kalm does not dare to give a single observation to the Academy or to others. . . . He has changed completely by talking a lot and performing little, as all he has written is empty words."[18] Linnaeus could only be appeased by the arrival of packages of seeds from Kalm, accompanied by numbered lists specifying their botanical names. The first of these consignments was announced by Kalm as a "kind of bribe," to mitigate Linnaeus's dissatis-faction.[19] Linnaeus subsequently began to praise Kalm's discoveries, as when he wrote to the academy in December 1750 that "Kalm's discovery about *Lobelia* [used by Indians to cure venereal diseases, or "French-men," as Linnaeus put it] is great and greater than I dare to tell."[20] And when he learned about Kalm's return to Stockholm in June 1751, he wrote to his close friend Abraham Bäck (1713–95): "Dear brother, bring Kalm here immediately. I long for him endlessly."[21]

What inspired Linnaeus's enthusiasm on Kalm's return? It can have been neither the prospect of detailed descriptions of North American nature nor practical results in terms of transplants. As to the former, a first volume containing extracts of Kalm's travel journal was published in 1753. Later volumes were published in 1756 and 1761, and the last one, containing Kalm's trip to Niagara Falls, never appeared in print. The delays were caused by doubts about how to arrange the material contained in the journal—whether day by day or object by object. Equally, a planned *Flora canadensis*, containing the botanical descrip-tions of plants in technical Latin, would never appear in print, probably because Kalm, once settled at the University of Åbo, lacked the necessary facilities; Åbo had no university garden.

What Kalm did publish immediately after his return was a list of the names of plants collected in North America, arranged alphabetically and numbered serially. This list contained barely any information on the plants or their geographic distribution, into which Kalm had gained certain important insights (summarized in a letter to Linnaeus dated 5 December 1750).[22] Rather, the contents of this list were restricted to the species' names, their natural environments, and their "use" and "care," that is, the measures to be taken in cultivating them. The short introduction stated that the list was primarily intended to inform others of the plants brought from America and to offer seeds to those who wished to attempt cultivating these plants.[23]

Kalm's hopes for extensive cultivation of North American plants in Scandinavia were not fulfilled. He engaged in agricultural experiments after his return to Åbo, for which purpose a piece of land had been placed at his disposal by the Reichstag. In terms of large-scale cultivation, these experiments were unsuccessful. North American plants, after all, were not so easily acclimatized in Scandinavia. Indeed, acclimatization would remain an unsolved scientific and technological problem throughout the eighteenth century (see Marie-Noëlle Bourguet's chapter in this volume).

Why Linnaeus continued to be enthusiastic about Kalm's travel is clear from the conclusion of the former's letter to Abraham Bäck. After reporting the progress on his *Species plantarum*, he wrote: "God give me health at least as long as Kalm is here, so that I may eat my fill of nice and delicate plants."[24] One of Kalm's chief clients for North American seeds was indeed Linnaeus, who grew them in the university garden of Uppsala. And in botanical gardens, where diligent care was invested in single plants, it was possible to cultivate such varieties as Kalm supplied. Already in 1751 a dissertation that presented eight new plant genera that had been identified by Kalm was published under Linnaeus's supervision.[25] Moreover, Kalm's name appears in some ninety of the seven hundred definitions of North American plant species in the *Species plantarum*, an indication that these species were derived from specimens gathered by Kalm. Of these, Linnaeus claimed seventy-five to be "new" (new to European professional botanists, that is). One genus, newly described by Linnaeus, would even bear the name "Kalmia" to honor its discoverer; another was named "Gaultheria [*sic*]" to honor Gaultier, the Quebecois physician who had helped Kalm so greatly in building up a collector's network.

What is ironic about these prominent, if scattered, acknowledgments of Kalm's results is that all the care he had invested in recording the morphology, geographic distribution, natural habitats, and uses of North American plants seems to have been in vain. The *Nova plantarum genera* offered only the most important morphological characteristics of the newly detected genera, and the *Species plantarum* listed only the following information for each numbered entry (see figure 1):

1. The name of the species—both in the traditional fashion, extending the genus name by a shorthand statement of the specific difference, and in the newly introduced fashion of "trivial" or binomial names;
2. A list of "synonyms," that is, a list of citations referring to other literary sources on the species;
3. Short remarks on geographical origin and symbols indicating habitat and life span (information important for cultivation).

Thus the plants Kalm collected in North America entered a system of representation reduced almost completely to mere reference by number, name, and citation. This system not only resulted from the work of collectors such as Kalm but also provided the foundation for their botanical activities. Trained in Linnaean "demonstrations" and "herborizations," Kalm worked to build a network of collectors spanning the social strata of colonial North America; and like Linnaeus in his *Species plantarum*, Kalm kept a record of his specimens through the use of referential terms such as serial numbers, literary references, or names. Through multitudinous "cycles of accumulation" set in motion by trained agents, Linnaeus's *Species plantarum* was built on the one element that remained constant throughout these cycles—namely, pure reference, established by numbering, naming, or citing exemplars.

Botanical knowledge was thus purged, so to speak, of all context-dependent meaning, resulting in a taxonomic system that consisted of little more than extensional relations of signifiers to signifieds, of names to exemplars. These exemplars, moreover, were no longer restricted to the particular specimens Kalm had brought with him. The synonym lists also contain exemplars that grew out of previous "cycles of accumulation," such as the specimens mentioned in J. F. Gronovius's *Flora Virginica* (Leiden, 1739) or Mark Catesby's *Natural History of Carolina* (London, 1730–47). The diversity of botanical knowledge, as it had been uncovered in botanical exploration, was thus made accessible from the vantage point of the species entries in the *Species plantarum*.[26]

What the story of Kalm's travel thus reveals is that two social systems have to be recognized as operative when we speak of "botany in colonial connection." At a fundamental level, Kalm's travel was clearly dependent on the political infrastructure and economic interests of the partic-

KALMIA.

1. KALMIA foliis ovatis, corymbis terminalibus. *Gen. latifolia.*
nov. 1079. *
Andromeda foliis ovatis obtufis, corollis corymbofis
infundibuliformibus, genitalibus declinatis. *Gron. virg.*
160.
ChamæDaphne foliis tini, floribus bullatis. *Catesb. car.*
2. *p.* 98. *t.* 98.
Ciftus ChamæRhododendros mariana laurifolia, floribus expanfis fummo ramulo in umbellam plurimis.
Pluk. alm. 49. *t.* 379- *f.* 6.
Habitat in Marilandia, Virginia, Penfylvania. ♄

2. KALMIA foliis lanceolatis, corymbis lateralibus. *angustifolia.*
Gen. nov. 1079. *
Azalea foliis lanceolatis integerrimis non nervofis gla-
bris, corymbis terminalibus. *Gron. virg.* 21.
ChamæDaphne fempervirens, foliis oblongis anguftis,
foliorum fafciculis oppofitis. *Catesb. car.* 3. *p.* 17. *t.*
17. *f.* 1.
Ciftus fempervirens laurifolia, floribus eleganter bulla-
tis. *Pluk. alm.* 106. *t.* 161. *f.* 3.
Anonyma. *Cold. noveb.* 100.
Habitat in Penfylvania, Nova Cæfarea, Noveboraco. ♄

Figure 2.2. Species entries for the genus *Kalmia* in Carl Linnaeus, *Species plantarum* (Stockholm, 1753). The specific epithets, or "trivial names," as Linnaeus called them, show up in the right-hand margin. From Landmarks of Science microfiche collection, Max Planck Institute for the History of Science, Berlin.

ular culture he inhabited; his travel would never have been possible without them. And yet the starting and end points of his travel—its inspiration and its ultimate motivation—belonged to an altogether different social system, which constituted the discourse and tradition of botany as a scientific discipline and consisted in the circulation of specimens and names among professional botanists. In this system the specimen grown in the garden served neither local economic needs nor locally circumscribed tradition. Rather, it was invested with the power to establish new social relations across divides in subjective opinion. As many of the essays in this volume also suggest, specimens acquired authority in matters of identity and authenticity; and they did so by way of social relations across cultural divides. How objectifying specimens (and the numbers and names that refer to them) mobilized social relations can be observed in the botanical networks built up by Linnaeus and Kalm: "tributes" and "briberies" were paid in the forms of seeds and specimens; rewards were given by coining names; revenues were counted in terms of "new" specimens; and scriptural techniques were developed to keep records of discovered specimens (Anke te Heesen discusses such techniques in her essay in this volume). Specimens, numbers, and names, which restricted representation to mere extensional reference, mobilized social relations (rather than amplifying preexisting ones) to overcome differences in language, culture, and knowledge frameworks.[27]

Conclusion

Looking at botany as a scientific discipline, binomial nomenclature was thus designed to operate in or, rather, to mediate between different cultures rather than to serve the interests of a particular one. It is this "anachronism" of science—this tendency toward being ahead of its time in the sense of not being fully determined by the cultures that support it locally—that made it possible for binomial nomenclature to come into existence in advance of the colonialist conquest for which it later proved to be instrumental. Linnaeus did not envision a colonial empire but rather an economic autarky for his nation. Yet the scientific tradition he reformed had in other places and at other times already developed a dynamic in terms of specimen and information exchange (see the chapters in this volume by Daniela Bleichmar, Harold J. Cook, and Claudia Swan) that would carry it beyond his idiosyncratic visions. Rather than instituting incommensurability between metropolitan and peripheral knowledge frameworks, science *makes* knowledge frameworks commensurate through symbolic representations.[28] This was not, to be sure, an unencumbered process of open communication, as Londa Schiebinger points out in her essay. The case studies presented by her, Emma Spary, and Andrew J. Lewis, as well as the one I have presented, amply demonstrate that this process was loaded with potential for friction and conflict. However, this very potential attests to the power that science possesses in upsetting established social relations and in creating new, often unexpected ones. Science, with its dialectical ability both to upset and to objectify social relations, constitutes one of the motors of colonialism writ large, rather than merely serving as one of its instruments.

Chapter 3
Mission Gardens
Natural History and Global Expansion, 1720–1820

Michael T. Bravo

When the French navigator Jean François de Galaup, comte de Lapérouse (1741–88), bade farewell to the missionaries and the garrison at Monterey Bay (in present-day California) on 22 September 1786 to begin his long voyage across the Pacific Ocean to the China seas, he had every reason to be grateful to his Spanish hosts. The ships had arrived in need of fresh food, after sailing from Concepción in Chile north to Alaska and back down the coast. The *Astrolabe* and *Boussole* were now setting sail laden with fresh vegetables, wood, water, cattle, sheep, and poultry. For Lapérouse, like Louis-Antoine de Bougainville (1729–1811) and James Cook (1728–79) before him, the success of a voyage around the world depended on maintaining adequate provisions for the long oceanic crossings. Having access to colonial victualing stations around the globe was therefore crucial.[1]

Following defeat at the hands of the British in the Seven Years' War (1756–63), France's power overseas had been severely curtailed. In a strategic bid to explore new imperial economic opportunities, Louis XVI's Marine Ministry had commissioned Lapérouse to sail around the world under the banner of disinterested science. His official instructions from the French Academy of Sciences specified his route and hundreds of tasks for the expedition's scientists to undertake. Among these were orders from the king's chief gardener, André Thouin (1747–1824), detailing precise botanical procedures for procuring, transporting, and sowing European plants in new places and climates, and for returning newly discovered species to the king's garden in Paris.

When Lapérouse landed at Monterey Bay, he was returning to a frontier of the Spanish American Empire. Although the backing of the Academy of Sciences entitled him to safe passage and courteous treatment, he knew that the Spanish garrison, the nearby Franciscan missionaries at Carmel, and the Rumsen (indigenous) people were under no legal compulsion to grant him access to local resources.

Initially his visit had not looked promising. The local plants had "entirely dried up" in the heat of the California summer, "their seed being scattered on the ground." All that Jean-Nicolas Collignon (1762–88), the expedition's gardener, had been able to collect and identify were two varieties of wormwood, the male southernwood, Mexican tea, Canadian goldenrod, Italian starwort, millefoil, deadly nightshade, spurry, and water mint. Fortunately, however, the governor and the missionaries kept gardens that "were filled with an infinity of plants for culinary use." So rarely were these minor Spanish officials able to welcome illustrious visitors to this distant margin of the Spanish Empire that they generously showered on the departing ships all that they could "in such abundance, that our people had in no country been better supplied with vegetables." Though it was difficult to reciprocate adequately, Lapérouse presented gifts, and Collignon, charged by the Jardin du Roi with the job of introducing plants around the colonies, made a gift of Chilean potatoes "in perfect preservation," certain to "thrive in the light but fertile soil of the environs of Monterey."[2]

Offering up the mission's horticultural produce was indeed a sacrifice. Missions were seldom fully able to meet their own needs, requiring them to release captive Indians from their garden duties to gather wild plants further afield, as had been the Indians' traditional practice.[3] Lapérouse, though grateful for the supplies, was distinctly unimpressed by the missionaries' lack of skill and capacity to improve the conditions of the colony. Their education at the San Fernando seminary in Mexico, remarked Lapérouse, had taught them to be "more attentive to their heavenly than their earthly concerns" and meant that they "have greatly neglected the introduction of the most common arts." The consequence of this unworldliness was visible in the Rumsen population, who had been reduced to a state of destitution. Their de facto imprisonment in the colony reminded Lapérouse of Caribbean slave plantations.[4]

Lapérouse's reflections call into question the notion of economic self-sufficiency and its dependence on particular kinds of botanical skill and forms of social organization. In this essay I examine assumptions about skill and self-sufficiency in the context of mission gardens. In marked contrast to the Franciscans, some mission societies—the Jesuits in particular—placed great emphasis on specialized forms of natural knowledge as a criterion for selecting or training prospective missionaries. The Jesuit missionaries were, of course, renowned for their higher learning in natural philosophy and cartography. To provide their South American missions with vital medicines they recruited pharmacists, normally among the laity of their own ranks, to cultivate medicinal herbs in mission gardens.[5]

This chapter focuses on the missions of the Moravian Brethren, which, though much less well-known than the Franciscan missions, gave artisanal skill a much more prominent place in the process of converting and civilizing natives. The Moravian Brethren emphasized practical, artisanal skills, though the missionaries were in fact self-selecting volunteers. Curing illness and finding reliable sources of food were integral to the planning and realization of conversions. Hence the skills required to name flowers, vegetables, trees, and animals and to cultivate them, harvest them, and transform them into useful forms for everyday life were highly valued. Yet, peculiarly, with the exception of the Jesuits, scarcely any attention has been paid to missionary contributions to early modern imperial natural history, botany in particular.[6]

Missionary Artisanship: An Overview

By way of introducing my argument about missionary botany, I will focus on the eighteenth-century Moravian Society for the Furtherance of the Gospel among the Heathen. The Moravians considered skill an essential personal attribute for mediating between missionaries and prospective converts. It figured prominently in their conversion method as set out by August von Spangenberg (1704–92), who explained that missions established themselves as collectives near their prospective converts in order to attract them by example. Building missions required artisanal skills such as carpentry. To enjoy a reliable subsistence, the ability to grow vegetables in the mission garden was crucial.

My argument is that artisanal skills in field sites such as mission gardens were "place-responsive" in four ways. First, mission gardens were places where missionary and indigenous botanical traditions intersected, as though they were a practical space where the incommensurability between cultures could be overcome. Second, the range of possible botanical practices was contingent on local political, physical, and climatic conditions. Third, becoming economically self-sufficient at mission stations required the missionaries to make local links to landscapes and peoples. Fourth, where inadequate colonial botanical practice made self-sufficiency impossible, the missionaries were forced to reconsider the social organization of the colony.

Missionary botanical skills could take on a multiplicity of meanings, in ways not limited to those who were able to observe the missions firsthand. Seen from the social and geographical distance of powerful figures at the center of natural history networks, skill was an ingredient in generating useful knowledge. For example, Joseph Banks (1742–1820), the autocratic botanist who played a near-omniscient role in matters relating to the conjunction of British botany and empire, acquired for his private library those newly published mission travels that showed an awareness of natural history and opportunities for economic improvement. He teamed up with William Wilberforce (1759–1833), the famous evangelical and abolitionist, to encourage mission societies to make available useful natural knowledge from mission stations. For Banks, skilled missionaries represented a potential source of collectors and new information about indigenous *materia medica*.[7]

Although missionary botany was usually tied into large-scale, long-distance botanical networks, it was not always organized along the lines of empire. The Moravians offer a case in point. Although they received imperial patronage from the Danish and British Crowns, their strongest ties were with the natural history networks of Saxony—then an unimperious Germanic state with close economic ties to Poland. The tiny agricultural village of Herrnhut was the administrative headquarters for all Moravian settlements in Europe and had its own pedagogical network in Saxony: a high school (*gymnasium*) at Nietsky, a theological seminary at Barby, and natural history collections in Leipzig and Görlitz.[8]

Conversely, attentiveness to collecting at mission stations was a viable means for showing influential patrons that a mission society was deserving of the goodwill or protection of the colonial state and could help to combat the "trouble-maker" image that colonial governments, settlers, and traders often had of missionaries. For the missionaries, access to natural history networks promised potentially valuable institutional and political uses of their knowledge and skills. Mission gardens played an important role in formalizing the economic structure of a colony. The consumption of plants could never be entirely divorced from the skills of those who cared for them. Aside from providing a colony with sustenance, a garden gave missionaries a legitimate point of entry into global natural history networks. Participation in botanical patronage required reasonable competence in identifying, collecting, and preserving specimens. After all, wrongly named specimens, false claims to novelty, damaged cuttings, or specimens desiccated or drowned in the course of travel were hardly likely to impress patrons in Europe. Missionary access to natural history networks could not therefore be taken for granted and carried with it rigorous standards without guarantees of success.

The export of botanical knowledge was clearly a two-way process between missions and European botanical gardens and collections. The significance of receiving familiar Old World plants and animals in new locations (often tiny outposts on imperial peripheries) should not be underestimated. Seen from the missionaries' perspective, the relocation of European plants and animals at unlikely stations could ameliorate missionaries' sense of dislocation and self-understanding. Familiar foodstuffs could reinforce a sense of being needed by those in their home countries, as well as a sense of connectedness with other parts of the world. Grains and grasses, for instance, were part of an agrarian or pastoral material culture through which the meaning of many parables in the Bible could be brought alive in sermons and readings. Botanical products such as hay or bread were loaded with symbolic meaning and could therefore be used to introduce Christian religious concepts, convey moral values, and help to explain conversion. Translating the Scriptures for converts required finding vernacular equivalents for animals and plants. In other words, the presence as well as the naming of plants could signify differences between the material and spiritual lives of missions and the cultures they were seeking to transform.

If plants served to mediate between missionaries and their patrons, they could likewise mediate between missionaries and their local indigenous neighbors and colony members. Botanical knowledge can therefore also be seen in terms of skilled practices that helped missionaries to position themselves strategically in relation to those they sought to convert as well as to establish social and spatial proximity, or what the anthropologist Paul Rabinow in the context of fieldwork has called "adjacency." Skilled practice can be understood as a phenomenological set of relationships that emerge through performing an activity in a particular environment; an extension of the sensual reach of the human body that is exemplified by the early modern artisan. It therefore lends itself to a historical understanding of missionaries, particularly those for whom "heart and hand" were a touchstone of self-understanding, as well as a license to perform practical, technically demanding work. Skill is likewise exemplified in the person of the hunter-gatherer, but through a different set of relations between person and environment than in the case of the artisan.[9]

The study of the Moravian missions' contributions to natural history holds a particular appeal for historians of science, by virtue of the fundamental role the Moravian Mission Society assigned to artisanal practices. It engendered colonies where knowledge was productive, valued, and taught to apprentices. The centrality of mission gardens in the design of Moravian colonies also suggests that they played an important symbolic role as places where the natural world was tamed and made virtuous through skilled labor. It is tempting to conjecture that there was something peculiarly Protestant about this emphasis on artisanship. Was the organization of Moravian missions an expression of the values of "hand and heart" of early modern pietist artisans? Going one step further, one might wonder whether Max Weber would have viewed the missionary artisans as ideal protoindustrialists in his essay on the Protestant work ethic.[10]

Such broad generalizations about the skilled practices of missionaries, however, need to be tempered. Among eighteenth-century Protestant churches, Moravian doctrine was frequently viewed with suspicion and sometimes attacked in a manner approaching persecution. Natural history practice as a whole was rife with theological controversy, so that inferring from a general notion of missionary artisanship to a type of inward, pietist, spiritual disposition is interesting but should be approached cautiously. A more promising tack is to review the circumstances in which the Moravians began their missionary work.

Moravian Natural History Around the Globe

In the Protestant world of northern Europe, the Lutherans had established a permanent mission in Lapland as early as 1632 and would in future generations play a significant role in supporting the state's efforts to incorporate Lapland and its Sàmi people into the Swedish nation. A century later Linnaeus incorporated the missions into his peculiarly Lutheran-inspired "cameralist" vision in which botany served as an instrument for managing the natural resources of the crown. He idealized the Sàmi as a people living a rural nature, eating local foods, and hence enjoying good health and moral virtue. To achieve his "cameralist" ends, he recommended in 1747 that Uppsala theology students also attend his natural history lectures and that the Lapland Ecclesiastical Bureau require their trainee clerics to learn the art of agricultural improvement. Sàmi, properly trained, "could perform thermometric experiments when shooting snow grouse in the highlands," while the missionaries could make observations for the Royal Academy of Sciences.[11]

The Moravian mission movement emerged in the shadow of state Lutheranism, and in Denmark rather than Sweden. Frederick IV (1671–1730), sympathetic to the cause of pietism, actively recruited missionaries from Halle, the center of pietism in Saxony, and established the first Danish Lutheran missions in Tranquebar, India (1706), and in Greenland (1721). His son, Christian VI (1699–1746), embraced pietism wholeheartedly and extended to the Moravians the license to evangelize with overseas missions (the Moravians were a displaced people living in relatively small settlements across Protestant Europe after fleeing persecution in Moravia and Bohemia during the Counter-Reformation). Most Moravian missionary volunteers were German-born, reflecting the formative influence of August Hermann Francke (1663–1727), a professor of theology and oriental languages, on Nikolaus Ludwig, count von Zinzendorf (1700–1760), who sheltered Moravian refugees and established a new headquarters for them nearby at Herrnhut.

Christian VI first granted the Moravians permission to establish missions in the Danish West Indies (Saint Thomas, Saint Croix, Saint John [1732–34]) and in Greenland (1733).[12] These two settlements soon grew into a global missionary network, in the British and Danish Empires, with stations spanning distinct geographic and economic regions: Surinam (1735), the Cape of Good Hope (1737), the British West Indies (Jamaica, Antigua, Barbados, Saint Kitts [1754–77]), and Labrador (1771). Where possible, the Moravians made land purchases from the respective colonial states to ensure their missions' long-term access and autonomy. A widespread and lasting perception arose in Europe in the 1730s that the Moravians were emerging as a separate denomination (for example, ordaining their own bishops). This elicited many accusations—of enthusiasm and heresy, for example—and as a result the patronage of friendly states would remain crucial to the Moravians' legitimacy and even their very existence. Consequently their land tenure was closely linked to the question of their trustworthiness, especially in places that were politically sensitive. Relations between the state and plantation owners in the West Indies were paramount in shaping Moravian access to slaves and plantation lands, and in Labrador geopolitical rivalry between the British and the French also played a determining role.[13]

The Moravian strategy for recruiting missionaries placed faith first before other considerations. Missionaries were unpaid volunteers required to support themselves as far as possible through their own collective labor. This was a tall order that often placed great demands on their physical, mental, and spiritual strength. The moral and economic viabilities of the missions were intertwined. To satisfy their need to sustain themselves economically in the face of highly contingent and variable political, cultural, and physical realities, the Moravians adopted a set of universal principles that they then adapted to local circumstances. In economic terms, the Brethren exported a model of limited-scale, agrarian production aimed at collective self-sufficiency. The many artisanal skills required for a colony to thrive ultimately depended on a reliable source of food and medicine, and ideally this meant a well-managed, fertile garden. Successful missions required gardeners who were not only knowledgeable about transplanting foodstuffs but also capable of acquiring practical knowledge of the local flora: how to identify and cultivate edible plants; when to plant and harvest them; how to work different woods; and how to transform plants into food, cloths, fuel, and building materials. Mastering the use of local flora and fauna therefore involved the skills of carpenters, tailors, cobblers, blacksmiths, cooks, and gardeners—as well as indigenous cultivators and harvesters, as my examples will show.

The importance of artisanship to the Moravian ideology of land tenure figured in their strategies for recruiting converts. The importance to Protestants of free will in guaranteeing genuine conversions meant that coercion (where, for example, the Franciscan authority was supported by a local garrison) was to be avoided at all costs. For the Moravians, religious conversion was always the principal goal. Artisanship and natural knowledge were seen as moral and practical instruments for conversion, and hence subordinate to the teaching and acceptance of the New Testament. Artisanship allowed the missionaries to attempt to lead by example and to bridge the language and cultural barriers separating them from those they sought to convert. In some cases converts were trained to generate income for the missions. In the West Indies and Surinam, for example, the Moravians "trained and employed in some instances quite significant numbers of slaves" as tailors.[14]

Just as the Moravians were great champions of church music, believing hymns to be something akin to a universal language of the spirit, they believed that work promoted pious contemplation. Piety could therefore be imparted to converts through shared work while the missionaries were still acquiring the rudiments of the local languages. The Moravian missionary ideal incorporated a vision of natives and Europeans working quietly and harmoniously side by side.

These agrarian ideals were put to the test at each mission station. In the West Indies the land had long been heavily capitalized by plantation owners. The dream of the first Moravians—to give themselves into slavery so as to gain access to the slaves—was abandoned before the mission was begun. Instead the Moravians reached an accord with some of the more evangelically inclined plantation owners, who appreciated that obedient pietist slaves were less inclined to be rebellious. The Moravians in turn were prepared to preach that the duty of slaves to submit to their masters was entirely consistent with biblical teachings (unlike the Lutherans in the Danish islands, who opposed slavery).[15]

Moravian gardens in the West Indies followed two models of land tenure. Where the missionaries were entirely dependent on the plantation owners, their own food requirements relied on the plantations' systems of gardens. In addition to the cane fields, large plantations could have several different gardens, growing different kinds of produce differentiated according to social function, plant type, and space. The most detailed descriptions and analysis of the Moravian economic response to the prior existence of plantation gardens were recorded by Christian Oldendorp (1721–87), who visited these settlements for eighteen months between 1767 and 1769. In a panoramic account of plantation gardens, he wrote, "around the plantation buildings, there is generally a courtyard or garden" for fruit such as "bacove, bananas, and similar plants." Vegetable gardens, by contrast, were planted in separate courtyards. Plantations, catering to the taste of their owners for luxuries, typically had "coffee courtyards" for the owners' own individual crops. It was rare but not unheard of to have a "pleasure garden" that featured mainly flowers.[16]

Moravian principles of land tenure varied depending on whether the missionaries were reliant on the existing infrastructure of a plantation owner or whether they were successful in purchasing plantation land, in which case they enjoyed greater autonomy. The normal practice of plantation owners, according to Oldendorp, had been to give slave families their own garden allotments, delegating to them the responsibility of growing their own food and tending to these plots during their time off, normally on Saturday afternoons or Sundays. When they could produce a surplus, slaves were able to trade for other goods. Slave societies on plantations were divided along protoindustrial class lines into three hierarchical groups: house, craft, and field slaves. House slaves were "entrusted with the supervision of the food supply and similar activities." The craft slaves (for example, blacksmiths, tailors, carpenters) closely fulfilled the Moravian ideal of a self-sustaining population of prospective converts. Among these were the overseers such as the *bomba* (foreman in the cane fields), the sugar cook, and the rum distiller.[17] Field slaves worked the cane fields, skilled but brutal and exhausting labor.

With the missionaries trying to grow their own food and stave off disease, and the slaves enduring long hours of hard labor and strict curfews, finding opportunities to preach to them was frustrating. The missionaries responded in various ways, such as creating formal opportunities for "speakings" with the slaves.[18] Oldendorp recognized that winning over the powerful, hardworking *bombas* was crucial because they held authority over many other slaves and because they acted as intermediaries between slaves and the plantation managers. This strategy can be seen as an effort to forge alliances between artisanal classes (rather than between elites, as was often the case under colonialism). Opportunities to make conversions sometimes emerged en masse, perhaps demonstrating the roles of leaders or key figures in the slave society hierarchy.

When the Moravians were allowed to settle on plantation land, their mission house and garden were not necessarily located anywhere near the slaves. For example, the house constructed by Brother Schenke on Saint Croix was "a half-day's journey from Christiansted in a low-lying, entirely wild and desolate region on the boundary of the plantation." Transportation was difficult, and the thick bush allowed "numerous vapours rising from the soil" to cause many diseases. Rats were "present in great numbers, doing a great deal of damage to the vegetables," and the missionaries' "batatas and other kitchen vegetables" were perpetually vulnerable to guerrilla-style raids from runaway slaves who retreated to outlying areas.[19]

However, after acquiring the trust of many plantation owners (delivering intelligence to preempt slave rebellions was a factor), the Moravians were able to acquire plantation lands on both Danish and British islands, such as the garden land at Friedensthal in Saint Croix (see figure 3.1). Although their plantations were comparatively small, those on Saint Thomas and Jamaica were large enough for the Moravians to worry that they might be perceived as running a venture for profit rather than salvation. In that sense, the economics of Moravian land tenure differed in degree rather than kind from that of the large plantation owners. Plantation ownership allowed Moravians much more autonomy in defining their own way of life, but it also put them in the position of be-

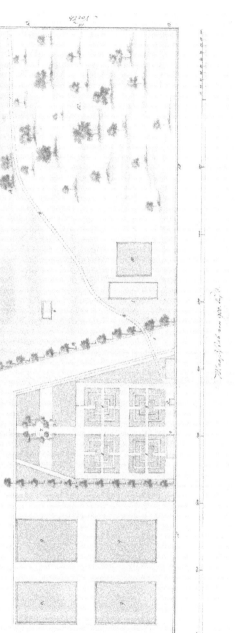

Figure 3.1. The Brethren's Garden at Friedensthal in Saint Croix (1757). In 1751 the Brethren were able to acquire this garden land, where Brother Ohneberg built his residence (building marked 2). He supervised and worked alongside slave converts to develop the coffee groves (8), vegetable gardens (9), and orchards (12). By 1757 Brothers Kremser and Töllner were organizing the construction of sugar mills. Reproduced with the permission of the Moravian Archives, Herrnhut.

coming slave owners. The missionaries transplanted and superimposed their ideal of collective self-sufficiency onto the global economy of sugar cane. The Moravian synod, when reviewing the morality of plantation ownership in 1769, found the connection between cane and conversion appealing, drawing a comparison between "Husbandry in Europe" and plantations as intrinsically "one of the most innocent and Natural Employments."[20] In order to finance their own presence, the Moravians, unable to fulfill their desire to be self-sufficient, turned to cane as a source of revenue. Faced with their own frequent bouts of ill health and periodic threats of famine, the missionaries chose to rely on *lamina*, the labor of field slaves.[21]

Oldendorp was among a small but growing number of mission visitors to appreciate the plantation "pleasure gardens" as well as being able to indulge his passion for walking and collecting around the islands. Unconcerned that the Danish West Indies, with "only a few educated people," could hardly be considered "a center of learning," he reflected that, for those who shared his taste for such pursuits, "the profusion of rare natural history phenomena provides them with sufficient means for their amusement."[22] Although Oldendorp was writing as an amateur, his interest in the islands and their botany bordered on obsession. He returned to Saxony with numerous notebooks and botanical specimens. So deeply concerned was he with the magnitude of his task ahead that his six-hundred-page, two-volume *Geschichte der Mission* (1777) would probably never have been published without the interventions of his editor and keeper of the natural history cabinet at Barby (near Magdeburg), J. J. Bossart, whom Oldendorp regarded as insensitive and heavy-handed. Nevertheless, the first volume, devoted to natural history, lent credence to the notion that Moravians were disinterested and hence trustworthy students of the natural world, having common interests with savants or physiocrats in Europe's centers of learning. The descriptions of the slave societies are expressed by making normative generalizations (for example, "Agriculture is by no means the favorite activity of the Caribs"[23]) that feign objectivity. The descriptions of the animal, vegetable, and mineral worlds are laid out clearly, with one section devoted to "Useful Plants," another to "Medicinal Plants, Building-woods, and Dyewoods," and a third to "Some Curious Plants and Flowers." The first volume also provided testimony to the Moravian presence in the Danish West Indies in a politically acceptable way to European audiences, at a time when the abolitionist movement was placing the islands under close scrutiny.

The world of eighteenth-century natural history in Saxony is still relatively unexplored territory, although the United Brethren have carried out considerable research. The end of the Seven Years' War in 1763 was clearly an important event for the Moravian Church, which could see that its interests could be well served by earning the patronage of the British state. The return to peace enabled the church to assert more direct control over its missions, send inspectors to check each colony's progress, plan further colonial expansion, and hold synods. During this period the synod planned the Moravian expansion from the Danish to the British Empire. Although the Moravians' linguistic and cultural knowledge had prepared them well to expand from Greenland to Labrador and from the Danish to the British West Indies, there were political and religious obstacles to be overcome in Britain.

At this time the Moravians had developed a regional natural history network that engaged in activities characteristic of other European networks: collecting rare plants, sketching them in their native habitats, transporting specimens and other curiosities back to Europe, organizing them in museum collections, and publishing natural histories as companions to mission histories. David Cranz's *Historie von Grönland* (1765), based on a six-month stay on the island, had established a literary model that would be followed by Oldendorp's *Geschichte* (1777).[24] The Moravians, encouraged by the success of Cranz's history, hoped that its timely two-volume translation into English would win popular support for their bid for a land grant in Labrador for new missions. Sadly for them, discerning British readers lapped up the first volume devoted to Greenland's natural history but largely ignored the second volume, devoted to the history of the missions.[25]

Cranz's Greenlandic history shows us the limits of the Moravian artisanal-agrarian model in a very different way from Oldendorp's history of the Caribbean colonies. Ironically the Greenlandic situation was better suited to the Moravian model because the missionaries enjoyed greater freedom to choose strategically the locations of their mission stations in relation to Inuit camps, trade routes, and the colonial fishing factories.[26] Although the gardens' positions within the colony varied, they were almost always located within the colony, in contrast to the undomesticated economy situated beyond the colony boundaries. The contrast between the artisan community at New Herrnhut, complete with its garden on the inside of the colony, and the mobile Greenlandic hunting and fishing culture on the outside neatly fit the Moravians' assumptions about the superiority of Christian agrarianism. Consequently, Cranz's depiction of the colony clearly identifies the garden and specifies its location in its layout (see figures 3.2, 3.3).

The Moravians had initially been optimistic about the potential for making Greenland productive, just as Linnaeus had been about Lapland during his tour in 1732. However, the Greenlandic soil and climate made its limits clearly known from the first season of their arrival. Cranz's study, reflecting this, was predictably short on botanical materi-

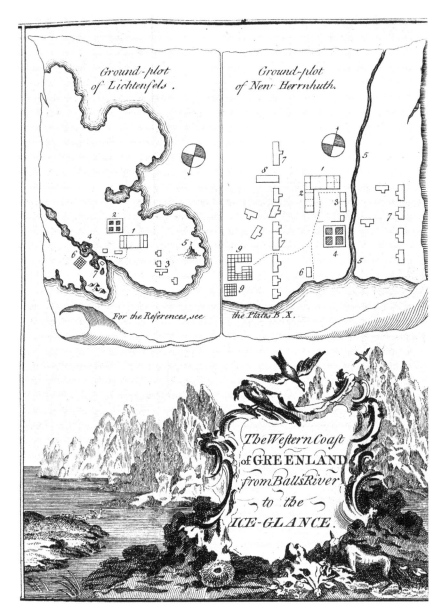

Figure 3.2. "The Western Coast of Greenland from Ball's River to the Ice-Glance," in David Cranz, *History of Greenland*, 2 vols. (London, 1767), vol. 2, 397. The use of space within the mission stations reflects the Moravian desire for order and civility, whereas the stations' location on the coast (protected from the interior by rugged mountains) mimics the maritime orientation of the West Greenlandic culture. Reproduced courtesy of the Scott Polar Research Institute, University of Cambridge.

Figure 3.3. "View of New Herrnhuth" in Greenland, in Cranz, *History of Greenland*, vol. 2, 397. Although the iconography of the garden is given prominence in the composition of the scene, the boathouse and the two people with spears or harpoons point to the greater importance of the seal fishery. Reproduced courtesy of the Scott Polar Research Institute, University of Cambridge.

als. The Greenlandic flora, he remarked, was, as might be expected, lacking in fertility.[27] Nonetheless, he was still able to describe "a few of the indigenous plants of this country, many of which are not common elsewhere." And if the soil was generally barren around the Greenlandic houses, it had been "dunged for years by the blood and blubber of seals," with the result that "plants of every kind flower most copiously, and grow to a large height." In addition to an abundance of mosses, lichens, and fungi, diminutive plants with practical benefits grew as well. Scurvy grass and cloudberries were plentiful (the missions were in southwestern Greenland) and reliable remedies for scurvy. Birches, where well sheltered, could grow to the height of a man, and willows could grow well along the ground, but neither burned nearly as well as turf, driftwood, or European coal.

Given the ubiquity of sea mammals and fish around Greenlandic camps, as well as kayaks, harpoons, and spears, it was made evident to readers that these were a maritime people who derived most of their food from the sea. These observations, however, also serve to expose a common myth about the Arctic regions—namely, that they are barren and of little botanical interest. The enthusiastic missionaries, undeterred, carried out trials to grow oats and barley in their new garden. According to Cranz, these crops "gr[e]w as fine and as high as in our countries," in spite of the short growing season between mid-June and September, but "seldom advance[d] so far as to ear" and even under the best conditions were prevented from ripening by the early onset of "frosty nights."[28] Settlers discovered techniques such as screening the garden from the wind and the seawater, and transplanting indoors come September everything except the hardy chives. In explaining the utility of plants, Cranz subtly defers to the experience of the Greenlanders who, for instance, used the moss to insulate their stone houses. A careful reading reveals that the most useful plants in the Arctic were gathered, whereas those that were cultivated with great difficulty had only limited use and could not easily be preserved.[29]

The missionaries' correspondence with Herrnhut shows that they and their families suffered greatly from hunger and cold, and depended on their supply ships and on the local Greenlandic hunting and fishing economy for sustenance. Given the inadequacy of cultivated plants, it is interesting to see how the garden (labeled 4 in the plan of New Herrnhut [figure 3.3]) was represented as a central feature of the mission, at least iconographically, while the boathouse (labeled 6 in figure 3.2 and 3.3) was introduced as a new architectural feature in mission iconography. The significance given the boathouse is heightened by attention to the relevant skilled practices and tools. Cranz is one of the first European naturalists to celebrate the kayak for the ingenuity of its design and the skills required to use it effectively. Of the two engravings in figure 3.4, the one on the bottom shows an array of Greenlandic hunting tools laid in the foreground alongside a kayak under construction, which reveals its normally hidden skeleton. The top engraving shows a hunting scene in which every instrument is stowed in its correct place. The hunter is deftly holding his paddle in one hand to balance the kayak, while taking aim at a seal with the harpoon in his other hand.[30] The scene aims to please Cranz's audience, but in its use of ethnographic aesthetic codes (such as the boathouse in figure 3.3) it domesticates the technologies of the sea in terms that complement and preserve those of the mission garden.

Conclusion

Moravian mission gardens reveal a great deal about the economic practices of the missionaries and the conditions in which they carried out their work to convert native peoples and slaves to Christianity. For the majority of missionaries, plants were important either as sources of food, as medicines to combat disease, or as building materials. In the cases examined above, while the missionaries aimed to cultivate their own sources of food, more often than not to meet those needs they were dependent on those whom they either wished to convert or had converted.

The Greenlander in his Kajak.

Figure 3.4. "The Greenlander in His Kayak," in Cranz, *History of Greenland*, vol. 1, 140. The tools and hunting implements emphasize the skill required to construct, maneuver, and use a kayak. For the missionaries, the kayak became an emblem of Greenlandic artisanship, disciplined and self-sufficient. Reproduced courtesy of the Scott Polar Research Institute, University of Cambridge.

Many missions, such as those at Carmel or Saint Thomas, were officially or unofficially places of incarceration. Loss of mobility and personal freedom among the converted was a general condition of most missions. Oldendorp was quick to admit that the black slaves "seem to have the most extensive knowledge about the healing power of these plants" and that "even European physicians do not hesitate to learn as much as they can from them."[31] Hence the missions' dependence on their target populations for their infrastructure—in particular their food and medicines—contributed to the complexity of hierarchies and the differentiation of privileges among social groups.

The importance of mission gardens had a strong geographical dimension. The depictions by Cranz and Oldendorp showing the precise layouts of the model Moravian colonies reflect the significance attached to the differentiation of space as a means of organizing the colonies' activities. The visual code, however, used to identify the gardens and other features as static and enclosed spaces, belies the importance of mission gardens—like other botanic gardens—as interfaces between a colony's interior and exterior spatial relations. In spite of the Moravians' desire that the gardens embody ideals of self-sufficiency, they were part of a much more spatially extensive and socially complex economy that linked the area surrounding missions to economies spanning other parts of the globe.

The distinction between the interior and exterior of missions was powerful for missionaries and reflected ways of thinking about the inner life of the soul and its relationship to the outer world. That the Moravians in particular idealized the struggle between internal and external worlds through the figure of the artisan is worth reflecting on. Artisanship, as the Moravians conceived it, was a normative, moral category rooted in a religious culture that embraced the values of pietism. However, this romanticized image of the artisan, epitomized by a pietistic desire to achieve self-sufficiency through collective labor, led the Moravians to co-opt and even incarcerate the indigenous other. Keepers of indigenous gardens, though recognized as skillful, were too often stripped of the ownership of their labor. With that in mind, historians cognizant of the great importance of artisanship in early modern social history should resist the temptation to treat it as a universal category, even though skilled practice, construed in different ways, has arguably been a conspicuous characteristic of all societies. The desire to recognize the contributions of artisans often subsumed by the cultures of gentlemanly science creates new problems for writing the history of colonial botany. Artisanship was in fact an instrument of missionary ideology, as well as a category of self-understanding, which helped to justify the Moravians' ownership of slave plantations in the West Indies and the transformation of hunting cultures such as the Inuit of Greenland.

Chapter 4

Gathering for the Republic
Botany in Early Republic America

Andrew J. Lewis

In January 1814 Benjamin Smith Barton (1755–1815), professor of natural history, *materia medica*, and the theory and practice of medicine at the University of Pennsylvania, received a letter from H. B. Trout (dates unknown), a frontier resident of western Pennsylvania, asking him about poppies. Trout, a correspondent whom Barton did not know, praised Barton's botany textbook, *Elements of Botany*, "wherein it is stated that the *peiparer sommiferum* [*sic*] might be cultivated in the countries of the Unighted-States [*sic*] with much pecuniary profit." Trout asked Barton for seeds, directions for their cultivation, and gardening tips in time for the following season: "What climate of the Unighted-States [*sic*] would be most favourable to the growth of the poppy—what sort of manner [*sic*] would be best calculated to put on the ground in which the poppy is to be sowed?" Trout believed that he could capitalize on a growing American demand for opium, a medicine frequently prescribed by doctors at the time and imported at great expense. For Trout, poppy cultivation provided an opportunity to correspond with an eminent naturalist and represented to this frontier resident a means to become an entrepreneur.[1]

Trout was not the only citizen of early republic America who hoped to make natural knowledge profitable. Unsolicited letters from ordinary Americans to eminent naturalists, letters that sought information about which plants to cultivate and how to tend them, are commonplace in archival collections. References to the receipt of these letters and discussions of them are, likewise, frequent topics in the correspondence of naturalists. Historians have overlooked these exchanges and their larger meanings, in the interest of other narratives of botanical practice in early republic America: transatlantic exchange, the beginnings of botanical gardens and botanical societies, and classificatory expeditions into the interior. These letters offer a window into the inner workings of early

American botany, a set of practices that closely resembled those pursued by European botanists and their colonial counterparts but that took place at a historical moment when ordinary and elite Americans were self-consciously crafting a national identity, reconfiguring individual identities to fit new political allegiances, and cataloging the new nation's flora. The following examination will reveal how unbalanced relationships—colonial relationships—shaped and continued to shape American botanical practice in the forty years following the end of the American Revolution in 1783. Moreover, it will suggest some of the ways that American citizens, elite and ordinary, adapted, or at least attempted to adapt, a hierarchical, mercantilist, and imperial botany practice to a rapidly democratizing nation.

A Revolution in Botanical Practice

Early republic American naturalists, the botanically inclined especially, imagined themselves at the center of a vast network of North American plant collectors, a web that stretched from the urban centers of Boston, New York, Philadelphia, and Charleston to the ever-expanding frontiers of the American West in the Ohio and Mississippi valleys and, for a few, into the Caribbean and the West Indies. Former colonial collectors for Europeans and now citizens of a republic, American naturalists regarded the American Revolution as a watershed for science that provided an opportunity to reinvent and reform scientific inquiry and methodology. Naturalists worked to replicate—but alter when necessary—the knowledge-gathering patterns of European botany and natural history. The changes they advocated and made, however, were more rhetorical than substantive. Many early republic naturalists claimed to champion a more democratic natural historical practice. As Francis Hopkinson (1737–91) reasoned to Philadelphia's preeminent scientific organization, the American Philosophical Society, in 1784, "the language of nature is not written in Hebrew or Greek," and a correct understanding of the natural world "is not involved in the contemptible quirks of logic, nor wrapt [sic] in the visionary clouds of metaphysical hypothesis"; rather, "the great book of nature is open to all—all may read therein."[2] But revolutionary enthusiasm went only so far. Just as Benjamin Franklin (1706–90) modeled the American Philosophical Society in 1743 after London's Royal Society, so early republic naturalists modeled their epistemology on European precedent.

In the years following the end of the American Revolution, American botanical enthusiasts enlisted and encouraged the assistance of domestic naturalists, physicians, military men, travelers, and gardeners as collectors for their own projects. American botanists requested samples of plants from throughout the new nation. Collectors would be rewarded with classificatory identifications and suggestions for potential uses, and they might be credited with the discoveries, and thus granted a form of immortality. Barton and the few other naturalists with his skill and European education, social standing, and contacts, both domestic and foreign, actively advocated a scientific practice that for years had been sustained through patronage and driven by a mercantilist and imperial logic. Barton imaged himself an American scientific savant. With assistance from the American Philosophical Society, he encouraged western physicians, military officers, and ordinary frontier residents of the new nation to collect samples for him and other Philadelphia naturalists. The specimens would be added to the national plant catalog, and botanical authorities would judge submissions individually for their worth and value to the nation: How to classify the plant? Was it a source of food? Might it have medicinal qualities? Could it become an exportable product? In short, Barton and other elite naturalists attempted to re-create the methodology of colonial botany in a democratic republic.

Barton and his fellow botanists confronted a society of artisans and farmers, however, who sought economic and political freedoms at odds with the dreams of their social betters and who possessed their own ideas about plants derived from experience, gardening, folk knowledge, Indians, and local usage. Individuals such as H. B. Trout looked to botany and to botanists to settle matters of local dispute as well as for economic advice; they were interested in the national flora catalog but ambivalent about becoming unpaid collectors for eastern elites. If fortunes could be made from the landscape—and the promises of naturalists suggested that they could—then those who possessed the land containing the plants should garner the rewards, ordinary Americans thought, rather than distant naturalists with specialized knowledge. In short, botany enthusiasts encountered a democratizing republic reluctant and at times resistant to adopting a hierarchical botanical practice. Embedded in letters from individuals such as Trout and in other early national discussions of plants are ideological and economic debates of the highest order: who were the rightful owners of the knowledge of plants, and to whom would the seemingly untold profits of an unexplored continent belong?

Naturalists were confident that they had the answers to these questions: the owners of this knowledge were those who classified the plants definitively, rather than the "vulgar," who confused species, misunderstood identifications, embellished stories, or dissembled about the plants they possessed. Furthermore, naturalists believed that the profits belonged to all Americans, as long as those who identified the resources got a cut. Naturalists considered themselves uniquely positioned—

situated in regional economic centers and intellectually prepared with botanical training—and offered the new nation their identification skills to classify its flora and fauna in the years following the 1783 Treaty of Paris. They reasoned in private and in public that their classificatory expertise would assist the republic in the effort to catalog its natural resources. In turn, these resources—various plants, minerals, and other raw materials—would provide the economic foundation essential to the fragile nation's political success. To hasten the acquisition of information and to establish a base-line catalog of the nation's nature, natural historians encouraged ordinary Americans to communicate with individual naturalists and scientific societies and invited submissions. In return, frontier Americans were promised, but did not always receive, putative lucrative "useful knowledge": classifications and information that would transform unidentified natural resources into valuable commodities.

In books, literary magazines, newspapers, broadsides, and personal letters natural historians such as Barton appealed to individual acquisitiveness and economic nationalism to generate interest in their practice. Characterizing North America as a continent of unrealized profits, they expanded on preexisting assumptions concerning nature's potential, portraying its imagined limitless resources as an individual's springboard to prosperity. If the intoxicating admixture of aspiration for wealth and confidence in natural plenty enticed ordinary Americans to practice natural history, the utility and efficacy of natural history and botany were demonstrated and reinforced as the practice proved a reliable economic predictor for a few commodities. When natural historical prognostications turned lucrative—as they did with a few objects, such as fossil remains and mastodon skeletons—natural history's authority increased. In a few cases individuals who located objects of extraordinary value were catapulted into local and national celebrity, generating for the discoverer social capital as valuable as the money received. However, many early republic Americans appear to have questioned the natural historical enterprise when its assessments challenged botanical folk knowledge, when it resisted settling matters of local dispute, or when it failed to confirm the scientific importance of curiosities or wonders. The easy calculus of natural knowledge leading to individual and national wealth floundered as botanical determinations exposed economies operating on different terms: a local economy of traditional medical practice, folk knowledge, curiosities, and marvels; and a broader, rapidly maturing capitalist economy determined by consistent supply. As a result, the intimate association between natural history and commodification enlisted to generate interest in and spread the practice, embroiled its authority in disputes over the definition of worth and conflicting ideas about value. Early republic Americans found themselves caught between old and new economies, modes of exchange exposed by different epistemologies and scientific practices and determined by the cycles of supply and demand.

Methods for Promoting Botany

Natural historians of early republic America attempted to advance a Baconian epistemology that comprised the gathering of individual facts that might lead to a discovery of the laws determining natural processes. Naturalists of the era were not mere fact collectors, however; they were compilers of knowledge. Like Bacon, who saw science and commerce working together, naturalists also considered themselves vital players in the economic success of the nation and believed that their knowledge and expertise were essential to establish its fiscal independence. More than one naturalist agreed with Nicholas Collin (1746–1831), a Lutheran minister and member of the American Philosophical Society, that although they might be "citizens of the world," American natural historians were duty-bound "to cultivate with peculiar attention those parts of science, which are most beneficial to that country in which Providence has appointed their earthly stations"; they may be curious about plants and animals in foreign lands, Collin wrote, but "Patriotic affections" required attention to objects closer to home.[3] Cursory examinations, anecdotal reports, and two centuries of European colonization convinced Collin and other native-born Americans that North America possessed the resources necessary to support and sustain the new nation economically. Yet many believed that more resources remained unclassified and underutilized. It was the naturalists' responsibility to gather information about these resources and learn how to bring them to market. First, they worked to assemble a base-line catalog of American nature, collecting information from distant and varied sources about the nation's plants, animals, and minerals. Second, they aimed to educate their fellow citizens in the identification and utilization of the products local to them.

Natural historians wrote simple keys to plants and trees as well as tracts aimed at improving agricultural methods, perhaps the most famous of which prior to 1800 was Humphry Marshall's *Arbustrum Americanum* (1785). Marshall (1722–1801), a Philadelphia resident, collapsed the two dominant natural historical goals into one text, which opens with a list of mid-Atlantic native trees and shrubs according to Linnaean classifications and later provides a description of their uses in "Medicine, Dyes, and Domestic Oeconomy." Marshall consulted "Botanical Authors" for information but intended his book for an audience without formal training. He explained that botany was the branch of natural history comprising "the right knowledge of Vegetables, and their application to the most beneficial uses." Botany was an object of study that merited the "attention and encouragement of every patriotic and liberal mind," particularly those interested in the nation's welfare. His countrymen were aware that the "continual enormous expense" of imported botanical products drained the national coffers. There was plenty of plant folk knowledge in circulation, Marshall admitted, but he explained that "observations and researches founded upon, and directed by, a knowledge of Botany" would meet a degree of success higher than what Americans "may gain by tedious experience, or stumble by chance upon . . . respecting the uses and medicinal virtues of plants." The "more general knowledge we obtain of the character and appearance of plants," Marshall wrote, "the more likely we shall be also to encrease [sic] our knowledge of their virtues, qualities and uses." Aware that his readership would be more "embarrassed and confused than profited" by technical language or terms, Marshall adopted the "most plain and familiar method . . . to render the work as generally useful as possible." His vernacular descriptions of plants and trees likely made botany more inviting to the residents of the new republic.[4] Marshall's dual attention to botanical classification and *materia medica* was typical of the period's natural history literature. Generally, however, the literature's efforts to classify and educate aimed also to eradicate alternative epistemologies. In particular, natural history literature distinguished itself from almanacs or collections of folk knowledge. Whereas almanacs might provide useful information, they collected it unsystematically and without design. The method he was championing, Marshall argued, would reveal truths obscured by older collecting techniques.

Natural history and botany enthusiasts considered the subject barely broached, Marshall's efforts notwithstanding, and early republic naturalists called for further action. "It is our duty to study [Nature's] ways, in order that we may know what is meant for our particular benefit," Charles Willson Peale (1741–1827), the artist and Philadelphia museum owner, told listeners in a public lecture in which he outlined his philosophy of natural history. The study of natural history is, he argued, "not only interesting to the individual," but a national priority; it held the potential to propel the nation toward economic independence, every citizen and profession benefiting from the pursuit.[5] To maximize these benefits and to understand their country more fully, natural historians cataloged locales—some as small as the countryside surrounding a city, others as large as individual states. Thomas Jefferson's *Notes on the State of Virginia* (1787) was the most famous and typical of these local natural histories. In it, Jefferson (1743–1826) described Virginia's boundaries, rivers, ports, mountains, "cascades," animals, vegetables, minerals, and climate. He plodded though subcategories of navigable and nonnavigable rivers, caves containing or without minerals, and plants sorted by their uses: medicinal, "Esculent," ornamental, and those "Useful for fabrication."[6] These lists provided evidence of Virginia's natural bounty and suggested that the state, and by extension the nation, could support itself on its own natural resources. When other authors turned their attention to New Hampshire and Vermont they followed Jefferson's example. "Few persons in this country, have studied natural history as a science . . . to the extent which is desirable," wrote Jeremy Belknap (1744–98), author of *The History of New-Hampshire* (1792); "it would be unpardonable not to take notice of its natural productions."[7] Both Jefferson and Belknap considered an examination of the natural resources vital to the success of the young nation and regarded natural history as a means by which to decipher nature's instructions for use.[8]

Natural historians' cataloging efforts sought to legitimize their knowledge and practice by demonstrating their utility to their fellow citizens. If natural historians could identify new sources of income for individuals and the nation, the endeavor was more likely to be regarded as valuable. Americans of all stripes would be drawn to it, and its spread would take advantage of innate acquisitiveness. Exposure to and enlistment in the cataloging efforts would generate the governmental and popular support they sought, natural historians believed. It would also spread natural history techniques and knowledge abroad, facilitating assessment of the country's natural history.

In many of the state and local natural histories, Jefferson's *Notes* and Belknap's *History* included, authors drew on accounts provided by fellow residents, soliciting and receiving information about areas unknown to them. Authors scrutinized and then included, qualified, or discarded descriptions of animals, plants, and trees. Reports that came from men of established authority were included without qualification. By contrast, anecdotal reports that were unverifiable or related by those who had "interests"—financial or otherwise—in the stories were cast aside or treated with caution. In addition, reports of Indian plant use were frequently dismissed as superstitious, ineffective, and often as representing knowledge already lost, "the combined effects of warfare, civilization, and amalgamation with the whites" having decentered Indian *materia medica* and *materia alimentaria*; in short, reports by the "vulgar" of plants' medicinal properties were to be doubted, and more often than not, the properties they described were thought to exist only "in the imagination of the credulous people who employ them."[9]

Enlisting Ordinary Americans

Natural histories of place and the reports they contained provided a foundation for a catalog of American natural resources, but the most energetic advocates of natural history called for greater effort. "Little [attention] is paid to the study of nature in the United States," Barton wrote in his *Fragments of the Natural History of Pennsylvania* (1799). Instead, educators spent time on "languages which are withered or dead." Barton lamented that natural history was not considered an "indispensable branch of polite or useful knowledge." Were it pursued "with a portion of that innocent and useful zeal with which it is cultivated in Europe . . . in less than twenty years, the animal, vegetable, and the mineral productions of the United States would be pretty well investigated." Instead, the "pursuit of gain" occupied the majority of American people; natural history was left to the "labours of two or three individuals, unaided by the public, and trammeled by professional engagements and pursuits."[10] Faced with discouraging prospects for natural history, practitioners emphasized forcefully the commercial benefits of their practice. Barton and others aimed to meet those interested in gain on their own terms.

To facilitate that process, natural historians and scientific societies solicited natural history samples from ordinary Americans, coaxing information with lucrative promises. Philadelphia's short-lived Linnaean Society issued a public notice for Americans to send "any plants, ores or any mineral substance whatever" to the organization, where they would be examined by the "botanical and mineralogical departments of the society." Members of the Linnaean Society believed that their knowledge would "assist in obtaining a full knowledge of the medicinal and dying drugs indigenous of our soil." It would "expedite the discovery of useful metals; to aid the manufactures of their country," all the while removing "the inconveniences and disadvantages of individuals, not possessing an acquaintance with natural knowledge." In return, the examination results would be communicated to the individual transmitting the specimen, "together with such information relative to its nature and uses."[11]

Whether any individual submitted specimens to the Linnaean Society is uncertain; its records no longer exist. Other natural history organizations made similar appeals—among them, Philadelphia's American Philosophical Society, the Lyceum of Natural History of New York, and Boston's American Academy of Arts and Sciences. The lofty goals of these appeals contrasted with the recorded results because ordinary Americans either misunderstood what the solicitors sought or differed over what commodities possessed value.

Some ordinary Americans saw these invitations as a way to satisfy their curiosity. One Gabriel Crane submitted an account of a grub to the American Philosophical Society. He reported that he had found a worm in a "state of vegetation" and that his examination had caused the worm's destruction; moreover, he submitted, his theory was that the worm was the sperm of a much larger animal. Crane related his discovery after "having seen an advertisement containing a general invitation to make communication to the American Philosophical Society."[12] Members of these organizations regarded requests such as Crane's— among them submissions of beans with "unusual qualities," worms pulled from inside children's ears, and a treatise purporting to overturn Isaac Newton's "Theory of Universal Gravitation"—of little value to the process of nation building.[13] Still, these submissions suggest the degree to which early republic Americans differed over the interpretation of natural history. Ordinary American writers sought to make their objects of curiosity known, but the natural historians reviewing the submissions considered them anomalous and unimportant. Naturalists imagined themselves using their skills and knowledge to make the natural world predictable, determine patterns, and locate sources of national wealth; they were not interested in identifying curiosities.

American natural historians of the late eighteenth and early nineteenth centuries had embraced the European recategorization of intellectual and emotional responses to wonders. Like their European counterparts, they pushed curiosities to the margins of natural history practice and collection and concerned themselves more with representative samples and objects illustrative of larger groupings.[14] American natural historians' cataloging efforts were part and parcel of this shift, though disputes and individual episodes of strange phenomena demonstrate that curiosities and wonders continued to occupy their minds. Still, natural historians doggedly pursued an intimate knowledge of America's landscape and its contents, particularly in those regions where one might find plants and minerals essential to national prosperity. More than one early republic natural historian lamented that fellow Americans have "as yet but little taste for Natural History," even the "most simple observations," one investigator wrote, "leaving but few traces upon their minds."[15]

A persistent interest in unique objects and unusual natural processes dominated the correspondence of those writing to the elite naturalists. The divergent emphasis in the understanding of natural history practice helps to explain the persistence of early republic curiosity stories and unique objects. It also suggests why natural historians so infrequently responded to stories of magical beans or frogs found inside rocks: this was an epistemology and practice they were attempting to eradicate, each submission forcing natural historians to admit the remoteness of success. Conversely, what those who wrote to natural historians hoped to gain is difficult to determine. Likely they expected a more complete understanding of the objects they possessed, legitimization of their efforts to preserve them, perhaps a valuation of their objects, and confirmation of their roles in the larger cataloging effort. In all probability, solicitors received no response, natural historians dismissing these requests out of hand.[16]

Ideals and Realities

Naturalists may not have succeeded in cultivating the sort of national network for compiling botanical information they wished for, but in matters of mining and ores they came closer to attaining their goals. Most Americans knew little about geology or mineralogy and contacted natural historians to have deposits or discoveries assessed. Since there was little anecdotal speculation about rocks and ores in circulation, naturalists judged themselves able to establish an authoritative voice on these matters. One such writer was Charles Creswell, a resident of western Pennsylvania. In 1809 he wrote to the American Philosophical Society soliciting their "deliberations and judgment on a matter that has excited a Curiosity." After heavy rains, Creswell wrote, "a large mist resembling in appearance and smell of a coal pit," would rise from the earth. Creswell thought that this was likely a naturally occurring oil deposit and wondered how he might exploit it. Likewise, a Mr. Brown wrote to Barton from Camp Legionville, Pennsylvania, in 1792 to report the discovery of iron ore and "black lead" in a hillside that "appears to have suffered a violent degree of heat," indicating larger subterranean deposits. He asked Barton's advice as to whether the find might yield more iron. In 1808 D. R. Patterson of Virginia wrote to Barton, "I own the quarry of Marble which Mr. Jefferson has described in his *Notes*"; the only one in Virginia and never "worked," its quality was "infinitely superior to any American marble hereto discovered." Patterson claimed that "several good judges say it is equal to the Egyptian & Italian marble." He contacted Barton because a "Gentleman" sought to purchase it but Patterson claimed ignorance of its value. He asked Barton to inform him "of the value of the quarries in your neighborhood, or the prices at which they usually sell, from which I could estimate the value of mine." In these cases naturalists drew on the collected knowledge of American geology to predict the viability of various mines and resources, fulfilling their hope that natural knowledge would be used to establish the economic foundation of the young nation.[17]

Natural historians similarly welcomed the opportunity to authenticate claims of lucrative gold deposits, specifically requested by those with stakes in North Carolina. William Thorton, an individual who owned "very extensively" in North Carolina, was of the "opinion that my Purchase contains an immensity of Gold." In 1805 he wrote to the American Philosophical Society with an account of his mines, hoping that they would assure him and potential shareholders that he was "without any danger of deceiving myself or others" that his stake possessed "incalculable value." Thorton enclosed a pamphlet from the North Carolina Gold Company describing the geologic characteristics of the area and awaited the society's assessment of his deposits. Joseph Richardson transmitted "statements of the quality of some native gold found in the State of North Carolina" for the society to consider. He wished to learn from them any information concerning the quality of future excavations. He was concerned that a "large piece being inferior in quality to the smaller grains" indicated that "the larger lumps are not so pure as the smaller grains." He hoped that the society could confirm that this was a "groundless suspicion." Thorton and Richardson legitimized their value claims by relying on the geologic expertise of natural historians; they enlisted the authority of natural history to stabilize the volatile and speculative economy surrounding gold. Natural historians, in turn, welcomed such solicitations, which authenticated their expertise on matters of natural resource value and provided them detailed information about the geology of regions unknown to them.[18]

Coal, ores, and gold were certainly valuable to nation building. Nonetheless, in his *Collections for an Essay towards a Materia-Medica of the United-States* (1798), Benjamin Smith Barton wrote that the "man who discovers one valuable new medicine is a more important benefactor to his species than Alexander, Caesar, or an hundred other conquerors." "All the splendid discoveries of Newton," he continued, "are not of so much real utility to the world as the discovery of Peruvian bark, or of the powers of opium and mercury in the cure of certain diseases." Barton urged young physicians, naturalists, and gardeners to search for new *materia medica*; the prospect for its discovery was "particularly happy." Barton wrote that "the volume of nature lies before you: it has hardly yet been opened: it has never been pursued." Botany and, more specifically, *materia medica* he called "the *punctum saliens* of science in our country," and he encouraged his students and his countrymen to explore the fields and forests of North America for discoveries that would result in the "happiness of one's country" and "add luster to your names." Specifically, Barton encouraged the search for New World analogues to Old World medicinal plants used in the tonics and emetics to treat yellow fever and other fatal epidemiological diseases.[19]

Few medicinal plants received more attention in early republic

America than American columbo. The columbo root, an expensive import from Asia, was an emetic and purgative important to the dispensar-
ies of all physicians. Rumors emanating from the trans-Appalachian
West convinced Barton that the New World possessed an analogue or an
American columbo. However, as Barton and others quickly became
aware, more than one plant was masquerading as the American columbo, eventually classified as *Frasera Walteri*. Two others, *Hydrastis Canadensis* and *Zanthorhiza apiifolia*, were also said to be the American columbo root. Excitement over American columbo was linked directly to
its purported uses. Among western residents columbo was used "in the
heat of the summer, [to] put a stop to a wide and fast spreading gan-
grene"; it was also rumored to be used by Cherokee Indians to treat cancer. Some used it to alleviate inflammation of the eyes; others told of its
use in cloth dyes.[20]

Identifying the American columbo frustrated some of its investigators.
After receiving a small specimen of *Zanthorhiza*, a plant thought "less
pure than the Columbo," Barton believed that it was "in certain cases,
to be preferred to that celebrated bitter." The *Zanthorhiza apiifolia*, commonly known as the parsley-leafed yellow root, grew commonly in the
Carolinas and Georgia but rarely, if ever, in the Ohio valley. *Zanthorhiza*
was a species often conflated with *Hydrastis Canadensis*; the fact that the
latter was often called "Yellow-root" in the vernacular undoubtedly
added to the confusion. But *Hydrastis* grew in "various parts of the
United-States; particularly in the rich soil adjacent to the Ohio and its
branches, in the western parts of Pennsylvania and Virginia; and in Kentucky," the region being settled by Americans most rapidly during the
period. Barton considered the *Hydrastis* a "powerful bitter: perhaps not
less so than that of the *Zanthorhiza*"; initial research into the plants' utility was promising, and Barton urged Americans to search out and to
send him examples.[21] Individuals answered his request, transmitting
specimens accompanied by urgent pleas to identify their submissions.
More than one correspondent was eager to be the first to harvest the
root and reap the material rewards. Frontier Americans anticipated
finding and harvesting the columbo, becoming local if not national heroes, and receiving the material rewards consistent with the discovery.
Barton's rulings on these matters disappointed most, however, dashing
their hopes of quick wealth.

John Beatty of Cambridge, Ohio, was frustrated to learn from Barton
that "the root I sent you is called *Frasera*[,] a term I don't understand
nor can I find any Doct[or] that does." (Unbeknownst to Beatty, he had
located American columbo). Undeterred by the discouraging information, Beatty wished Barton to be more precise: was his "the real Columbo or not?" He tried to cajole a rapid reply, "as now is a good time

to gather it." Should Barton think it columbo root, "pray let me know [how] much it is worth a pound as I intend to gather and cure a quantity if I find it [and] turn it to my advantage." One of the best doctors in the western country told Beatty that his sample was the "true Columbo." However, with misinformation swirling and sharp dealers ever present, "I depend more on your judgement," Beatty wrote. In a postscript Beatty once again underscored the urgency: "be explicit about the root as it is a good time to gather it now the tops are just showing above ground as my son was that way and brought me some."[22]

In the years surrounding Beatty's letter, Barton was one of a few naturalists engaged in an effort to determine whether various roots growing in the Ohio valley might be the lucrative columbo and part of a larger effort to categorize the *materia medica* of the United States. Yet records show that rumors, exaggerated stories, and confusion characterized and impeded that search. One correspondent enclosed "seeds of the plant found in the Western country which is said to possess the vertues [*sic*] of the Columbo root." The writer found it in "swamps" but also "on the tops of hills." Paradoxically, it grew both "in the richest bottoms and on the poorest ridge," and he observed that it "grows best in rich loose soil, & that no animal will eat it." The writer wished Barton to give his opinion and awaited a reply.[23] There is no record that Barton responded to this correspondent, but confusion waned for metropolitan naturalists once specimens replaced anecdotal reports. Still, distance and a dearth of trained naturalists in the West could do little to reduce frontier uncertainty over plants.

Richard Brown, a former medical student of Barton's who lived in Louisville, Kentucky, sent Barton a specimen "of what the people of this neighborhood call the Columbo root. It grows here in great abundance." He described what he knew of the plant but admitted that he had never seen one and apologized that "my entire ignorance of botany renders it impossible for me to give you a better description." The people of the region asked him frequently to identify the root and to authenticate their finds. Unfortunately, Brown had been "obliged to answer doubtingly," not possessing the skills to say whether it was or was not the true columbo root. "It would give me much pleasure to hear your opinion of it," he wrote. He enclosed a packet that contained a specimen he believed to be *Orybanche Virginiana*. The people of Kentucky, Brown wrote, "call it the cancer root and Beech drop" and used it to treat various diseases. Again he asked Barton if he would "do me the favor to let me know your opinion of these points. They have excited a great deal of enquiry in this country and I should be happy in having it in my power to pronounce an opinion respecting them with confidence." Though he was a physician, Brown admitted, "It is subject of great mortification to me that I have never studied botany." The western country "is such a field for research" that knowledge of the science, Brown mused, "would be to me a source both of profit and pleasure."[24]

Conclusions

Brown, like many others in early republic America, embraced the fusion of natural knowledge and economic profits proposed by natural history advocates. He and his neighbors in Kentucky looked to the fields and forests as landscapes where natural knowledge and entrepreneurial zeal could blend seamlessly. For some of those who gathered plants, fossils, and ores, the location of lucrative resources resulted in financial windfall. For others, success proved more elusive, with confusion rather than clarity characterizing the transactions over putative discoveries. Stories such as that of the columbo root suggest that appeals to the most prominent adjudicators on these matters were only one means to determine the identity of plants; and evidence suggests that these prognostications were often disregarded or manipulated, raising questions and possibilities about the efficacy and limits of botany and natural history that surely troubled its advocates. These moments of dispute, perhaps imagined as the optimum opportunity to establish epistemological authority, instead made manifest how tenuous was botanists' ability to settle these controversial matters. The authority of natural knowledge, Barton and others came to realize, was compromised and challenged by the authority of the early republic marketplace—an economy that was rapidly transitioning toward capitalism even as older forms of exchange persisted. If a purported root sold as "true columbo," natural historians could do little to dissuade the possessor that he purchased something other than the authentic article. Local knowledge, medicinal efficacy, and faith that a plant was what one believed it to be were powerful means of shaping belief. Misinformation, misunderstanding, and manipulation came to typify early republic natural history transactions as natural historians faced the results of so closely aligning their practice with commerce. Suspect botany bolstered by the authority of the marketplace was an alliance difficult to undo.

As others in this volume have shown, indigenous knowledge of plants was valued as Europeans brought new worlds and new flora under their control. This chapter suggests that the value of local knowledge was limited in early republic America, a historical time when a colonial scientific practice encountered a society in which unbalanced relationships in politics and governance were increasingly unwelcome and challenged. Certainly, naturalists received information about American nature and disseminated their techniques and epistemology throughout the continent; eventually their methodology identified plants that were useful to the nation and to the world. Yet this process was neither easy nor predetermined; naturalists in early republic America encountered resistance from ordinary Americans who, at times, transferred their disdain for hierarchical relationships in government to their dealings with naturalists. Frontier Americans were hesitant to establish and to accept colonial relationships with eastern naturalists if no personal benefit were obvious. Ordinary Americans might correspond with naturalists and describe specimens in their possession, but they were unwilling to provide them plants without compensation, either informational or fiduciary. Early republic Americans were acutely sensitive to the inequalities of colonial relationships, and ordinary Americans were unwilling to permit themselves to be manipulated by elite naturalists. When early republic Americans sensed that their knowledge was to be used by others for financial gain, they wanted to share in the promised profits of American nature.

Elite naturalists were frustrated but circumspect about this state of affairs. Naturalists continued to preach the economic benefits of botany, especially for Americans in an acquisitive society. An individual who could identify lucrative trees and plants was more likely to succeed economically than a neighbor who could not; the marketplace prompted and reinforced this equation. As the worth of many commodities increasingly rested on natural historical valuations, those with broader botanical knowledge could better judge claims of value, ignoring competing claims when they believed the information to be wrong. As the United States matured, the close relationship between science and market economics promoted an intimate understanding of the natural world, its products, and the ability to place both in larger scientific and medical contexts. Those who remained attuned to botany's colonial overtones were left behind.

II.
Translating Indigenous, Creole, and European Botanies: Local Knowledge(s), Global Science

Chapter 5
Books, Bodies, and Fields
Sixteenth-Century Transatlantic Encounters with New World
Materia Medica

Daniela Bleichmar

Early modern European exploration and colonization of the Americas have been described as a series of successive encounters between peoples, religions, languages, and cosmologies characterized by misunderstandings, appropriations, and syncretic reinterpretations.[1] As regards medicine and natural history, the notion of encounter can be fruitfully supplemented by examining the relationships among knowledge, trade, and consumption. Recent research has demonstrated the importance of trade for early modern science and the links between empiricism and commerce.[2] This is particularly significant in the case of New World plants, which entered the European repertoire as natural commodities. In natural history and medicine, as in other domains of colonial activity, the trajectory of encounter, reinterpretation, and appropriation implied the redistribution of political, economic, and social power. Interest in New World nature was inextricably linked to interest in its commercial exploitation. Political domination, profitable trade, and competition among European powers were the starting points for the colonization of this new natural world.

An emphasis on trade also focuses discussions of colonial science on the local experiences of the agents involved. The great majority of Europeans came in contact with the New World in Europe: colonial science was often enacted at home, not abroad. Encounters were conducted not only secondhand, through reading and conversation, but also firsthand, as Europeans acquired *materia medica*, foods, spices, and numerous other imports. Colonial botany was practiced not only in the Americas but also in courts, gardens, battlefields, consulting rooms, and pharmacies throughout the world.

This chapter examines the book that did the most to publicize New World *materia medica* in the late sixteenth and early seventeenth centuries: the *Historia medicinal de las cosas que se traen de nuestras Indias Occidentales que sirven en medicina*, written by the Sevillian physician Nicolás Monardes (*ca.* 1508–88).[3] This work reveals what Europeans knew about New World *materia medica* at the time of its publication and how they learned about it, both in Europe and in the New World. This chapter highlights two themes embodied by Monardes's work: the connections between learning, commerce, and printed books; and the impact of local conditions on the production of knowledge and the uses to which it can be put. This approach complicates an important historiographical assumption: that botanical and medical knowledge were to be found, like specimens, in the field, where they were readily available for appropriation. As the *Historia medicinal* shows, Monardes became the foremost European authority on New World *materia medica* without ever traveling there; it was easier for him to obtain information than it was for a European in the New World. At a time when Europeans with the most experience of the New World were those who had participated in its conquest, Monardes could plausibly present a soldier as a reliable witness and expert on the subject.

Exploring American Nature in Europe

Although the Sevillian physician Monardes never crossed the Atlantic, his life and work revolved around the exploration, use, and sale of New World plants and animals (see figure 5.1). The son of a bookseller and grandson of a physician, Monardes studied medicine at the University of Alcalá de Henares, where his education centered on commentaries and new translations of Greek texts by Dioscorides, Galen, and Hippocrates, the pillars of medical education. Soon after his return to Seville in 1533, Monardes established a successful practice. His patients included the archbishop of Seville and members of the aristocracy and local government. At the time Seville was quickly becoming the most important city in Spain, as the only port from which ships were authorized to sail to the New World or return to Europe. Since 1503 the Casa de la Contratación (House of Trade) documented and controlled the flow of merchandise and people between Spain and its American colonies and functioned as the seat of the royal cosmographer, a training center for pilots, and a repository for nautical information and charts. Given the availability of ships, information, and a market, many prominent Sevillians became involved in the transatlantic trade.[4] Monardes did so in 1553, establishing a partnership with the Spaniard Juan Núñez de Herrera (?–1563), who was settled in Tierra Firme (Central America). Over the next thirty years Monardes's three sons and one of his four daughters immigrated to Tierra Firme and New Spain and were involved to different extents in a commercial venture that made Monardes one of the foremost European experts on New World products and, for some time, a wealthy man. Almost no information about the operation of this enterprise survives, though records indicate that Monardes and his partners imported medicinal products, dyes, and hides and exported cloth and African slaves to work in New World mines.[5]

In addition to selling products from the New World to customers throughout Spain, Monardes also incorporated these products into his medical practice, experimenting over the years with preparation, use, and dosage. He collected natural history specimens and kept a garden of foreign plants that he used for experiments and distributed to correspondents across Spain and Europe. He also exchanged information and specimens locally with at least three other Sevillians who owned gardens and cabinets: the celebrated poet Gonzalo Argote de Molina (1549–96?); Rodrigo Zamorano (1542–?), the royal cosmographer and piloto mayor at the Casa de la Contratación; and the physician Simón de Tovar (?–1596?).[6]

Monardes, physician and entrepreneur, was also a skillful navigator of the world of book learning, in which he had been born and educated. Monardes published several short monographs before the *Historia medicinal*, undoubtedly his major work. The latter consists of a series of entries on New World *materia medica*, providing similar information for each item: name, aspect, and place of origin when known; how it is prepared and administered; its uses; and anecdotes about Monardes's experiences with it over the years, which often include the story of his first encounter with the product, as well as European encounters with it in the New World. Each entry concludes with a description of the product's *complexión*, or Galenic attributes.

Although the *Historia medicinal* is not well known today, it was extremely well received when it first appeared. Translated into Latin, Italian, English, French, and German, nineteen editions were printed during the author's life and another fourteen after his death. The text was frequently summarized, cited, and plagiarized—the true mark of an early modern best-seller.[7] It became widely known through a revised Latin translation by the renowned Flemish naturalist Carolus Clusius (1526–1609), *De simplicibus medicamentis ex Occidentali India delatis, quorum in medicina usus est* (Antwerp, 1574 [parts I and II]; 1579 [Part III]). Likewise, Clusius's *Exoticorum libri decem* (Antwerp, 1605) included material from Monardes as well as Garcia da Orta's *Coloquios dos simples e drogas e cosas medicinais da India* (Goa, 1563) and Cristóbal Acosta's *Tractado de las drogas y medicinas de las Indias Orientales* (Burgos, 1578).[8]

Not only was Monardes's book widely read, but at the time it was also practically the only European source of information on New World *mate-*

Figure 5.1. A portrait of Sevillian physician and author Nicolás Monardes, age fifty-seven, was included in his *Primera y segunda y tercera partes de la historia medicinal de las cosas que se traen de nuestras Indias Occidentales que sirven en medicina* (Seville, 1574). By permission of the Biblioteca Nacional, Spain.

ria medica. Although the sixteenth century was a golden age for printed illustrated herbals—important resources for both naturalists and physicians, with their descriptions of the "virtues" or medicinal properties of plants—these rarely contained information on the *materia medica* of the Americas. And while natural histories of the New World, such as Gonzalo Fernández de Oviedo y Valdez's *Historia general y natural de las Indias* (Seville, 1535), contained descriptions of American diseases, botany, zoology, landscape, native customs, and history, these texts did not provide the specific information on preparations and treatments required by physicians to treat patients or apothecaries to advertise their wares.[9]

This unavailability of practical medical information is not an index of a lack of interest in the matter. In 1552 the Aztec healer Martín de la Cruz (dates unknown) produced an illustrated manuscript describing indigenous botanical medicine, which was translated into Latin by his collaborator Juan Badiano (1484?–?). The *Libellus de medicinalibus indorum herbis*, better known as the *Codex Badianus*, was commissioned by Francisco de Mendoza, son of the viceroy of New Spain, and intended as a gift to Charles V, king of Spain and Holy Roman Emperor (1500–88, r. 1519–55). Neither published nor circulated, it disappeared soon after arriving in Europe and resurfaced only in 1929. Spanish physician Francisco Hernández (1515–87), named "Protomédico de las Indias Occidentales" (Chief Doctor of the West Indies) in 1565 by King Philip II, traveled through New Spain between 1571 and 1577 under royal orders to gather information on the medicinal uses of plants from European and native sources. His descriptions were first published, in abridged form, by Francisco Ximénez as *Quatro libros de la naturaleza y virtudes de las plantas y animales* (Mexico, 1615) and in fuller form by the Accademia dei Lincei as *Rerum medicarum Novae Hispaniae thesaurus* (Rome, 1651). In New Spain, Hernández probably met the hermit Gregorio López (1542–96), whose treatise on New World medicines suffered a similar fate and was published nearly a century after its author's death as *Tesoro de Medicinas para todas las enfermedades* (Mexico, 1672). Hence, for fifty years after the *Historia medicinal* appeared, Monardes was the major European authority on the uses of New World products to treat a wide range of medical conditions.[10]

That Monardes could achieve this position without traveling highlights both the fragility of information in motion and the fact that it was possible for him to produce it locally in Europe. In Seville, Monardes was in frequent contact with travelers returning from the New World. One of his first encounters with New World *materia medica* took place in the early 1530s, when he visited a sick man recently returned from New Spain. When Monardes prescribed a course of treatment, the patient suggested he might instead use a product he had brought back with

FLOR DEL MECHOACAN.

Figure 5.2. Monardes's *Historia medicinal* included woodcuts illustrating some of the medicinal products described, particularly those with higher potential economic or curiosity value. These images served less to depict objects realistically than to highlight their importance: the "Flower of Mechoacan," for instance, is described in the text as a root. By permission of the Biblioteca Nacional, Spain.

him. He claimed that this purgative, unknown to the doctor, was superior to those being used in Europe and that in the New World he had used it repeatedly and with great success. In his account of the encounter in the *Historia medicinal*, Monardes portrayed himself as initially skeptical: "I reprimanded him for using such new medicines about which nothing was written or known, and persuaded him to purge himself with the medicines we had here, about which there was so much experience and knowledge, and much had been written by wise men." But when it became necessary to repeat the treatment, the man refused to use any product other than the suspect import, which he called "Michoacan root" (likely *Ipomaea jalapa*) after the region from which it came (see figure 5.2). It proved to be extremely effective. Nonetheless, Monardes harbored reservations about this new medicine, preferring to exercise caution until he had gathered sufficient evidence of its worth: "Although I was pleased with its effects, I was not satisfied until many others who arrived in a similar state and fell ill were purged with this Michoacan. It worked very well on them, since they had the habit of using this purgative in New Spain. When I had seen how good it was for so many, I started using it and purging many with it, now believing in its good effects."[11]

On the basis of clinical experiences such as this, Monardes became an enthusiastic prescriber of the product as well as its advocate and distributor. He claimed to have sent "great accounts of it to almost all of Europe, both in Latin and in our language," noting that as a result of those written reports and of word-of-mouth publicity from travelers returning to Europe from New Spain, the use of Michoacan root had become widespread throughout the New World, Spain, Italy, Germany, and Flanders. By the time Monardes published the *Historia medicinal*, the product had become so popular that, he wrote, "there is no town where it is not used, as a medicine that is very safe and has great effects," and apothecaries boasted of selling it profitably throughout Europe.[12] This popularization depended on commercial distribution of the import and on the written and oral publicity by doctors and apothecaries who profited greatly from New World merchandise.

While firsthand experience forms the basis of Monardes's book, his investigations often began through correspondence and conversation. Between the 1530s and the 1560s Monardes seems to have talked to everyone who had anything to say about medicinal products. Many entries in the *Historia medicinal* mention such conversations and highlight the privileged circles in which Monardes moved: he notes talking with the archbishop of Cartagena de Indias during the latter's visit to the archbishop of Seville, Monardes's patient, and interrogating travelers he met in visits to the Casa de la Contratación.[13] He also learned a great deal from travelers whom he saw as patients, as in the case of the Michoacan root.[14] In addition to these conversations with travelers, Monardes mentions talking with other doctors and apothecaries who shared his interest in New World *materia medica*. This expansive network included doctors and apothecaries as well as royal administrators, nobles, merchants, and household employees such as servants, children's caretakers, and slaves. The talk is copious and ongoing, and Monardes conveys the impression that, no matter how much had been found and discussed, there was always more to learn. Roots, plants, woods, barks, leaves, seeds, resins—all merited attention; and the more Monardes learned, the more he suspected remained to be discovered about the natural richness of the Indies.

Stories and conversations were merely starting points for Monardes, to be supplemented by his own experiences with the products he describes. "What is written here," he explains, "we have learned partly from those who have come from those parts bringing news of them, and partly from their complexion and qualities—what it is that they do—and partly from our experiments."[15] Monardes complements *historia*, the narrative description of an experience, with *autopsia*, direct observation or manipulation. Monardes depended heavily on anecdotes and reports about new plants since they were not mentioned in classical sources. The style in which these reports are told is an *autoptic* one: emphasis is placed on the traveler's experience. Details—names, places, dates, situations— promise certainty through specificity. This *autoptic historia* is in turn complemented by Monardes's personal experiences, which he translates into descriptions of the individual products in which he speculates on their attributes, as per classical medical theories, as well as accounts of experiments and experiences with patients. In this way Monardes achieved firsthand knowledge of New World products without leaving Europe and was able to represent himself as the authoritative expert on the matter. Thus his narration of a conversion experience regarding the Michoacan root, in which his early disbelief gave way to enthusiasm, is rhetorically useful to establish his reliability.

Monardes concluded the first part of the *Historia medicinal* with a passage that suggests clear reasons for his interest in New World products. "How many trees and plants," he wondered,

with great medicinal virtues are there in our Indies . . . leaving no need for the spices from the Moluccas, and the medicines from Arabia and Persia, given that our Indies yield them spontaneously in the untilled fields and in the mountains. It is our fault not to investigate them, not to look for them, not to be as diligent as we should be in order to profit from their marvelous effects. And I hope that time, which is the discoverer of all things, and diligence and experience will demonstrate them to us, to our great profit.[16]

With those words the first part of the *Historia medicinal* ends on a particularly persuasive note: the imagined natural riches of the fertile Indies promise material gains. The word "profit" appears twice at the end of the paragraph, linking the new knowledge of the medicinal effects of New World products to the possibility of a lucrative western spice trade through which Spain could compete with the eastern trade.[17] But the financial optimism contrasts with frustration that it was difficult to obtain information about these products. Information traveled more easily within Europe than between the New World and the Old. Monardes mentioned, for example, his eagerness to find out what the Michoacan plant looked like since he knew only a dried root that provided no clues about the plant's appearance. He repeatedly questioned travelers returning from New Spain, asking them for information about "the fashion of the plant that has this root, and what form and figure it has."[18] Such information would have been of interest to Monardes as a potential source of further medicinal products, since different parts of a single plant were often used to treat different ailments, and also as an indication of whether the plant could be brought to Europe and grown locally, eliminating the costs and difficulties of transatlantic commerce. Unfortunately, most travelers had no inkling of what the plant looked like since "the only thing that concerns them is their interest and profits, and so they are careless and know nothing about it. The Indians in Michoacan sell them the roots, dried and clean like they are when brought here, and the Spaniards buy them and send them to Spain as a type of merchandise."[19] (On the production of commodities in the context of colonialism, see also the chapter by Harold Cook in this volume.)

The privileging of commercial interest over knowledge provoked an outburst from Monardes, who added a revealing note that captures both his expectations for New World richness in natural products and his exasperation at the difficulties of learning about them:

And truly in this matter we deserve great reprehension: given that there are in New Spain so many herbs, and plants, and many other medicinal things of such importance, and yet there is no one who writes about them, or knows what virtues and forms they have, so that we might compare them with ours. If they had the will to investigate and experiment with the many kinds of medicines that Indians sell in their markets or *Tianguez*, it would be of great utility and profit to see and know their properties, and to experiment with their varied and great effects, which the Indians make public and manifest through the great experiences they make of them among themselves. But our people discard them without consideration, and even in cases where they do know their effects, they do not want to send us relation, nor any news, nor describe their effigy and fashion.[20]

For Monardes, European failure to investigate this rich medicinal tradition represented a great loss, both in information and potential profit. So, while he was able to provide some detail in the *historia* of Michoacan root—anecdotal descriptions of how the medicinal properties of this plant became known to Europeans in the New World and of how he learned about the plant in Europe—he had almost nothing to say about New World understanding of the plant. This omission has a great deal to do with how Europeans in the New World learned about *materia medica* and medicinal practices.

Reading about American Nature in America

While it is fairly clear how knowledge of medicinal properties of New World plants was obtained within Europe, it is more difficult to learn what happened overseas. Many entries in the *Historia medicinal* include anecdotal accounts of how Europeans in the New World learned about local medicinal products. The entry on Michoacan root, for example, states that when a group of Franciscan missionaries in that region of New Spain fell ill, the local indigenous leader was moved by the priests' suffering to offer the counsel of his personal healer. The principal friar, "seeing how little he had there in the way of doctors and resources," agreed to consult the native doctor and was given the powder of a root mixed into wine. This medicine worked such wonders that the friar mentioned it to his superior in Mexico City, who in turn relayed the story to other priests in the city. Through this transmission mechanism the root came to be "used by many, and noted for its marvelous works," and it was soon taken to Europe "as a highly valued merchandise."[21]

The *Historia medicinal* is peppered with this sort of anecdote of initial European contact with New World medicinal plants. The content and tone are expedient, glossy, and suspiciously unproblematic; such an account does not address the uneasy relations between Europeans and Americans. Only rarely do these stories provide telling glimpses into a postconquest world of fighting and servitude: Spaniards became aware of the curative properties of balsam by observing Amerindians using it to heal war wounds; a plant with a similar function was named *yerba de Juan Infante* after the Spaniard who popularized the remedy, not the Amerindian servant who demonstrated its uses.[22]

The most detailed account of New World encounters in the *Historia medicinal* appears in the second part of the book (1571), where Monardes inserted a lengthy letter he claimed to have received from a Spanish soldier, Pedro de Osma (dates unknown), writing from Lima in late December 1568. The author's motive for ceding the narrative voice for twelve of the total ninety-four pages is clear, given the letter's laudatory tone and the way it dramatizes the importance and utility of the book. It begins by celebrating Monardes's (first) book, explaining that by reading it soldiers in the Indies had learned "how we should use the remedies we have here. Before, we would use them without rule or system, and they would have no effect, nor would people be cured. But now it is just the opposite. Through the books written by your grace people have recovered who thought they would never regain their health, nor be cured."[23] The letter goes on to describe how Osma, a model reader, followed Monardes's exhortation to investigate native knowledge. After reading in the *Historia medicinal* that bezoar stones, brought to Europe from the Portuguese East Indies, were found inside animals that looked "much like sheep or goats," Osma attempted to obtain them in Peru. He organized a hunting party, and after killing an animal he cut it open, book in hand. When no stone could be found, "we asked some Indians who had come along as our servants where the animals had these stones. But since they are our enemies, they did not want us to learn their secrets, and said they knew nothing of such stones. Eventu-

SEGVNDA PARTE

Del Libro, delas cofas que fe traen de
nueftras Indias Occidentales, que firuen
alvfo de medicina. Do fe trata del Tabaco,
y dèla Saffafras, ỳ del Carlo fanto: y de otras
muchas Yeruas, y Plantas, Simientes, y Licores:
que nueuamente han venido de aquellas partes, de gran-
des virtudes, y marauillofos efetos.

¶ *Hecho por el Doctor Monardes: Medico de Seuilla.*

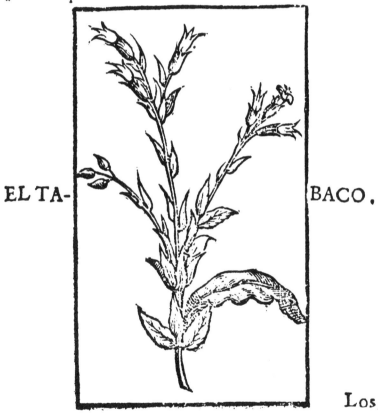

EL TA-   BACO.

Los

Figure 5.3. The frontispiece of the second part of Monardes's *Historia medicinal* advertised its inclusion of plants that had already aroused much interest in Europe, such as tobacco and other plants of "great virtues and marvelous effects." By permission of the Biblioteca Nacional, Spain.

ally an Indian boy who was also there, ten to twelve years old, seeing that we wished to know this, showed us the secret of the matter. He showed us where the dead animal had the stones." This angered the Amerindian men to the point that they threatened to kill the boy, whom the Spaniards protected, albeit briefly. Osma wrote that later "we learned that they had sacrificed the boy. While hunting we had lost track of him, and they took him into the mountains: we never saw him again."[24]

Such secrecy and violence stand in marked contrast to the openness of the Amerindian doctor who introduced the Franciscans to Michoacan root. Occasionally Osma did find native healers who were forthcoming as sources. In Peru a man had shown him a plant that could be chewed to alleviate headaches, and Osma claimed that this plant became known by his own name since he had popularized its use.[25] Osma also sent Monardes a fruit used to treat skin sores; his description of how he learned about it exemplifies the ambivalent combination of mistrust and resignation that characterized European encounters with native healers in the New World. Osma explained that when one of his black female slaves developed very bad skin sores he had allowed a native healer to treat the sores, "seeing that . . . she had no other remedy."[26] Osma witnessed the procedure and was able to describe in some detail the fruit used and the preparation of the remedy. The account contains no indication, however, that the healer volunteered any information. Another episode in Osma's letter suggests that healers tended to be as secretive as hunting companions were. Osma mentioned a man who used an unidentified plant to treat both Amerindians and Spaniards. No Spaniard was able, however, to learn exactly which plant he used, where it came from, or how it was prepared, and efforts to glean this knowledge from the healer were completely unsuccessful. "Regardless of our kind words and gifts, of our fierce words and threats," bemoaned an exasperated Osma, "he never told us what herb it was, nor showed it to anyone."[27]

Even if Osma had spent twenty-eight years traveling in the New World, local information was largely inaccessible to him, making Monardes's book extremely useful. "We have begun," Osma wrote to Monardes, "to use these [bezoar] stones as your grace indicates in your book, giving them in the amount you indicate, and against those diseases for which your grace mentions they are beneficial." The reader thanked the author profusely, "since through your book you gave us the knowledge to find, discover, and remove them from these animals, in which they were so hidden. Surely much is owed to your grace, because you have discovered to us a treasure as great as this, which is the greatest that has been discovered and found in these regions."[28]

The paradoxical implication of Osma's letter is that to obtain informa-

tion about his local *materia medica* a European in the New World depended on a printed book written by an author who had never been
there. Given the unavailability of doctors or healers, European or native,
travelers in the New World often had to take care of themselves when
they had physical problems.[29] With little access to indigenous knowledge
of healing products and practices, they turned to Monardes as a source
of both inaccessible local knowledge and European experiences with
this *materia medica.*[30] Monardes, back in Spain, was hailed as the discoverer of local, American secrets. The circulation of knowledge from the
New World to the Old and back to the New was dependent on native
knowledge yet unable to access and credit indigenous populations as
sources. American natives were at the center of this cycle and at the
same time excluded from it.

Transatlantic Botany and the Local Problematics of Power

Juxtaposing Monardes's and Osma's experiences, it becomes clear that
European reactions to New World *materia medica* depended greatly on
one's location: not only did the two men's experiences differ, but their
interests and preoccupations were dissimilar as well. Monardes wrote of
cooperative healers and helpful Amerindians who aided Spaniards,
while Osma described himself as surrounded by secretive and hostile opponents. Both had great faith in the potential riches of the New World;
but if Monardes faulted Spaniards in the New World for not investigating native medicines, Osma blamed Amerindians:

> I write your mercy about these things so that you may consider how many more
> herbs and plants possessing great virtues, similar to these, our Indies must have.
> But they are out of our reach and knowledge because the Indians, being bad
> people and our enemies, will not reveal to us a secret, not a single virtue of a
> herb, even if they should see us die, or even if they be sawed in pieces. If we
> know anything of the matters I have treated, and of others, we learned it from
> the female Indians. Because they get involved with Spaniards, and reveal to them
> all that they know.[31]

This difference between Monardes's and Osma's attitudes was a result
of the kind of contacts each had with information on New World *materia
medica* and of what this information represented. Unlike Osma, Monardes did not mention active or passive resistance to the transmission
of information around Seville—and it does not appear that he encountered hostility that he did not mention. Most of his sources were European and interacted with him in contexts of social, economic, or professional exchange. On the rare instances in which Monardes mentioned
contact with non-Europeans—such as a conversation on tobacco dosage

held with East Indian dock laborers, which led him to compare ethnic differences in physical constitution—he reported the experience as entirely unproblematic.[32] Nor was he troubled by non-European knowledge of *materia medica*. The New World and its medicinal products did not challenge his beliefs about the functioning of the body, the authority of classical writers, or the validity of his theories or practices (see figure 5.4). He was clearly aware of substantial differences between European and New World medicine, even if he did not possess many details about the former, but was not troubled by them. In a description of a village by a river near Guayaquil where people visited healing springs, much like European spas, Monardes mentioned that patients "neither purge at the beginning of the cure as we do here, nor in the middle of it nor when it is finished, because in that place there is no doctor or medicines. But there are certain women who administer this water, who are experts in its use, and they provide and arrange as they see fit." Monardes approved of these women's work, finding it "very good, and in accordance to good medicine." His positive opinion was based on these women's practice—how their preparation of a remedy managed to preserve most of the properties of a plant, for instance—and completely overlooked potential theoretical conflicts regarding conceptions of how the body operated or how the medicine functioned.[33]

Even those differences in medical practice that challenged the authority of European classical authors did not temper Monardes's optimism. He explained these disagreements as representing advancement of knowledge rather than catastrophic disruption. After extolling the efficacy of Michoacan root as a purgative, for instance, Monardes addressed the apparent conflict that this enthusiasm for purgatives introduced vis-à-vis ancient sources, which maintained that bloodletting constituted a safer practice than purging. This was not at all surprising, Monardes wrote, since while ancient doctors or barbers could control the amount of blood they let from a patient, once a purgative was administered they could not intervene until it had run its course, which could at times be too taxing on the patient. With the introduction of Michoacan root, however—and here Monardes's text adopts the manic enthusiasm of the advertising pitch—doctors now had recourse to a purgative they could control since its action was immediately halted with the ingestion of food.[34] A potential conflict of authority was justified as innovation impelled by new products. New World medicines were not a cause for concern but for celebration, and the ancients were not threatened but rather improved on. "While time," Monardes concluded, repeating a favored humanist topos, "has taken from us the true myrrh, and the true balsam, and other medicines that the ancients had, about which we know nothing in this age because they were lost in the course of time, it

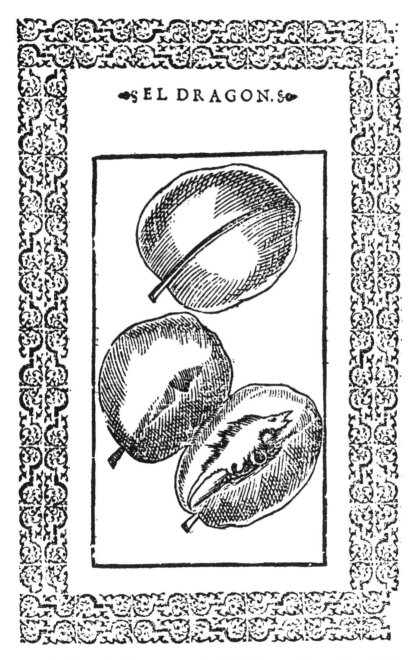

Figure 5.4. Images had important symbolic significance: the woodcut of "The Dragon" plant shows the imprint of that animal on one of the three nuts (Monardes, *Historia medicinal*). By permission of the Biblioteca Nacional, Spain.

is also time who in their stead has discovered and given us so many and such diverse things as we have written that our Occidental Indies send us, especially the Michoacan."[35]

Monardes used European theories to explain non-European practices, describing the properties of new products in terms of Galenic and Hippocratic conceptions of the four humors of the body and the four qualities of matter. New World medical theories went unmentioned; whether Monardes ignored or disregarded them is unclear. But it was precisely this neat split between approved foreign practices and unmentioned foreign theories that allowed Monardes to embrace New World *materia medica*. Products were carried from the New World to Europe, and at times some information regarding their use traveled with them, but the web of beliefs and associations that made these objects meaningful to Amerindian societies were left behind. (See, on the dissociation of beliefs and meanings from objects that entered the European colonial market, the essay by Claudia Swan in this volume.)

Such categorical dissociation was impossible for Osma. In his letter he described the practical difficulties of obtaining information on medicinal products and asserted that these problems arose from the natives' desire to preserve the secrecy surrounding ritual and religious beliefs. The boy who helped the Spaniards find the bezoar stones was murdered for revealing a religious secret, not a medical one. "The Indians," Osma explained, "care deeply for these stones, and offer them to their *Guacas*, which are the shrines where they keep their idols."[36] Osma did not expand on these religious beliefs and was ultimately as little preoccupied with them as Monardes was. In the New World as in Europe, native religion was of less concern to doctors or soldiers than to priests and philosophers. Amerindian medical knowledge and practices incorporated religious and cosmological beliefs that contradicted the tenets of Christianity, thus threatening European religious and political authority. Missionaries found it necessary to be informed about native medicine in order to discredit it; they also needed to appropriate and justify the efficacy of remedies by attributing their effectiveness to either natural causes or demonic intervention. Medical explanations that contested European authority were replaced by others that demonstrated the superiority of Christianity and aided conversion projects.[37] Catholic attacks on native practices stemmed less from the supposition that indigenous beliefs were superstitious, irrational, or wrong than from the fear that they were efficacious and dangerous.[38]

The need to convince and convert was not expressed by Osma. New World missionaries, in close contact with religious practices, urgently needed to provide alternative explanations that would further instead of compete with evangelization; soldiers had neither as much access to native beliefs nor as great a need to attack them. Whether or not Osma knew and understood more about native religion and cosmology than he described, he helped to erect the barrier between New World practice and belief that made it possible for these products to be incorporated into European medicine. Objects and practices that in the New World were inextricably linked to ritual and religious beliefs were cleaved from that context and thoroughly cleansed before being shipped across the Atlantic. Stripped of their local connotations, New World natural substances resurfaced as global goods to be enthusiastically investigated and promoted by European physicians, apothecaries, and traders.

Chapter 6

Global Economies and Local Knowledge in the East Indies
Jacobus Bontius Learns the Facts of Nature

Harold J. Cook

Besides, every Malayan woman practices medicine and midwifery with facility; so (I confess that it is the case) I would prefer to submit myself to such hands than to a half-taught doctor or arrogant surgeon, whose shadow of education was acquired in schools, being inflated with presumption while having no real experience.

—Dr. *Jacobus Bontius*, De Medicina Indorum

One of the most important aspects of the so-called scientific revolution in Europe was the accumulation of detailed information about natural things and natural events. Numerous early modern accounts stress that before speculating about the underlying processes of nature, one had to know the things experientially. Such careful description often went by the name of "natural history," a rubric under which even famous astronomers in later centuries placed their work.[1] Recently historians have paid renewed attention to the importance of the "big science" that was descriptive natural history,[2] and to the "matters of fact" that became so important to the understanding of nature.[3] Europeans wanted to assemble accounts of all of nature so as to know it completely, for the sake of truth and benefit. To find things out, they made inquiries among local informants wherever they went, including craftsmen, alchemists, medical empirics, herb wives, engineers, shipwrights, and many others closely involved with things and their uses.[4] After all, the accumulation not just of things but of information—accurate information—was essential to commerce. People who promoted trading ventures depended on finding, gathering, and redistributing the products of nature and on knowing all about them and their uses. As one economic historian of the European trading companies put it, "the supply of accurate information must have been one of the first things one expected of a clever merchant."[5] The letters they sent home to their associates while abroad indicate just how active they were in accumulating information about people, places, prices, and ways of evaluating the quality of the goods available for purchase. (See Anke te Heesen's essay in this volume, on the adaptation of financial accounting practices in natural history.) For the hardheaded merchant and other men trying to plan to their advantage, the foundation of true knowledge lay not in debating general premises or conclusions but in accumulating precise and accurate information.[6] Thus, in the first period of globalization, which linked the silver mines of Peru with the spice trade of Asia and the gun foundries of Europe, a worldwide natural science rooted in descriptive natural history also developed for the first time. Nonetheless, how people composed accounts of these matters of fact is not always clear, requiring us to reassess "the process by which authorship is attributed to matters of fact in science."[7] What follows is an examination of one such author's work, noting the importance of local informants in supplying him with knowledge, his debt of respect to them for doing so, and the kinds of information he drew from what he learned.

The patterns of knowledge acquisition germane to early modern global natural history are worthy of close study. On the one hand, authors had to interact with other people both at home and around the world in order to acquire knowledge. On the other hand, when they acquired it, they took only what they wanted, mainly information about description and use. In both respects, authors acted in ways similar to those of merchants who accumulated and exchanged goods. (See also the essays in this volume by Claudia Swan on collecting *naturalia* and E. C. Spary on the difficulties of transforming knowledge of particulars into universalized science.) Many of the best examples of contributions to descriptive natural history can be found among the activities that occurred under the aegis of the Dutch East India Company (Verenigde Oostindische Compagnie, or VOC). The Gentlemen Seventeen—the governors of the VOC in the Netherlands—sometimes even encouraged such efforts. Despite the violence of the VOC's efforts to monopolize the spice trade (see the essay by Julie Berger Hochstrasser in this volume), employees of the company produced some of the most important works on early modern medicine and natural history of the New World, Africa, and Asia. Several members of the company became famous for the publications that issued from their work under the VOC. The most renowned of the Asian natural historians included Hendrik Adriaan van Reede tot Drakenstein (1636–91), who investigated the Malabar coast of South Asia; Paulus Hermann (1646–95), who studied Ceylon and nearby regions; and Georgius Everhardus Rumphius (1628–1702), who wrote on Ambon and places nearby in the Indonesian archipelago.[8] Other writers with more explicit interests in medicine, such as Jacobus Bontius (1592–1631), Andries Cleyer (1630s–late 1697 or early 1698), and Willem Ten Rhijne (1649–1700), also produced landmark works of descriptive natural history.[9] These authors' efforts deploy similar rhetorical devices as well. They share the quality of conveying matters of fact as if newly discovered, although careful examination reveals that their accounts were written on top of erasures, as in a palimpsest. While European authors often represented their observations as unique, personal experiences garnered independently of any help by agents of other knowledge systems, it seems that the most important means for acquiring new information actually involved contact with other people and familiarity with their experiences and accounts. In their various publications these European authors all similarly reinscribed conversations with local people in the language of commensurable matters of fact.

The work of Jacobus Bontius, who died in Batavia (now Jakarta) in service to the VOC, is exemplary of this process, a process we may come to understand as exemplary of natural history at the time of the so-called "Scientific Revolution." The Gentlemen Seventeen appointed Bontius physician, apothecary, and surgical inspector of the VOC territories in August 1626. The VOC had previously employed ships' surgeons, and a few physicians had also been sent out to certain large stations, but Bontius was given a general remit to oversee all of the VOC's medical affairs in Asia. This appointment served Bontius's ambitions in turn, for as he later remarked in a letter to one of his brothers, he expected that his travels, writings, book collection, and exotic botanicals would earn him a professorship in the Leiden medical faculty, a position both his father and his eldest brother had held.[10] Bontius set sail with the outbound fleet on 19 March 1627 accompanied by his wife and two sons, and reached his destination on 13 September, having lost his wife en route.[11] During his stay in the Dutch East Indies, Bontius suffered many bouts of illness, especially during two sieges of Batavia in 1628 and 1629, when he contracted dysentery, beriberi, and other serious diseases. Bontius also suffered many personal losses in Batavia: his second wife, whom he had married shortly after his arrival, died on 8 June 1630 of "a vehement cholera"; his eldest son died at the beginning of 1631 of "kinderpoxkens" (perhaps measles); and the deaths of friends and acquaintances are mentioned several times in passing in his works.[12] Not only were his personal sufferings great and his medical duties relentless, but he was assigned additional responsibilities as well. Between the sieges of 1628 and 1629 he became a member of the Court of Justice, the highest judicial body in the Dutch East, and in 1630 he assumed the mantle of chief law officer for the Dutch Indies, *Advocaat Fiscaal.* He also served as bailiff of Batavia from 15 October 1630 to 18 January 1631, finally giving up the ghost on 30 November 1631.[13]

Despite heavy duties and considerable suffering in body and mind, Bontius assiduously investigated the medicine and natural history of the region. His *Methodus medendi qua In Indiis Orientalibus oportet* (On the Proper Treatment of Diseases of the East Indies)[14] was, according to the 19 November 1629 dedication, completed immediately after the lifting of the second siege. It describes nineteen major diseases of the belly, chest, and skin observed in the East Indies but unknown in the Netherlands. At the same time the work celebrated the ways in which local diseases found their remedies in local plants, a common medical point of view that emphasized the beneficence of nature: "Where the diseases . . . are endemic, there the bountiful hand of Nature has profusely planted herbs whose virtues are adapted to counteract them," he commented.[15] By the time Bontius composed his dedication he had also begun a work on the natural history of the region, which he seems to have considered his major task. According to a later letter to his brother (apparently written on 18 February 1631 and printed as a preface to his first four works, published in 1642), he set out to acquire knowledge of the plants and especially the spices of Java immediately on arriving. The dedication addressed to the Gentlemen Seventeen in his *Methodus medendi* expressed his continuing devotion to their service, which would be even more evident, he promised, when he had finished his "commentaries on the shrubs, trees and herbs which grow in Java." In the same work he lamented, "And would [that] this disease, which has laid me low for about four months . . . have permitted me to travel around the countryside to freely explore the delightful woods of Java and gain an exact knowledge of the many most noble herbs that are here!"[16] Perhaps even from the start, then, but certainly no later than his recovery from the effects of the second siege, Bontius kept a record of his observations on natural history, in both words and pictures.

Some of Bontius's observations on natural history appeared in a second work, his *De Conservanda Valetudine. Seu de diaeta sanorum in Indiis hisce observanda Dialogi* (On the Preservation of Health: Or Observations on a Sound Way of Life in the Indies in the Form of a Dialogue), which he finished on or before 18 January 1631.[17] The *Conservanda Valetudine* is modeled after the famous work of a physician who had lived and worked in Portuguese India more than half a century earlier, Garcia da Orta (c. 1501–68), who published *Colóquios dos simples e drogas . . . da India* (Colloquies on the Simples and Drugs of India) in Goa in 1563.[18] (On the influence of da Orta's text, see the essay by Daniela Bleichmar in this volume.) Da Orta's work took the form of a dialogue, in the course of which he discussed the uses of what he considered to be the most important medicinal plants of Asia. As Guy Attewell has shown, da Orta had taken the opportunity while in Goa to learn some Arabic, and in his book he criticized classical Greek and Latin sources sharply because their authors were ignorant of most of the medicines and spices of Asia.[19] While few copies of da Orta's book appear to have made it back to Europe, a young Flemish naturalist picked one up during his travels in Portugal and in 1567 brought out a heavily edited and annotated edition of it in Latin.[20] This edition, by Carolus Clusius (1526–1609), made Clusius's reputation as a botanist and remained the standard work on Asian botany for several generations—until Bontius's work appeared. (For more on Clusius, see the essay by Claudia Swan in this volume.)

Like da Orta's *Colóquios*, Bontius's "On the Preservation of Health" takes the form of a conversation organized around the familiar medical theme of the six non-naturals, or environmental and personal habits that can support health or cause disease.[21] Bontius is almost as critical of da Orta as he had been of the ancients. For asserting that the Javanese and Indians attribute to pepper a cold quality, Bontius disparages da Orta as "again ridiculous." In the same dialogue Bontius writes that da Orta mistook the uses of *calamus aromaticus* (sweet flag [*Acorus calamus*]), for although he "knew no other use for it or the sweet smelling reed in India than as bedding for horses, if he had truly been as diligent in investigating the qualities of aromatics as he was discerning in reading Arabian physicians, he would not have been ignorant of the uses of that plant, for throughout India both fish and meat are cooked with a bit of *calamus aromaticus* or the sweet smelling reed, both to improve their flavor and to invigorate the stomach."[22]

Immediately after finishing "On the Preservation of Health," Bontius undertook a more systematic review and critical commentary of the work of da Orta, in which his views are somewhat tempered. He completed the study by February 1631; it was later published as *Notae in Garciam ab Orta* (Notes on Garcia da Orta).[23] When he surveyed da Orta's work as a whole rather than select excerpts from it with which he disagreed, Bontius was much less critical of his predecessor. Ultimately, his *Notae* offers gentle correctives or supplements to da Orta's findings. For instance, da Orta remarks that those who use opium appear drowsy, and Bontius tempers his implied criticism of the drug, for "if we did not have this opium and opiates the prospect in this very hot region of making medicines to treat dysentery, cholera, ardent fevers, or other bilious diseases that swell the organs would be frustrated." Bontius also added considerable information. For example, da Orta confessed that he had not seen *assa foetida* (*Ferula assa-foetida*), "called 'Hin' by the Javans and Ma-

'T AMSTERDAM. by IAN TEN HOORN 1693.

Figure 6.1. The title page of the 1693 Dutch translation of Willem Piso's work *Oost- en West-Indische warande* (East and West-Indian Veranda) depicts a European physician (holding the urine flask) and a surgeon (with his instruments) in conversation, with several Asian people asking for help and, in the background, a view of the interior of a hospital. The engraving appears to refer to the "Dialogues" between Bontius and his surgical friend Durie, held in front of the hospital at Batavia, and may be based on a contemporary drawing or painting now missing. L. S. A. M. von Römer, *Historical Sketches*, trans. Duncan MacColl et al. (Batavia: Far Eastern Association of Tropical Medicine, 1921), 28–30, argues that the upper part of the physician's face shows a strong family resemblance to extant portraits of Bontius's brothers, Regnerus and Willem. By permission of The British Library.

layans," so Bontius described it. Likewise, when discussing ivory, da Orta confessed that he had never seen a rhinoceros, while Bontius had "not only seen them a hundred times hiding in their lairs, but also wandering in the woods," which gave him an opportunity to give the reader an account of one fearsome encounter with the beast.[24]

Bontius's last work, however, better represents his methods of gathering information. This was the promised natural history of the region, which remained unfinished on his death in 1631, although he had worked on it for two or three years and it was beginning to come to completion.[25] In a letter of 18 February 1631 to his brother Willem he told him to "expect next year, if the power of life remains, a full description of plants, shrubs and trees, with a delineation of each drawn from life."[26] Unfortunately, Bontius did not retain his power of life. But many years later, in 1658, Willem Piso (1611–78) published Bontius's medical works in conjunction with his own natural history material and work on Dutch Brazil by himself and Georg Marcgraf (1610–43) in a composite, voluminous, illustrated edition on the medicine and natural history of both Indies.[27] Piso's *De Indiæ Utriusque Re Naturali et Medica* thus preserves Bontius's final efforts in published form. The natural history material attributed by Piso to Bontius contains information on thirty-three animals and sixty-two plants. A large portion of Bontius's original material, on which Piso's account is based, survives in manuscript in Oxford, where I rediscovered it among the papers collected by the early eighteenth-century lawyer and keen botanist, William Sherard (1659–1728).[28] These manuscript materials contain information on sixteen animals, birds, and fish and forty-two plants, in random order. Most of the information on particular specimens includes illustrations together with descriptions and commentaries on the facing or following pages; one long textual description (on the tea plant) lacks any illustration. Presumably, yet another manuscript volume, still missing, contained similar information on the additional animals and plants. A comparison of the printed version and the surviving manuscript reveals that Piso reordered Bontius's material, did some light editing of the Latin, added introductory poems and occasionally additional information, and even introduced a few new items, some based on new information from witnesses who had been in the East Indies.

Bontius's descriptions of animals are full of interesting anecdotal as well as morphological information. Some specimens he had to hand: Bontius kept the skin of a thirty-six-foot snake he killed in the woods at home; his observations of the chameleon were based on one he kept "in a case [*cavea*] at home"; and he also had a flying lizard, "which measured three quarters of an ell" (probably meaning a Flemish ell, which would make the lizard about twenty inches long) and which "can fly,

GULIELMI PISONIS
MEDICI AMSTELÆDAMENSIS
DE
INDIÆ UTRIUSQUE
RE NATURALI ET MEDICA
LIBRI QUATUORDECIM.
Quorum contenta pagina sequens
exhibet.

AMSTELÆDAMI,
Apud Ludovicum et Danielem
ELZEVIRIOS.
X. cIɔ Iɔ c LVIII.

Figure 6.2. The title page of Piso's 1658 edition, which printed Bontius's natural history text and illustrations for the first time. Some of the latter (of the dodo and the rhinoceros) are depicted prominently in the center. By permission of the Wellcome Library, London.

Figure 6.3. A drawing of a tiger in Bontius's notebook by his friend Governor-General Specx. In the entry on the animal Bontius mentions that a "tremendous" one was caught and killed just outside the city wall in May 1630 in the presence of Specx. The annotations are probably by Bontius and include a note for the printer on which chapter the illustration should accompany in Piso's reordering of the material (it appeared in print as chapter 2 of book 5 but occurs randomly in the middle of Bontius's notebook). The annotation reads, "Tygris, quam Radja Outang / Hoc est regem Sylvis indi vocant / Cap. 2" (Tiger, called in India "Raja Outang" or "King of the Woods," Chapter 2). By permission of the Plant Sciences Library, Oxford University Library Services.

but does not persist in flight for long . . . reaching forty paces, or in turn thirty, just like flying fishes." He kept birds of various kinds in his back garden and thought it a "great pleasure" to observe the speed of house lizards when they pursued flies and ants.[29] He mentioned going into the woods when possible and even being squirted in the face with black ink by a cuttlefish he picked out of the water, much to the amusement of onlookers.[30] A few animals he dissected. As for the plants, the ones he described were almost entirely medicinals, along with a few cooking herbs; Piso added a general chapter on flowers, in which he cites a source who comments that the Javanese and all Muslims are fond of fragrant flowers and perfumes.[31] According to Bontius's letters, most of the drawings were executed by a cousin, Adriaen Minten (dates unknown), whom he employed with the permission of the governor-general (as would have been necessary to transfer a VOC employee to another job), but since Minten did not work hard, Bontius arranged to have him sent back home.[32] Presumably, some of the drawings were done by tracing

T I G R I S.

Figure 6.4. In the print in Piso's book made after the drawing, the background is eliminated although some shading under the feet is retained, while the engraver's rendering of the tiger's stripes looks more like the shading of musculature. Reproduced by permission of the Wellcome Library, London.

the contours of a specimen on a piece of paper, after which the details were filled in: Bontius describes this method going awry in the case of the gecko on account of the animal's sticky feet.[33]

Perhaps the most important description was of the tea plant. Acting Governor-General Specx (c. 1588–after 1638), who had spent a long time in Japan for the VOC and had seen it growing there, told Bontius tea was a shrub; until then, Bontius had heard conflicting opinions as to whether it was an herb or a shrub from Chinese sources. In Bontius's manuscript, however, this is one of the few accounts that are not illustrated. Bontius explains why: "I could never manage to see the green leaves here," since it was grown elsewhere and imported in dried form. "It is certainly true that it undoubtedly encourages good health and as a medicine acts not unhelpfully to rid the chest of thick phlegm." With its "excellent diuretic properties," tea also "acted as a fine remedy against bladder and kidney stones."[34] By the time Piso published Bontius's natural history, he had access to a drawing of the shrub provided by François Caron (1600–1673), who had been one of the commanders of the Dutch station in Japan following Specx, to which Piso also

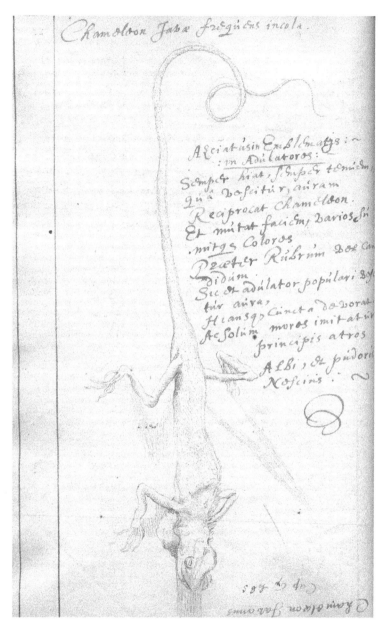

Figure 6.5. A depiction of a chameleon (the caption at the top reads, "Chameleon, a common inhabitant of Java") with a quotation from Alciati's *Emblemata* down the right side, also reproduced in the printed text; the illustration appeared in the book in a horizontal rather than vertical alignment. By permission of the Plant Sciences Library, Oxford University Library Services.

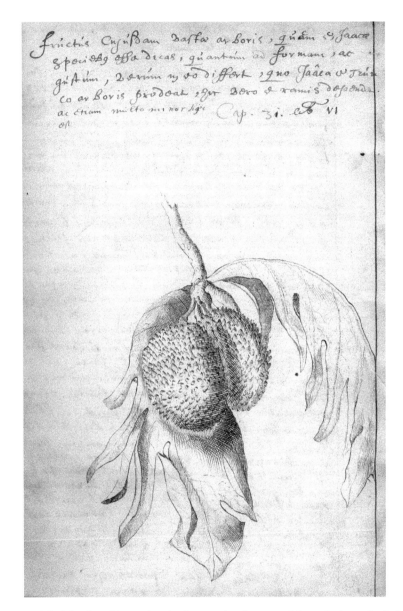

Figure 6.6. The first illustration in the notebook, a careful rendering of a fruit called the Durion, Jaaca, and Champidacca (now known as durian), drawn as if hanging from a tree. By permission of the Plant Sciences Library, Oxford University Library Services.

FRVCTVS DVRIONIS Minoris CHAMPIDACA.

Figure 6.7. The same fruit, in Piso's book; the illustrator has preferred the durian to be stretching upward. Reproduced by permission of the Wellcome Library, London.

added further valuable information he had collected on tea and its properties.[35]

Bontius proclaimed of all his works that they were based on experience alone. In the *Methodus medendi*, for instance, he wrote that all the treatments he recommended were affirmed by practice, seen with his own eyes, and judged to be true.[36] His claims for autoptic empiricism did not prevent him from drawing on previously published texts as well, however. As we have seen, he cited and commented on da Orta extensively. He also incorporated passages from the works of all conceivable predecessors in the domain of Asian medicine and natural history: he cites Pliny's *Natural History*; Pierre Belon's *Les observations de plusieurs singularités* (first published in 1553) on the natural history of the Near East; Cristóbal Acosta's *Trata de las drogas y medicinas de las Indias Orientales con sus plantas* (1578), which is almost entirely derivative of da Orta; Prosper Alpino's *De medicina Aegyptiorum* on Egyptian medicine and botany (first published in 1591); and Jan Huyghen van Linschoten's famous Dutch *Itinerario* (1596) on Portuguese Asia (see Claudia Swan's chapter in this volume for more on this). He also mentions Andrea Alciati's *Emblemata* (first published in 1531) and a book by Pedro Texeira on the history of the kings of Persia, and he quoted from Horace, Juvenal, Martial, Virgil and—several times—the Roman playwright Plautus. Bontius clearly had a small library with him and may even have brought along the whole of his sizable collection of over two thousand books.[37]

While Bontius regularly incorporated textual references, in his later writing he also acknowledged that he depended on informants. Especially when it came to discovering the uses of things, most of his information came from other people, although he merely hints at how he went about obtaining it. One source available to Bontius from the beginning was the information accrued over time by European medical practitioners, especially ships' surgeons. Moreover, Bontius might also have gotten information from the official midwife appointed by the council in 1625 and paid to serve the needs of poor Dutch women, although he does not mention her.[38] Bontius does refer to merchants as sources—as, for example, when he corrects da Orta's identification of *assa foetida*; Bontius writes that he had acquired the second of its two varieties, "given to me by an Armenian merchant friend, who brought them out of Persia." Writing of the origins of bezoar stones, he credits Persian and Armenian merchants who gave him information "completely faithfully." Of a medicinal substance called "tutty," Bontius writes that it was produced by calcining a "glutinous earth," which "Persian and Armenian merchants who come here to negotiate business" assured him was available in great quantities in Persia. Bontius frequently refers to how the "Indians" or, more particularly, the "Javans" or "Malays" did things. When reporting on "tutty," for example, he notes that women in the Indies use it to remove unwanted hair. In the case of *agallochum* (Aloes-wood; probably *Aquilaria* Thymelaeaceae), "called by the Indians *Calambac*," he informs his readers that "a powdered scruple of it" cures the cholera locally called "mordexi" and heals disorders caused by coldness in the stomach and intestines, "in children efficaciously countering tineae, and ascarides"; he could not resist noting that the Chinese and "Moors" used it as an incense sacrifice.[39]

Bontius suggests that he made every effort to acquire information by means of friendship, persuasion, and payment, but he was sometimes met with fear, or at least caution. In acquiring goods, and in its efforts to establish a monopoly over the trade in fine spices, the VOC sometimes resorted to extreme violence. The worst atrocities were perpetrated on the Banda Islands, where the world's supply of nutmeg grew, just a few years before Bontius arrived in Batavia. After years of attempts on the part of the islanders, by violence and by "smuggling," to resist the VOC's contractual monopoly on their produce, Governor-General Coen exterminated most of the islanders, replacing them and their villages with Dutch-run plantations worked by slaves (see Julie Berger Hochstrasser's essay in this volume). It is perhaps not surprising, then, that one of the few stories Bontius tells about his relationships with local people is permeated with fear. An elderly Javanese neighbor, a female slave of a Chinese gardener, possessed a "woodpecker" that could speak even more like a human than a parrot could (it was undoubtedly a mynah). Bontius tried several times, to no avail, to purchase the bird from her in order to observe and draw it. He then tried to convince her to lend it to him. After many such attempts she finally agreed, stipulating that he not feed it any pork. When Bontius and the artist got it home, the bird began saying "Orang Nasarani Catjor Macan Babi," or "Dog of a Christian, eater of pork." He implies that he neither fed it pork nor injured it.[40]

Bontius relates much information about how local women used plants in the kitchen or in medicine; this too he must have acquired from female sources. For example, the "hog stone" or Malacca stone (a soft and fat stone "that feels like Spanish soap"), which came from the gall of hogs and the stomachs of porcupines, was infused with wine to treat mordexi, but it was also dangerous to pregnant women and caused abortions; "I have been told by Malayan women that it is certain to provoke an abortion, and if their menstrual purgations do not come at the right time, if they only hold this stone in their hand they are rejuvenated." About Indian saffron, or turmeric [*Curcuma Zingibereceae*], locally called "borbory," Bontius wrote that "throughout the Indies no plant was more frequently used." It was taken internally and applied topically for obstructions of the bowels and mesentery and for urinary complaints. Moreover, "in diseases of women nothing is so much celebrated by the Malayan women than this borbory. It has a divine effect in easing childbirth, in cases of difficult urination, and in kidney problems. For problems of the uterus it is a specific." (That is, it had powers against these complaints in particular.) "And to make sure of this truth, among all medicines I have myself found nothing better in these afflictions than this remedy."[41]

This last remark suggests that Bontius practiced medicine on women. Although it is likely that his patients were exclusively European, Bontius may have learned from them about indigenous remedies. European women would have mixed more easily with Asian women in the early days of the Dutch colonies than in the later years, when they often behaved in a high-and-mighty way. It is also possible that Bontius obtained local information through his own (European) wives. In addition, a growing population of people of mixed heritage and multilingual abilities, many of whom became crucial information brokers, was beginning to take shape in Batavia and other VOC settlements.[42] Moreover, masters could obtain information from household servants and slaves. Bontius owned slaves, as is clear from his reference to "my Moorish slaves" as well as the human "chattel" mentioned in his will.[43] Perhaps slaves were the main informants on the uses of herbs in the kitchen to which he frequently referred; they certainly told him about tigers.[44] Bontius may also have obtained information about local plants and their uses from women selling vegetables in the Batavian markets: still today women tend to be the main retailers of local produce, not only bringing things to market that university-educated ethnobotanists and others have not seen before, but also providing information about the uses of them.

Bontius's praise for local women extended to the people of the region more generally: "And here, by the way, I note that these nations, though many among us call them barbarian, are superior to the Poles and Germans in pickling fish, who nevertheless are awarded these plaudits without blushing."[45] Elsewhere he notes that the people who came from Surat and the Coromandel Coast must be followers of Pythagoras since they were vegetarians who even abstained from red beans and herbs because of the color. "Thus it is that those who in other things are illiterate have an exact knowledge of herbs and shrubs, such that if the most learned [Pieter] Pauw,[46] prince among the botanists of our age, came back from the dead and traveled here, he would be surprised that these barbarian peoples could instruct him." Bontius even noted, with reference to Clusius's annotations of da Orta, that "they write so elegantly as to excel us by a long way; and when they draw the characters they delineate on these [palm] leaves (which are Arabic) then my indignation rises against those of our Europeans, and especially our compatriots, who admire nothing unless it is their own, even calling these peoples barbarian who, of a more laconic mind, can express more of their meaning in only a few significant characters than ours can with long phrases and useless multiplicity of words." He even praised the Mataramese, who had besieged him: "Although it appears that the kingdom of Java is despotic (*Tyrannicum imperium*), they exercise their authority in light of the condition of the people so that everyone, unless blinded or completely thick, will quickly realize that their political system benefits them, the government ruling well and the people obeying even better." "I often marvel at the carelessness of our people, who without respect call these people barbarians," he reiterated, who "not only in their knowledge of herbs but in all aspects of their economic system (*oeconomica administratione*) leave our own far behind."[47]

Bontius's positive estimation of local knowledge may have a great deal to do with the fact that the relationships he entered into often required a modicum of mutual respect. He met many people who gave him objects and information, whose gifts in turn solicited personal obligation on his part. The exchanges between Bontius and his informants might therefore be said to fall into the category of "gift exchanges." The French sociologist and anthropologist Marcel Mauss offered a well-known model for thinking about the function of gifts in "archaic" societies as an index of their difference from commercial societies of twentieth-century Europe and America.[48] Gift exchange requires reciprocity: the given embodies the character of the giver, and the giving and receiving take place in a face-to-face exchange through which the parties are linked to one another personally. In commercial exchange, transactions are voluntary and involve impersonal and alienable objects; and seller and purchaser are (or can be) completely unknown to one another. Mauss's model does not correspond exactly to the complicated worlds in which people actually live: even highly commercialized exchange economies involve gift relations, just as societies based on giving also transact commercial exchanges.[49] Societies along the island coasts of the Dutch East Indies had been commercialized for some time before the Europeans arrived, even though they were not as highly commercialized as Dutch society. But the simple message of Mauss is that gifts are special and personal, obliging their recipients to the giver and weaving together the lives of the participants in the exchanges. "It goes without saying," comments one of Mauss's interpreters, that the gifts "are not necessarily 'things' in the sense of material objects having a cultural significance. The 'thing' may also be a dance, a spell, a human being, support in a dispute or a war, and so forth"[50]—or even information. Bontius was awarded a position by the Gentlemen Seventeen of the VOC that he repaid by recording his observations on the medicine and natural history of their eastern lands; in the course of his efforts he also received gifts of knowledge from many other people, which he repaid in part by defending their reputations.

The economy of Bontius's information gathering went two ways: on the one hand, the personal relations he cultivated in the East Indies entailed obligations; on the other hand, he transformed what had been gifted to him into a form he could transmit to European sponsors, readers, and markets. Indeed, Bontius "took" something of what he learned from his informants by transforming it into public knowledge. He converted their utterances into information packaged in Dutch or Latin words and syntax, and aimed to transmit them via publication. Bontius transformed indigenous knowledge into those "matters of fact" so highly regarded by his European peers. One aspect of "matters of fact" that sometimes goes unnoticed is that they appear to be true without regard to time or place, just as commodities appear to be the same regardless of where they are consumed. (That is, just as a packet of cigarettes is not altered by its context despite the diversity of meanings its possessors might invest in it, so any simple and clear descriptive "matter of fact" is universally the same in any culture no matter what variety of implications it might acquire.)[51] While he appreciated local knowledge, Bontius dismissed out-of-hand the values Malays and others attributed to plants, and very seldom included any such information in his written accounts. In one case, in his comments on Indian verbena (*Cymbopogon citrates*), he revealed his assumptions by qualifying the inclusion of indigenous knowledge as follows: "this herb," he wrote, "is considered sacred among the old Indian women (which they have in common with our own old women)," but he mentioned this only to "demonstrate the foolish habit" of mind that considers such things to be true, for "I am not one of those who has a propensity to superstitious belief about the powers of medicines, which are [rather] from nature."[52] In other words, he privileged certain kinds of knowledge and favored those that had to do with things and the material uses of those things.

Translated to a European context, such descriptive statements acquired a common value and were easily exchanged, just as commodities were exchanged on the Amsterdam stock market. There merchants traded one thing for another by agreeing on a common value. It had come to be recognized that the ability to transform the value of one thing into another was not neutral but had a value of its own, prompting important philosophical investigations.[53] It has even been argued that early modern European mathematics developed a concept of "general magnitude," which emerged from attempts to calculate apparently incommensurable things in the marketplace.[54] The merchants of the VOC for whom Bontius worked were keenly interested in making the diverse things in which they dealt commensurate, and they needed detailed information in order to do so. In works such as Bontius's, then, foreign nouns, adjectives, and verbs that were tangible—simple morphologies that address the five senses rather than the mind's eye—were valued because they were readily transferable, while he ignored, misunderstood, or dismissed as superstition local interpretations of them. He (re)produced knowledge of objects, accumulated it, and exchanged it, just as the merchants who governed the VOC did with commodities. The author appropriated information as his own not just by learning about material things and practices but by transforming them into commensurable facts.

Long before the rise of nineteenth-century racism and orientalism,[55] then, much of the work of natural historians from Europe, including Bontius, reinscribed information developed by other people in other cultures, past and contemporary, making it objective and exchangeable. Whatever we call the method by which Bontius rendered reading, conversation, and observation into "his own" work, he owed enormous debts to others and recognized his obligations. One can detect beneath his words literate traditions that reach back to Greek antiquity and Islamic culture, as well as oral traditions reaching from Persia and Armenia, through South Asia to Java and on to the Spice Islands, China, and Japan. In reinscribing this knowledge Bontius universalized and objectified it, making it a kind of commodity. Nevertheless, he knew even better than we what debts this placed him under, eliciting his respect for those from whom he received knowledge, whether European or Asian. His acknowledgments should remind us that our own objective science is built from countless human interactions occurring around the globe.

Chapter 7

Prospecting for Drugs
European Naturalists in the West Indies

Londa Schiebinger

> *It is quite by accident and only from savage nations that we owe our knowledge of specifics [medicines]; we owe not one to the science of the physicians.*
>
> —*P. L. Moreau de Maupertuis, 1752*

Pierre-Louis Moreau de Maupertuis (1698–1759), president of the Berlin-Brandenburg Akademie der Wissenschaften, surely overstated his case when he claimed that European drug discoveries depended on either "accidents" or the knowledge of non-European peoples, whom the Europeans often characterized as "savages." How were new drugs identified in the eighteenth century? How did useful new medicines such as ipecacuanha, jalapa, and Peruvian bark arrive at London shops and Parisian hospitals?

Historians of the early modern period have begun detailing how botanical exploration—bioprospecting, plant identification, transport, and acclimatization—worked hand in hand with European colonial expansion. While Spanish conquistadors had entered the Americas looking for gold and silver, by the seventeenth century Europeans increasingly turned their attention to "green gold." The riches stored in plants would supply lasting, seemingly ever renewable profits long after the mineral wealth of the New World ran out. The very definition of *Amerique* in Denis Diderot and Jean Le Rond d'Alembert's *Encyclopédie* emphasized the business of plants—trade in sugar, tobacco, indigo, ginger, cassia, gums, aloes, sassafras, brazil wood, and guaiacum.[1] By the eighteenth century sugar was the largest cash crop imported into Europe from the Americas, but Peruvian bark (quinine) was the most valuable commodity by weight shipped out of America into Europe. Identifying profitable plants in the Caribbean was one way European naturalists contributed to the colonial effort.

Figure 7.1. Frontispiece to Jean-Baptiste-Christophe Fusée-Aublet's *Histoire des plantes de la Guiane Françoise*, 4 vols. (London and Paris, 1775), vol. 1. The native guide (portrayed as an effeminate male) guides Fusée-Aublet (cameo portrait) to the "different interesting objects relative to the culture and commerce of French Guiana." © Bibliothèque Centrale, Muséum National d'Histoire Naturelle, Paris 2003. Reproduced by permission.

European naturalists also contributed to Europe's expansion by providing new medicines to keep European troops and planters alive in the colonies. Colonial botany was central to the project of Europe's control of tropical areas where voyagers from temperate zones became sick and died in alarming numbers. Nicolas-Louis Bourgeois (1710–76), who served as secretary of the Chamber of Agriculture during his twenty-eight-year residence in Saint Domingue (Haiti), noted that finding new drugs was not merely a matter of curiosity but was necessary "to cure our maladies and provide new assistance." Europeans moving into the tropics encountered illnesses completely unknown to them; their standard pharmacopoeia proved largely ineffective against new diseases. Bourgeois complained that "apothecaries and surgeons buy their remedies from ship captains coming from France, but the vegetable remedies last hardly one year, and when they are old, they do more harm than good."[2]

This chapter explores European naturalists' field practices as they collected plants in the West Indies either for profit or for medicinal use. Was it scientific training that allowed for new drug discoveries? To what extent did Europeans in the West Indies depend on indigenous populations and African slaves for their superior knowledge in what later came to be known as tropical medicine, as Maupertuis suggested? How did naturalists from Europe wrest this information—on which their very survival often depended—from populations they had conquered and enslaved?

Drug Prospecting in the West Indies

Before the onset of rampant racism in the nineteenth century, many Europeans valued the knowledges of indigenous Americans, Africans, and peoples of India and the East Indies (see "Introduction" in this volume). In the eighteenth century, too, Europeans in the West Indies often took as their starting point for empirical investigations the drugs, dyes, and foodstuffs suggested to them by native "informants." Before leaving home, learned naturalists pored over documents from all across Europe seeking information concerning the flora they were likely to encounter in the course of their voyages. Hans Sloane (1660–1753), future president of the Royal Society of London, was typical in this regard. Before setting out for Jamaica, where he was to take up the post of physician to the governor there, Sloane collated all data available in Europe concerning plants of tropical areas (in both the East and West Indies) so that he would recognize new plants when he encountered them. On arriving in the West Indies, Sloane turned to locals for information, collecting "the best information" concerning the natural products of the country from "books and the local inhabitants, either European, Indian or Black."[3]

Who were the local informants to whom Europeans turned for information concerning useful foods and medicines? When Christopher Columbus (1451–1506) arrived in Hispaniola in 1492, the island was inhabited by approximately one million Tainos and Caribs.[4] By the eighteenth century native populations in the Caribbean had been decimated by conquest and disease. The Caribs had run the Arawaks out of the Lesser Antilles; the Spanish had crushed both peoples. A 1660 peace agreement between the English, the French, and the Spanish exiled the Caribs to the islands of Saint Vincent and Dominica, leaving the larger islands, such as Jamaica and Hispaniola, heavily inhabited by Europeans and Africans, with only small populations of Amerindians.[5] European physicians nonetheless gleaned what information they could from the few survivors.

In the eighteenth century, as earlier, most botanists were physicians. Medicine and botany were closely allied, and botany remained a required element in European medical education. Jean-Baptiste-René Pouppé-Desportes (1704–48), royal physician in Cap Français, Saint Domingue, from 1732 until his death in 1748, typifies doctors working in the Caribbean. In efforts to increase the efficacy of his cures, he supplemented his mainstay of remedies sent from the Hôpital de la Charité in Paris with local "Carib simples." Because the first Europeans who came to the Americas, Pouppé-Desportes wrote, were afflicted by illnesses completely unknown to them, it was necessary to employ remedies used by "the naturals of the country whom one calls savages."[6] In the third volume of his *Histoire des maladies de Saint Domingue*, he presented what he called "an American pharmacopoeia," offering an extended list of Carib remedies. Europeans had begun producing pharmacopoeia, official compendiums of drugs for major cities, in the sixteenth century in an effort to secure uniformity in remedies. Pouppé-Desportes's pharmacopoeia is one of the first to record Amerindian remedies. As was typical of these works, he cross-referenced plant names in Latin, French, and the vernacular Carib. Offering synonyms (across languages and cultures) was a common European practice, especially in the pre-Linnaean era, but only rarely were Amerindian names systematically included.[7]

Even more important than adopting Amerindian cures, Pouppé-Desportes urged that Europeans emulate the native Caribbean way of life; Europeans living in Saint Domingue would not need medicines, he wrote, if they lived as "frugally and tranquilly as the savages."[8] But, and his is a common refrain among medical men serving in the colonies, since Europeans (and he means males, who made up the majority of the European population in the West Indies) are "given to excesses in both their eating and liquor, they often require strong remedies."[9] During his years on the island Pouppé-Desportes continued to collect and test local remedies that he then classified according to their medicinal properties.

Native populations, however, continued to decline and by the 1780s Nicolas-Louis Bourgeois complained that "of the prodigious multitude of natives of which [the Spanish chroniclers] speak, not a single of their descendants can be discovered whose origin has been conserved pure and without mixture."[10] By the 1790s a chronicler of the revolution in

Saint Domingue bemoaned the fact that no trace of a "single native" remained.[11]

With the decline of indigenous populations, slave medicines took on an unexpected importance in the West Indies, even though Africans were originally no more native to the area than were Europeans. Unlike Europeans, however, Africans knew tropical diseases, their preventions, and their cures. The Scottish mercenary Lt. John Stedman (1744–97), for example, living in Surinam, worked alongside a number of African slaves. One old "Negro" named Caramaca had given him the threefold secret of survival: (1) never wear boots but instead harden your bare feet (which Stedman did by incessantly pacing the deck of his boat); (2) discard the heavy European military jacket and dress as lightly as possible; and (3) bathe twice a day by plunging into the river—something that distressed the water-wary Europeans.[12]

Bourgeois, a longtime resident of Saint Domingue, also appreciated slave medicines. Considering health a matter of state importance, he eulogized the "marvelous cures" abounding in the islands and remarked that *les nègres* were "almost the only ones who know how to use them"; they had, he wrote, more knowledge of these cures than the whites (*les blancs*).[13]

It is impossible to know with any precision how much African herbal knowledge was transferred into the New World. Displaced Africans must have found familiar medicinal plants growing in the American tropics, and they must have discovered—through commerce with the Taino, Arawak, or Caribs or through their own trial and error—plants with virtues similar to those used back home. Bourgeois confirms that there were many "doctors" [*médecins*] among the Africans who "brought their treatments from their own countries," but he did not discuss this point in detail.[14]

Bourgeois also praised the skill of slave doctors. "I could see immediately," he wrote, "that the negroes were more ingenious than we in the art of procuring health. . . . Our colony possesses an infinity of negroes and negresses [*nègres & meme des négresses*] who practice medicine, and in whom many whites have much confidence. The most dangerous [plant] poisons can be transformed into the most salubrious remedies when prepared by a skilled hand; I have seen cures that very much surprised me."[15] Confidence in African cures was so high among whites that when Sir Henry Morgan, lieutenant governor of Jamaica, became dissatisfied with Sloane's treatment of his disease, he sent for a "black doctor."[16]

Attitudes among Europeans across the Caribbean, of course, were not uniform. The French royal botanist Pierre Barrère (1690–1755), working in Cayenne off the coast of French Guiana from 1722 to 1725, did not think much of Amerindian cures. The good health the Indians (he named twenty-four peoples living there) enjoyed he supposed to be the result of their careful diet, frequent bathing, and moderate indulgence in pleasure. He wrote, "our Indians are completely ignorant of how to compound medicines. The few remedies they know they have learned from the Portuguese and other Europeans."[17] He nonetheless preserved several Amerindian plant names along with their medical uses. David de Isaac Cohen Nassy (1747–1806), a Jewish physician working next door in Surinam in the latter part of the eighteenth century, remarked that the "negroes" played a large role in the health of the colony with their "herbs and claimed cures"; however, he also noted that their cures were "more valued among the Christians than among the Jews."[18] Sloane had a similarly poor opinion of slave medicine. Although he took care to collect what the Africans living in Jamaica told him, he did not find their cures in any way "reasonable, or successful"; what they knew, he wrote, they had learned from the Indians.[19]

Even in this era when many Europeans valued Amerindian and African knowledge useful to them, mythologies of drug discoveries suggested that knowledge traveled up a rather anthropo- and Eurocentric chain of being, from animals (with their instinctive cures), to indigenous peoples, to the Spanish, and, according to the French mathematician Charles-Marie de La Condamine (1701–74), ultimately to the French. La Condamine, who traveled extensively in what is now Ecuador and Peru in an effort to procure Peruvian bark (*Cinchona officinalis*—the source of the therapeutically effective alkaloid quinine), recounted the ancient legend that South American lions suffering from fevers found relief by chewing the bark of the cinchona tree. Observing the curative powers of the bark, the Indians too began treating malarial and other recurring fevers with it. The Spanish then learned of the cure from the Indians, and the French, the self-appointed keepers of universal knowledge in the age of Enlightenment, learned of it from the Spanish.[20]

La Condamine's countryman Pouppé-Desportes offered two further examples in this genre—one from Martinique and another from Saint Domingue. In the first, a potent antidote to snakebite was discovered by the lowly grass snake. "Unhappy enough to live on the serpent-infested island of Martinique," wrote Desportes, the snake learned to employ a certain herb when attacked by a venomous serpent. The effect was so wonderful that the natives called the plant *herbe à serpent*, or serpent herb. In similar fashion the native hunters discovered the excellent qualities of the "sugar tree" by observing wild pigs, which are called "maroons," shredding the bark of this tree with their tusks when hurt and then rubbing their injuries with its sap; for this reason this sap was called "wild pig balm."[21] Edward Long (1734–1813) in Jamaica was very much a racist in how he looked at local cures. Although he agreed that Negro cures often worked "wonderfully" where even European art had failed, he did not admit to any creativity among Africans. Negroes, he charged, apply their herbs randomly and, like monkeys, whom he claimed they resembled, they received their skill only from "their Creator, who has impartially provided all animals with means conducive to their preservation."[22]

The process of colonial expansion opened Europeans' minds to new knowledge systems. From the fifteenth century onward, European naturalists underwent an epistemological shift away from relying solely on the "summa of ancient wisdom" (Dioscorides, Pliny, Galen) toward valuing (or at least appreciating) the knowledges of native peoples encountered through colonial expansion.[23] European physicians no longer defined their task as verifying the effectiveness of ancient medicines or identifying local substitutes for those remedies. Their desire—on the one hand, to identify valuable economic plants for themselves and their patrons, and on the other hand, to find useful cures for their countrymen who all too often succumbed to the ravages of tropical diseases— led physicians to rely on colonial locals for information concerning potentially helpful plants. Botanists in the eighteenth century trod a fine line between their nascent prejudices against the peoples of America and Africa and their need to survive in unknown environments.

Biocontact Zones

In her study of French natural history, Emma Spary has analyzed how botanists and gardeners at the Jardin Royal des Plantes in Paris, such as André Thouin (1747–1824), instructed and manipulated voyagers to speed new specimens into urban centers in order to enhance both the power and the wealth of nations. Successful acclimatization of plants for agriculture, medicine, or the luxury trade required that plants be sent into France with precise information about their cultivation, virtues, and uses. Following Bruno Latour's model, Spary has postulated that botanical men at the center needed a "rigidly structured, unvarying, universally agreed upon method of describing and inscribing" so that the useful plants could be known and successfully positioned within European frameworks of agri- and horticulture.[24]

What the naturalists at the Jardin du Roi, the Hortus Medicus in Amsterdam, or, later, at the Kew Gardens in London could not control and standardize was the "contact zone"—where Europeans negotiated with informants of all kinds and those with medicinal and natural historical expertise in particular. Mary Louise Pratt has defined this "space of colonial encounters" as "the space in which peoples geographically and his-

torically separated come into contact with each other and establish ongoing relations, usually involving conditions of coercion, radical
inequality, and intractable conflict."[25] Here I wish to introduce the notion of "biocontact zones" in order to set the contact between European
medical botanists and African and Native American healers in a context
that highlights the exchange of plants and their cultural uses.

The notion of a "contact zone," as it is often understood, is not unproblematic. To isolate the contact between Europeans and non-Europeans as the primary site of analysis improperly constructs an overly
rigid notion of non-Europeans as "others." Furthermore, contact is not
restricted to a bounded zone; Europeans had "contact" at many junc-
tures over the course of a voyage. There were the "encounters" among
persons of different classes and professions within Europe. There were
certainly the "encounters" of young men who, driven by hunger to seek
refuge in urban port cities, were kidnapped and shipped out as sailors
on trading company ships.[26] Abroad, encounters between Europeans
and other Europeans were often as troubled as those with non-Europeans. Maria Sibylla Merian (1647–1717), for example, who traveled to Surinam to study exotic insects, had unhappy relations with Dutch planters
in Surinam. "They mock me," she wrote, "because I am interested in
something other than sugar."[27] In her turn, Merian criticized the planters for failing to cultivate plants other than sugarcane and for their
harsh treatment of the Amerindians.

Here I will focus (despite my caveats) on the biocontact zones in the
West Indies and look closely at strategies Europeans deployed in their
efforts to learn about useful plants to entice information from their informants and how they were often, in turn, manipulated by them. Inside
Europe naturalists are known to have carefully choreographed complex
systems of patronage—coaxing favors while also disciplining one another with varying degrees of success. Europeans in the field often
found it difficult to manipulate their informants, whom they wished to
exploit and control. These "actors" in the network often simply refused
to cooperate. My account necessarily relies on European sources; neither Amerindians nor Caribbean slaves left written records document-
ing these encounters. As we shall see, botanical exchange in these transcultural zones was fraught with difficulties.

The loudest "noise"—or intellectual interference—in the biocontact
zone was the cacophony of languages. Europeans usually only scratched
the surface of local peoples' knowledges of plants and remedies because
they were often unable or unwilling to speak local languages. Edward
Bancroft (1744–1821—a Massachusetts privateer and double agent who
sent missives in invisible ink during the American Revolutionary War)
worked in Guiana as a young man in the 1760s. He bemoaned the fact

that he was "but little acquainted with the Indian languages," which he judged necessary for "acquiring that knowledge of their properties, and effects of the several classes of Animals, and Vegetables, which experience, during a long succession of ages, must have suggested to these natives." Though he endeavored to overcome these difficulties through the use of interpreters, he remarks that these efforts were largely "in vain."[28]

La Condamine, Pouppé-Desportes, and Alexander von Humboldt (1769–1859) were all keenly interested in local languages. La Condamine spoke what he called "the Peruvian language" and even owned a 1614 "quichoa" (Quechua) dictionary—which he used to study etymologies (of "quinquina," or quinine, for example). Concerning the language of the ancient Peruvians, La Condamine noted that by his day (the 1730s and 1740s) it was "strongly mixed" with Spanish.[29] The French mathematician judged the languages of South America harshly, writing that "many possess energy and bear traces of elegance, but they are universally barren of terms for the expression of abstract and universal ideas, such as time, space, substance, matter, corporeality. . . . Not only metaphysical terms, but also moral attributes are completely absent."[30] One wonders, however, if La Condamine had sufficient knowledge of the language to make such judgments. Humboldt, who traveled extensively in present-day Venezuela and Colombia, prepared a dictionary of the Chaymas language that consisted of only about 140 words, most of which depicted simple things such as fever, hammock, boy, girl, bridegroom, fire, sun, moon or were used in phrases such as "he likes to kill" or "there is honey in my hut."[31]

The problem of language in the West Indies was sometimes ameliorated by the slave and native populations there who, serving as active linguists, created Creole languages. The Caribs, for example, created "a jargon" through which they dealt with the French in Saint Domingue, which was a mixture of "Spanish, French, and Caraïbe pell-mell [all] at the same time."[32] African slaves in Surinam created a language that served as common currency there called "Negro-English," it was composed of Dutch, French, Spanish, Portuguese, and mostly English.

Problems of communication did not arise simply from the lack of knowledge. The Frenchman Charles de Rochefort (1605–83) as early as 1658 placed the problems of language in the context of war and conquest. "Some of the French have observed," he wrote, "that the Caribbeans have an aversion to the English tongue; nay their loathing is so great that some affirm they cannot endure to hear it spoken because they look upon the English as their enemies." He noted that the Caribs had, in fact, assimilated many Spanish words to their own language but that this was done at a time when there were friendly relations between the two nations. De Rochefort noted further that the Caribs shied away from teaching any European their language "out of a fear that their [own] war secrets might be discovered."[33]

"Noise" in the medical contact zone also resulted from the inflexible theoretical frameworks that made Europeans unable to absorb radically new information. Understanding of Carib, Taino, Arawak, or transplanted African medicinal herbs was limited to what made sense to Europeans within a humoral context of heating and cooling drugs.[34] European naturalists thus tended to collect specimens and specific facts about those specimens rather than worldviews, schema of usage, or alternative ways of ordering and understanding the world. They stockpiled specimens in cabinets, put them behind glass in museums, and accumulated them in botanical gardens and herbaria, but as specimens "stripped of narrative," supporting once again the notion that "travelers never leave home, but merely extend the limits of their world by taking their concerns and apparatus for interpreting the world along with them."[35]

In addition to problems involving language, conceptual frameworks, and physical hardships in biocontact zones, Europeans, Amerindians, and slaves each had their own economic interests and cultural aims—all of which further curtailed transcultural exchange. Humboldt complained that his Amerindian guides were interested in trees only as timber for building canoes and as a result paid little or no attention to their leaves, flowers, or fruit. Exasperated, he exploded: "like the botanists of antiquity they deny what they had not taken the trouble to observe. They are tired with our questions, and have exhausted our patience in turn."[36]

Encounters between Europeans and Amerindians or African slaves were not pure meetings between equals but antagonistic struggles full of exaggerations on both sides. Humboldt derided his "Indian" pilot, who spoke to him in Spanish, for exaggerating the dangers of water serpents and tigers. "Such conversations are matters of course when you travel at night with the natives," he wrote. "By intimidating the European travelers, the Indians believe, that they shall render themselves more necessary, and gain the confidence of the stranger." Humboldt attributed the exaggerations of the Amerindians to "the deceptions which everywhere arise from the relations between persons of unequal fortune and civilization."[37]

Even had conditions been ideal, the enormity of the task of understanding the medical value of New World plants was mind-boggling. La Condamine, already exhausted and in Quito, wrote that to detail the plants of the Amazon basin would "require years of toil from the most indefatigable botanist . . . and draughtsman. . . . I speak here merely of

the labor which a minute delineation of all these plants, and the reduc-
tion of them into classes, genera, and species, would necessarily require,
but if to this were superadded an examination of the virtues ascribed to
them by the natives of the country, certainly the most interesting part of
a study of this nature, how daunting would be the task."[38]

Conditions in biocontact zones were hardly ideal. The excesses of colonial rule brought a Carib rebellion on Saint Vincent and slave upris-
ings in Saint Domingue and elsewhere in the 1790s. Fear of poisonings
led planters to outlaw slave medicine in English colonies in the 1730s
and in the French islands in the 1760s. Even as early as the 1680s, Hans
Sloane reported the dangers of herbalizing in islands where "run away
Negroes lye in Ambush to kill the Whites who come within their
reach."[39]

Secrets and Monopolies

Europeans were often curious about Amerindian and slave medicines
and eager to learn, but the indigenes and slaves were less eager to divulge this knowledge to their new masters. Along with miraculous cures
came the silence of secrets. Bourgeois characteristically remarked that
though "the negroes treat themselves successfully in a large number of
illnesses . . . most of them, especially the most skilled, guard the secret
of their remedies."[40] A Dutch physician, Philippe Fermin (1729–1813),
confirmed that the "negroes and negresses in Surinam know the virtues
of plants and offer cures that put to shame physicians coming from Europe . . . but," he continued, "I could never persuade them to instruct
me."[41]

Naturalists in the West Indies devised various methods for wresting
secrets from unwilling informants. Bourgeois attempted to win slaves'
confidences with friendship; failing this, he offered money, but often
without success.[42] Fermin in Surinam, anxious to save the colony the cost
of foreign drugs and the malfeasance of ill-intentioned slaves, attempted
to learn from the "black slaves" their knowledge of plants, "but these
people are so jealous of their knowledge that all that I could do," he
wrote, "be it with money or kindness [*caresses*], was of no use."[43]

Europeans also threatened and coerced reluctant informants. Sloane
related the story of how Europeans learned about *contra yerva*, the potent antidote to the natives' poisoned arrows. The story was told to him
by an English physician named Smallwood, who had been wounded by
an Amerindian's poisoned arrow while fleeing the Spanish in Guatemala. Not having much time, he took one of his own Indians prisoner,
tied him to a post, and threatened to wound him with one of the venomous arrows if he did not disclose immediately the antidote. At this, the

Indian (of an unnamed people) chewed some *contra yerva* and placed it into the doctor's wound, and it was healed.[44]

The effort to secure secrets against enemies or competitors was not unique to the vanquished in the colonies. Europeans, of course, had many secrets of their own. Before patents, guilds kept the techniques of their trades secret. In the medical domain many physicians and apothecaries protected their remedies by keeping recipes secret until they could sell them for a good price. In a celebrated case the apothecary Sir Robert Talbor of Essex (1639–81) garnered fame and wealth from his "marvellous secret" that cured fevers: his *remède de l'Anglais* secured him a knighthood in England and an annual pension of two thousand livres from Louis XIV.[45]

The eighteenth century saw the rise of public health measures that celebrated the free exchange of life-saving remedies and procedures. Sir Hans Sloane, president of the London College of Physicians, walked a fine line between respecting the monetary security bound up in secrets and advancing the emerging ethic of making effective cures available to the public for the "welfare of mankind" when he kept secret his remedy for "eye soreness." Ever on the lookout for new drugs both at home and abroad, Sloane attempted to procure the secret of Dr. Luke Rugeley's eye balm. Avoiding direct contact with the doctor, Sloane sought out a "very understanding apothecary," a mutual acquaintance of both Rugeley and Sloane—but the apothecary either did not know or was not talking. After Rugeley's death Sloane acquired all his books and manuscripts, including his *materia medica*, but still "all in vain." Finally, a "person" who had made the medicine for Rugeley sold Sloane the secret on the condition that Sloane would not divulge it until after the unnamed person's death (because he or she was still making a living by it). Toward the end of his life Sloane published *An Account of a Most Efficacious Medicine for Soreness . . . of the Eyes*, which he sold for sixpence each.[46]

The great trading companies of the early modern period also guarded their investments through scrupulously protected monopolies. The historian Francisco Guerra has perceptively noted that "medical science has been the heritage of the human race, but the drug trade has been throughout history the object of economic monopolies."[47] The Dutch, for example, held a monopoly on trade with Japan from 1639 to 1854.[48] One of Linnaeus's many students, Charles Thunberg (1743–1828), who traveled to the Cape of Good Hope and on to Java and Japan for the Dutch in the 1770s, told how the Dutch East India Company continued in the eighteenth century to hold a monopoly on the spice trade and opium. "If anyone is caught smuggling," he warned, "it always costs him his life, or at least he is branded with a red hot iron, and imprisoned for life."[49] While many naturalists caught passage on company ships, the

East and West India Companies cautioned scholars not to reveal too much in their publications.[50] The French Compagnie des Indes, for example, urged naturalists to limit published information. The French company also blocked British efforts to buy Michel Adanson's papers treating the natural history of Senegal after the British took Saint Louis in 1758. Few French academic reports dealing with this area were published while Senegal was under British control.[51] Commercial exchange—the desire for profits by both individuals and joint-stock companies in this era—countermanded emerging practices of free scientific exchange.

Drug Prospecting at Home

Beginning in the late seventeenth century and throughout the eighteenth century, academic physicians prospected for drugs inside Europe, using techniques similar to those employed by their colleagues in the colonies. From Sweden, Linnaeus remarked that, "it is the folk whom we must thank for the most efficacious medicines, which they . . . keep secret."[52] In England, Joseph Banks's (1743–1820) interest in botany was kindled when, as a youth in the 1750s, he watched women gathering "simples" for sale to druggists.[53] In encounters strikingly similar to those in the West Indies, European medical men interrogated and cajoled their countrymen and women into disclosing the secrets of their indigenous cures. Sometimes persuasion and at other times the power of the purse yielded the secrets of their indigenous cures. As with colonial plants, physicians began testing on the basis of ethnobotanical clues and published the results. Thomas Sydenham (1624–89) provided a popular rationale for these new practices: any "good citizen," he wrote, in possession of a secret cure was duty-bound "to reveal to the world in general so great a blessing to his race"; medical experiments were to benefit the public good (not only physicians' purses) for, as Sydenham continued, "honors and riches are less in the eyes of good men than virtue and wisdom."[54] Not all would have agreed.

Physicians cast their nets widely in the search for effective cures at home. Published instructions encouraged travelers inside and outside Europe to question and learn from people of all stations and sexes— from statesmen, scholars, and artists as well as from craftsmen, sailors, merchants, peasants and "wise women."[55] The process of collecting information from old women or particularly successful women healers was strikingly similar to that of prospecting abroad, a kind of internal bioprospecting. The major strategy was to buy the cure, and often, as noted above, the government put up the money to purchase the secret of a useful remedy. One woman who did well for herself was the spinster

Mrs. Joanna Stephens, daughter of a gentleman of good estate and family in Berkshire, England. She was paid five thousand pounds on 17 March 1739 by the king's exchequer for her drug—an eggshell and soap mixture reported to dissolve bladder stones.[56] Her cure for this "painful distemper" was highly prized because the only other option was surgery, known as "cutting for stones." The London surgeon William Cheselden (1688–1752) was able to extract stones in less than one minute (a procedure he perfected on dead bodies); nonetheless, the operation was excruciating and dangerous.[57] Most women, however, whose cures were eventually adopted and published in the various European *Pharmacopoeia,* remained nameless (and probably unpaid), as was true also of the greater part of West Indian indigenes and slaves who offered cures.

As with informants in the West Indies, we have little access to the women's own reactions to their encounters with academic European naturalists. The historian Lisbet Koerner has highlighted an article in a 1769 Stockholm magazine purporting to give voice to "wise women." It was not uncommon in the eighteenth century for articles such as this to be written by men under female pseudonyms. Nonetheless, the wise women's assertions air complaints that appear in other women's writings from this period.[58] The women noted their "joy and pleasure" when a physician, standing by a sick child's bed, judged, "here no other help can be had than that of finding an experienced wise woman." They complained that physicians "exert themselves to both smell and taste our pouches, creams and bandages," attempting to divine the secrets of the medicines. These women, like so many others at the time, ended the article by asking to be admitted to professional training, in this case to the Stockholm Medical College, "for we are after all considered as highly as the gentlemen [physicians] in the homes into which we are called."[59]

The Swedish physicians offered to purchase the wise women's secrets, but, like other folk healers, these women would not sell them because their livelihood, and often that of their descendants, depended on monopolizing the cure. Mr. Ward, for example, who developed Ward's pills, would not divulge the recipe "because it would be worth so much yearly to him and his successors."[60] Thus a conflict arose between irregular healers (and Sloane classed some of his London colleagues in this category), who depended on their secrets for their livings, and the beginnings of academic research medicine with its ethics of sharing results for the "benefit of humankind"—or at least such was the ideology. Patents, developed to safeguard university and company investment in drug development, tempered this ethic.

Botany and Colonial Expansion

Since Columbus first set foot on the Caribbean islands that came to be known as the West Indies, Europeans scoured these tropical areas for useful and profitable plants of all sorts. Already by the sixteenth century "wonder drugs" such as quinine, jalap, and ipecacuanha had been found and were soon to be introduced into the pharmacopoeia of major European commercial centers. Mercantilists and cameralists all across Europe attempted to staunch the flow of bullion to foreign countries to pay for these new drugs. The French physician and botanist Pierre-Henri-Hippolyte Bodard decried the annual loss of revenue to foreign countries, a loss aggravated by the French upper classes' preference for "objects difficult to obtain."[61] He estimated the loss from Peruvian bark alone at 7,380,000 livres per annum. Other West Indian and South American exotics, such as cacao and jalap, similarly drained European coffers.[62]

Where Europeans could not import and acclimatize these tropical plants inside their own borders, they turned to their colonies to secure such valuable commodities for the metropolis. Bourgeois, speaking from France's richest colony, Saint Domingue, advocated not only fulfilling France's needs by cultivating medicinal plants in the colonies but also the continued search for new and profitable drugs. "Why have recourse to foreign drugs," he asked, "when it is believed that there are so many here [in the Antilles] of use?" There is "a mass" of simples, he continued, in Saint Domingue that need only be examined.[63]

The search for new medicines was fueled by the desire to supply markets back in Europe and also by the need to keep troops, colonists, missionaries, and traders alive in the tropics. Here colonial botanists had to rely on peoples indigenous to these areas to expand their knowledge of the multitude of new diseases they encountered daily. To this end, European-trained physicians and privateers in the West Indies turned to indigenous peoples and slave populations in their efforts to find new and effective drugs.[64]

Historians of science often present the creation of new knowledge as an unencumbered search for truth characterized by open communication of the best results. As we have seen, however, conditions were hard in biocontact zones. Mortality was high; informants were unwilling. The search for potential life-saving new drugs was mired in relations of conquest, commerce, and slavery. European colonial expansion depended on and fueled the search for new knowledge concerning tropical medicines. At the same time, colonialism bred dynamics of conquest and exploitation that impeded the development of this knowledge.

Chapter 8

Linnaean Botany and Spanish Imperial Biopolitics

Antonio Lafuente and Nuria Valverde

Science and empire are cause and effect of one another: they are not identical; each determines and is defined by the other. Empire requires that scientists and their patrons share the belief that the stuff of nature can be captured in words, figures, lines, shading, gradients, or flows. Eighteenth-century voyaging naturalists, for example, had to believe that the essential characteristics of natural organisms and processes could be sketched onto pieces of paper, entered into journals, and cataloged. For them and their successors, representations of nature were destined to become pieces of paper in a filing cabinet. Reducing nature to representation, however, is but one aspect of science. What distinguishes geographers or botanists from archivists is that the naturalists also serve as "witnesses," experts who study their objects in situ and who "bear witness" to their appearance in the field. All this lends authority to travelers' observations and representations, and convinces those who commissioned them of the truth of their representations. When travelers return home they, as well as those who sent them, share the belief that their papers—inscribed with representations of nature—can lead to exact knowledge of nature despite the distance separating nature from the metropolis: knowledge, in other words, becomes manifest in a network of signs, symbols, patronage, and conventions of representation.

A network extends the range of sight and the reach of the hand. It is a *technoscope* that depends on empire and at the same time underpins it: in short, it is a device used to know at a distance.[1] Distance, however, poses serious problems of scale since the diversity of places and the multiplicity of things are too great to be collected together. Problems of scale require that people invent principles of order to contend with the problem of data overload. Geographers have traditionally solved this problem with what we like to call "à la carte cartography," that is making maps to order by playing with scale.

Decisions about the scale of representation determine not only the data to be shown but also the conceptual matrix in which they are to be organized. Scale thus molds the relationships between selected data points; or, to put it more provocatively, we might say that geographers, rather than simply describing reality, first invent it and then depict it.[2] This scientifically produced network of data points is presented to readers as a coherent whole, preexisting and independent from the observer, although in fact it has been highly manipulated to suit the purposes of empire. What we have described here concerning territories also applies to plants: Carl Linnaeus (1707–78) saw a clear analogy between *Geographia Naturae* and *Historia Naturae*.

Our purpose in this chapter is to show how discussions about "scale" conditioned the type of botany deployed in Spain's New World colonies. We explore two conflicting biopolitics: Spanish imperial botanical policies emanating from the metropolis; and Creole political botany homegrown in the colonies. Botanists and policy makers in Spain championed an organizational model for botany focused on the production of data that allowed the metropolis to take advantage of the floral resources of its colonies. The political botany fostered by New World Creoles (elites of European origin born in the Americas) attempted, by contrast, to ensure that the plants being consumed locally improved the moral fiber of the community. Spanish imperialists demanded policies that organized and mobilized bioresources; Creoles in the colonies fostered a type of botany that supported civic life in the Americas.

Disputes of Scale

The concept of a "science-of-scale" is a useful tool to differentiate from among a multiplicity of cognitive practices those practices that are manipulated by a particular political entity—whether a royal court, an empire, or a village—until they become socially and culturally viable. Here we explore how varying scales lend authority to particular scientific practices.

During the first half of the eighteenth century the Spanish Crown hosted scores of official and quasi-official institutions. Scattered throughout the country and lying outside any large-scale network, these institutions tended to function locally and in line with age-old traditions. Things were to change when the Spanish court came to view America as a panacea—a place to export its domestic ills. With this change in perspective came a change in the scale of governmental administration. Control of such vast possessions meant massive military and administrative expenditure: all aspects of governmental apparatus—naval, sanitary, and educational—reached nearly global scales. This change of scale also affected the management of American resources, both vegetable and mineral.

Imperial expansion fostered new institutions designed to wrest the American territories from local control and implement instead a single administrative system joining the colonies to the metropole. This move toward administrative consolidation also promoted new scientific languages that allowed for coherent exchange of information between imperial botanical and cartographic centers. Institutions that understood these newly emerging requirements gained levels of influence and cultural profiles unthinkable a few decades earlier. Their function ceased to be courtly and became imperial.[3] Savants no longer resembled collectors or antiquarians but became instead agents in charge of governing nature.

Overseas expeditions served as the mainstay for the metropolitan science: between 1760 and 1808 Spanish kings sent fifty-seven expeditions to investigate the flora of their colonies.[4] A consistent botanical language was crucial to the success of these endeavors. From the time of the Orinoco Expedition in 1752, in which Linnaeus's disciple Pehr Löfling took part, the Spanish Crown imposed the Linnaean system on all botanical institutions, including New World expeditions.[5] In Spain botanists wrangled over rival classification systems. The Spanish Crown insisted, however, that all documentation, including manuals, inventories, and so forth, be expressed in the same code. Linnaeus's system was efficient since among its merits was its ability to disregard local circumstances, such as climate and soil conditions, without renouncing its claim to be describing a natural, or universal, order. Linnaeus saw nature as divided into "Kingdoms," each ruled by laws similar in kind to those that governed empires. For Casimiro Gómez Ortega (1741–1818), a strong supporter of Linnaean taxonomy, there was no room for argument: "All natural bodies form as it were an extended Empire, governed by the unalterable laws imposed on them by the Creator."[6] The Spanish doctor and academician Francisco Bruno Fernández put it more clearly, distinguishing between the universe as the collection of natural beings and nature as its sovereign.[7] For Spanish biopoliticians this was a convenient distinction because it ensured the viability of Spanish policies despite the great diversity in plants or peoples around the world. Gaston Bachelard was correct in pointing out that those who seek the exceptional or are moved by the beauty of nature are not on the side of science but obstacles standing in the way of its disciplinary institutionalization.[8]

Simply put, Spanish administrators and scientists in this period advocated the notion that there is no nature without empire. This conception of nature supports the ontological value of the concepts of *species*, insofar as *nature* captures notions of repetition and resemblance founding the notion of species. For this reason nature is a concept that should not be confused with the sum of the creatures and plants that inhabit the planet, or with the sensations that these multitudinous phenomena provoke in us. Rather, nature ought to be conceived to comprise the species that inhabit it, the meteorological phenomena that govern it, and the machines used to measure them. Nature exists only when technologies (that is, instruments, books, maps, tables) mediate between the sensations of the subject and the object toward which they are directed. Nature is the figures and charts produced by these measurements; it operates in such a way that weather becomes meteorology, water becomes H_2O, and parsley turns into *Petroselinum sativum.* Nature, then, is a world that distances itself from common experience. As things are geometrized, tabulated, and named, as order is given to data (soon called "facts" by the supporters of this cataloging system), scientists proclaim themselves the only reputable witnesses.[9] As Donna Haraway has pointed out, natural history and nature have been created to conform to particular notions of economy.[10]

Metropolitan botanists, then, reached New Spain satisfied by two simple ideas. The first was that of the species, a concept embodying the belief that the whole of the animal and vegetable kingdoms can fit neatly into a table. The other was the idea that species could be differentiated by morphology. A species concept permitted the unification of knowledge concerning flora and fauna, minimizing distances between near and far, between Europe and America. Nature could be depicted as a continuous mantle of plant life made up by forms that would be described in Linnaeus's binomial terms. Nature, in short, became a structure of data whose objective was not to appreciate but to process local peculiarities into information using the botanical system best able to homogenize diversity. To put it in modern parlance, the Linnaean system worked as an interface between morphology and nomenclature, between the shape of things and the language of systems. As Michel Foucault has shown, the grand project of the Enlightenment was to render objects part of a logical order.[11]

There were, however, scholars on both sides of the Atlantic wary of these conjuring tricks. The objections of Georges-Louis Leclerc, comte de Buffon (1707–88), could be heard in Paris and London as well as in Philadelphia, Mexico City, and Lima. Spanish Creoles from all parts of the New World found the Linnaean system lacking; they knew a lot about plants and little about scales. José Antonio de Alzate y Ramírez (1738–99), the ironic and brilliant agitator of public opinion in New Spain, shared several of Buffon's reservations concerning Linnaeus's system.[12] Alzate criticized Linnaeus's system for its insensitivity to local circumstances, considering it unnecessary to disregard what was known

about a plant's location, environment, flowering season, or soil characteristics simply to classify it. This stubborn Mexican priest wrote, "it is a
remarkable thing that the short-sightedness of one man, be he ever so
painstaking and observant as we suppose Linnaeus to be, should seek to
review the whole globe in order to index it, impose new names, and allot
them their proper place."[13] Alzate's remarkable insight reveals the gap-
ing disparities between the vastness of the world and the smallness of
the laboratory. Linnaeus's study in Uppsala, its brilliant occupant, and
its heaven-knows-how-many filing cabinets were too small to contain the
world. Alzate complained of a series of injustices that botanists sent from
the metropolis claimed not to understand: he lamented the high cost of
new instruments and the time wasted by not being able to collect plants
except in the flowering season; he bemoaned the obscenity of the sexual
system and its incompatibility with local knowledge.[14]

Linnaean botanical nomenclature also served as a political *nomenkla-
tura* insofar as the exclusion of native names from the field of science
defined new power relations. Linnaeus's nomenclature acknowledged
the authority of imperial botanists and belittled local herbalists and
herbal practitioners. Nomenclature, indeed, displaced traditional wisdom: Nahuatl came to be considered an unintelligible, garbled language, fit (as was remarked in 1788 by Vicente Cervantes [1755–1829],
the director of the botanic garden in Mexico) "to be spoken in public
places and small groups, with Indian women selling herbs and vegetables, but not in the academies of the learned." This had not always been
so: Martín Sessé, a doctor settled in Mexico and director of the Royal
Botanic Expedition to New Spain (1787–1803), cited his knowledge of
the native language as a qualification for his inclusion in several botanical expeditions.[15] In the hands of the imperial botanists, however, the
new Linnaean system also widened the split between botany and medicine: a worrying tendency for everyone would be affected if usefulness
were no longer the criterion for interest in plants. The Creoles turned
their bewilderment into a struggle to conquer public opinion. For them,
what was needed—what tradition and common sense dictated—was just
the opposite: to subordinate taxonomy to what was tangible, and what
was tangible to what was useful. A plant name, they argued with passion,
should express not a logical but a functional order.

Biopolitics: The Metropolis versus the Colonies

Creoles in New Spain disagreed with many aspects of Linnaeus's system.
The majority of those disagreements centered on the notion of natural
diversity. Alzate wrote in 1788, "in New Spain there are products of nature that refute and overturn all theories and botanical systems hitherto

devised.''[16] José Hipólito Unanue (1755–1833), an influential Peruvian physician with a profound knowledge of and admiration for Linnaeus's work, endorsed Alzate's anti-Linnaean position, writing that ''all the systems drawn up in Europe on this subject [natural history] require a thousand elaborations when applied here.''[17] The message was clear: America was luxuriant, too biodiverse for any scheme of classification. Nature would not fit into a simple tabulation. Just to imagine that it could was again to degrade a continent already stigmatized by Buffon, William Robertson (1721–93), and other Europeans as imperfect due to its immaturity.[18]

Metropolitan botanists accepted that America was rich in plant varieties, but they still believed that these varieties could be subsumed as specimens under larger taxa. They had no wish to argue against the importance of knowing the usefulness of plants, but they were horrified by the idea that a plant's properties could be conditioned by local circumstances, reasoning that ''the earth supplies no nutrition to plants, [but] merely serves as a means to support them and thus is not absolutely necessary for plant life. The smell, color, taste, lushness or other accidents of plants,'' these Spanish botanists continued, ''are of no use when setting out their specific differences; . . . the same applies to their uses or virtues, which therefore should not be taken into account for this purpose.''[19] These were serious arguments, for the foregoing were theses set for the students of the botanic garden in Mexico to be defended in public before their professors and the local dignitaries.

Similar theses were set in Guatemala, where students were asked to determine the properties of a plant from its morphological characteristics. In other words, the Linnaean system was credited not only with descriptive and denominative qualities but also with the capacity to make predictions. In 1769, at the opening of the Cabinet of Natural History in Mexico, José Mariano Mociño (1757–1819), a disciple of Cervantes and botanist to the Royal Expedition to New Spain, asked his student Ortiz de Letona, ''although *Nigella damascena* is not known to have any particular virtue, according to Linnaeus's system and in view of its morphological characteristics, what use or powers could it have?''[20] Again local criticisms were put forward, this time by the examiners, to the effect that botany should cease to be seen as a discipline of ''haughty savants.'' The examiners also denounced the unbridgeable gulf between the everyday use of certain plants and the theoretical proposals defended by those coming from Spain. They rebuffed the tendency of metropolitan botanists to ignore cultural experiences unique to New Spain.

''Deanthropologizing'' the knowledge of plants also involved ''delocating'' or, better still, ''deterritorializing'' it. Botany was emerging as a science separate from medicine but also from chorography as territory became a sort of carpet for plants to lie on. On this point, too, there was considerable local opposition. It is difficult to persuade a farmer that the land, his land, is not an important factor. The Sociedades Económicas de Amigos del País, private associations dedicated to the promotion of knowledge relating to agriculture, technology, and industry that emerged in Spain after 1763 and then spread to the Americas, were not groups of farmers but groups of patriots who were less interested in taxonomy than in production.[21] The priorities of these associations, made up of local elites committed to economic utilitarianism, had more to do with plant acclimatization and technical innovation than with botanical classification and the study of mechanics. They may not have known much about labels or calculations, but they did know that the land was temperamental, that general principles were for books only, and that crops were produced by miracles, not theorems. For these worthy people, being a patriot meant securing the stability of plantations and reducing the care of each new planting to a few indispensable and locally variable rules.

These concerns stood in direct opposition to those of Linnaeus's disciples. In the colonies floral identity was not defined in terms of botanical species but in terms of the particular characteristics of soils. It was thought, in other words, that the earth changes the properties of plants. José Antonio Liendo y Goycoechea (1735–1814), a Franciscan professor of philosophy at San Carlos University in Guatemala and an intensely active intellectual, wrote amid the controversy over the quality of American indigo and the possibility of acclimatizing it to Europe: "foreigners cannot take home the weather, the climate, the land, the water, and other precious assets."[22] As pointed out above, what was at stake was the prosperity of Creole lands, and while the Spanish members of various expeditions stood by Linnaeus, local Creoles propounded arguments that threatened the very structure of European botanical knowledge. Goycoechea, writing for practical people, spoke of production and the circulation of goods, not fixed rules. "Let dealers in indigo be wary of everything," he wrote, "first, of the many practices that are adopted blindly; second, of the theories that are not based on facts and manifold experiments; and third, of this very treatise and notes."[23] His was just one of a thousand ways of calling for common sense over authority, demanding experience in the face of artifice. And the word *artifice* could include any kind of artificial event, including, of course, experiments.

Why emphasize the importance of the soil? Here we have shown the main features of an eighteenth-century dispute over the place of local knowledge in a science scaled for Empire. This dispute dealt with two characteristic ways of symbolically appropriating America: the first, emanating from Spain, stressed common features and denied any value to

tradition and local sites; the second, championed by Creoles, highlighted native and practical knowledge as part of the peculiarities of
each region. Both held political and epistemological consequences insofar as they implied different understandings of botanical practices and
knowledge. The Creole pathway that valued details, to tradition, and pa-
rochialism we shall discuss later. Here we turn to the relationship between imperial politics and systematic botany.

As we have seen, imperial botanists disregarded diversity in their quest
for taxonomy. Botany, as developed in Europe over the course of the
eighteenth century, did not deal with things that change with time or
place but with organisms that are imagined to be perennial and stable:
that is, with fixed abstract forms that could be transformed into abstract
data. In this sense, Linnaean botany was a form of biopolitics, what we
might call "imperial biopower" devoted to turning diversity, local variation, and qualia into data. Systematics represented a useful simplification of reality that enabled expeditions to finish in a reasonable length
of time the immense tasks with which they were charged. Linnaean taxonomy and nomenclature provided an efficient way to create an economy of space founded on a logic of resemblance, shortening the time
and work needed to draw conclusions of political consequence. Mociño,
for example, noted extraordinary floral continuity between Mesoamerican and Guatemalan territories, which appeared to confirm the political
unity enforced from Mexico City: "Having visited nearly the whole of
the provinces of New Spain, we have examined the Kingdom of Guatemala, and on our last journey we have found regions not far different
from those found on our first expeditions. So then it can be of little
wonder to anyone that the flora discovered in New Spain contains the
whole catalogue of Guatemala with the sole addition of a very few species not observed in our first incursions."[24] It is worth noting that the
kingdom of Guatemala then included modern Guatemala, El Salvador,
Honduras, Nicaragua, Costa Rica, and Belize. And Mociño's expeditionary route was limited to the Pacific coastline because there was a passable
road as well as a village for supplies. Botanical homogeneity deduced
from a partial survey confirmed political unity. From this point of view
there was no perceptible break in floral types from region to region.

Linnaeus's system also helped the expeditions find shortcuts to limit
the hardships of their work. Finding new species often required long
journeys to out-of-the-way places. Working at the frontiers of unexplored
territories meant suffering torrential rain, bad roads, and shortage of
supplies. There were simply too many plants and too many hardships.
Diversity, in short, was an obstacle. And if, seen from the comforts of
viceregal offices or Jesuit classrooms, America seemed to be an orchard,
viewed from the depths of the forest or the loneliness of the mountains

this abundance seemed to inflict undue punishments. Hipólito Ruiz
(1754–1816), director of the Royal Botanical Expedition to Peru (1777–
88), described his stay in Chacahuassi, a small Andean enclave, as "a
very deep, narrow dungeon, where the sun hardly enters except at midday and at night, if there were ever a break in the clouds at that time of
year, the stars could [hardly] be counted, . . . we looked around this
oppressive and gloomy place, . . . examining those three steep and inaccessible hills, covered from the top right down to the banks of the roar-
ing rivers with tall trees, bushes, and scrub."[25] Insofar as species were not
defined by their surroundings, one could avoid such ordeals by establishing beforehand botanical identities between administrative units.
Considered in this way, the quest for new species could stop where regions became the most inaccessible: "On 30th July I returned from
Xauxa to the town of Tarma," recounted Ruiz, "confident that at no
time would the valley and hills of Xauxa present enough materials for
the members of the Botanic Expedition to work on . . . unless we went
into the Mountains, which we sought to avoid, fully informed by experts
and by the Missionaries of Ocopa that the Mountains of Tarma were as
lushly covered in vegetation as those of Xauxa and much easier and
safer to cross for the collection of botanical specimens."[26] The desire to
discover diversity, which had never been the purpose of the expedition,
ended as conditions became increasingly difficult.

This spatial strategy, devised by imperial botanists, tended to identify
certain regions with characteristic floral species. This allowed a selective
fostering of farms and forestry in those areas rich in species of particular
interest to the Spanish Crown. Plants ceased to be the business of experts and became the concern of politicians. In this way, botany came to
strengthen imperial politics.

Botanical policy developed into political botany, and just as we now
talk of environmental crises, there was much talk then of the decadence
of nature and the erosion of the empire. If political economy gave relevance to the historicity of resources and peoples, political botany ignored local features, while boosting monoculture and encouraging acclimatization mainly in the metropolis. The plants and not their
products became the stuff of trade, and what circulated through the networks was ideas on how to mobilize species and how to denature them.
Botany began as a *technoscope*—a way to see at a distance—but at the end
of the eighteenth century it was already a *teletechnique*—a way to act at a
distance. Its success as an imperial undertaking was linked to its ability
to set up an international network of professorships, gardens, expeditions, and publishing companies able to produce a version of nature easily put into words, and deducible from very little data.

Climate, Plants, Race, and Nation

An emphasis on individuality hindered the formation of political consensus. This was a problem for which the local Creoles sought a solution. Indeed, the complications posed by the Linnaean system and the political struggles over academic appointments associated with it made the obstinacy of its supporters hard to understand. Spanish scientific policy fostered not only scientific expeditions but also the founding of new scientific institutions, such as the Royal Botanic Gardens and Colleges of Mining, Surgery and Navy. Political struggle over appointments aroused great tensions because in many cases Creoles were passed over.[27] It did not matter what the subject was: Creoles started with species and ended up arguing about questions of space.[28] If scientists from overseas made plants a matter of politics, local savants politicized space.[29] Here we turn to a discussion of the controversies over woods, ponds, and diversity.

Europe's demand for timber grew so fast that there was soon an awareness of depletion, if not exhaustion, of resources.[30] These perceptions suggested two kinds of crisis in the American environment: the first centered on the decline (of floral diversity) and the second on mismanagement (of forest policy). Imperial agents were unwilling to recognize a difference between the two. They needed timber for ships and decreed protectionist measures and administrative regulations. When ordinances for the protection of the Guayaquil forests were approved, local officials complained that the measures implemented were out of proportion to the deforestation experienced and that the damage done by protectionism would wipe out any expected benefits. Local councillors suggested that instead of implementing draconian protectionism, scientists should tap into the deep knowledge of the natives, since "as long as the laws of Nature do not fail, there will be more than enough forests in Guayaquil without the help of decrees to provide for their conservation." Moreover, the decrees, they declared, revealed the "lack of knowledge of these lands and their climate"; such attitudes were so ignorant that they are comparable to those held by people who would "keep mosquito eggs for fear of losing the breed."[31] There were those who went further, declaring that "it is morally impossible [that depletion] should happen within the natural order of things."[32] The Creoles confidently argued that imperial policy was scientific but not at all natural. What was at stake here was the very idea of nature and underlying political schemes.

Creoles used the same arguments when discussing specialized land use and the Spanish Crown's monopolies on some monocultures. The *Gazette de Guatemala* declared that imperial policies would impoverish their lands and destroy biodiversity: "What wretched value is set on our produce," these patriots wrote on 16 October 1797, "for hardly has it come onto the market than it disappears like lightning. . . . I hear many people singing the praises of the riches contained in the natural products of this Kingdom; yet little or nothing is said about the conservation of those which, once a real inexhaustible treasure, now exist only as a pitiful memory."[33] The protection of certain native species considered useful and profitable (certain varieties of cinchona or rubber, for example) interested policy makers in the metropolis but not colonists. According to imperial logic, places were interchangeable because territory was no more than a floral *continuum*, and species could be transplanted, acclimatized, and restocked. Such thinking protected forestry business but not biodiversity. Those who lived in the colonies protested that they were the only ones who would regret the impoverishment of their land. For the Creoles, everything argued for a change of scale; for them, the goal was to abolish global viewpoints and recognize local perspectives.

If imperial agents detailed strategies and rationales for managing floral diversity, so too did the Creoles, who found ways to demonstrate the link between their history and their knowledge. This they did by invoking the symbolic topoi of mountains and forests as centers of fertility, topoi highly prized by Enlightenment thinkers. The New Granada naturalist Francisco José de Caldas (1768–1816) discovered what we call biogeography when he realized that each level of a mountain had its own different type of vegetation. In articulating this theory he gave scientific form to something that had long been familiar to peasants.[34] *Andes* is a Spanish word derived from *andenes*, a term used by the colonists to denote terraced and irrigated mountain country. In formulating his theory Caldas also seems to have been influenced by Incan land-use practices by which each administrative unit of their empire unfolded in a V-shape from the sea to the peaks of the Andean range and downward again to the depths of the tropical forest. To the Incas the mountains were a temple and also a living museum guarding all flora diversity. Moreover, the mountains were the greatest laboratory on the planet, for it was there that nature had experimented with all forms of hybridization and acclimatization. The mountains accommodated all types of land and all types of climate.[35] They also fostered diverse cultures: each place had a distinctive population, adapted to its local environment. Moreover, the mountains represented the link between the heavens (climate) and the soil (the land with its nutrients and its peoples). Caldas's biogeographical theory required a shift in thinking away from a view of the territory as two-dimensional flat space to a view of the land as three-dimensional stratified topography. This shift in thinking recapitulated the shift away from an imperial and botanical viewpoint to a Creole appreciation of biogeography.

Caldas recognized that the sheer vastness of the territory under discussion made it appear as though "plants have been sprinkled at random all over the surface of the Andes, and that confusion and disorder reign everywhere."[36] But these were appearances; mountains needed only to be deciphered, and the key was in the climate. In order to understand flora, he taught, one had to understand climatic fluctuations. And a change in the scale of thinking was necessary in order to appreciate the vicissitudes of climate. No one explained this more clearly than Caldas: "It is no longer about an ordinary map: reduced scales and all that has the appearance of smallness and economy should disappear. . . . Two square inches have to represent at least a league on the ground. Here are to be seen the hills, mountains, pastures, forests, fields, lakes, swamps, valleys, rivers, their twists and torrents, straits, waterfalls, fishing, all the settlements, agricultural concerns, minerals, quarries, in short every feature on the face of our land."[37] After all, "it is an error of judgment and reason," Caldas's colleague Unanue wrote, "to try to characterize a vast country by what is observed in one of its parts."[38] According to Caldas and Unanue, any attempt to generalize was an abuse, a display of ignorance that took no heed of the Creoles' call to appreciate local conditions and knowledge.

This controversy over scale makes sense of Unanue's statement that "plants are more sensitive in the tropics than outside them."[39] If plants are sensitive to local soil conditions and climate, then their locations are not interchangeable. Everything has its place and each country its own national forest. The same is true of all living beings, including people, for "although all humans on earth are descended from the same Father, the difference in climates, customs, and diet to which the first Diaspora reduced them has introduced such diversity into their features and attributes that, comparing several nations at the same time, they all appear to have sprung from different origins."[40] It was a commonplace in the eighteenth century that bodily differences arose from climate. But our illustrious man from Lima was not concerned with race alone. Like many of the enlightened, what interested him were nations, and what was unusual in his thinking was the inclusion of the diversity of vegetation among the fundamental features distinguishing nations. It is not that he was particularly keen on the scent of flowers or the taste of spices, but he imagined that human beings were created from the plants they ate. He only had to look out of the window to see that those who chew coca adapted better to the land and were resilient in the face of hardship. To deny this relationship was to be ignorant of the world of the Andes, to turn one's back on the evidence.[41]

What was true of coca, the "tonic of tonics," Unanue reasoned, might be true of other plants, such as cocoa, tobacco, cinchona bark, pineapple, and coffee. These were all remarkable products, and their effects were immediately obvious. It was also reasonable for Unanue to suppose that all plants had a beneficial, or beneficent, function since God had to fit into the explanatory machinery somewhere. But when the effect of any plant was not evident, Unanue, who knew nothing of biochemistry, turned to physical reactions. He agreed with the *foodist* claim that *we are what we eat!* He also included physical reaction (to taste, touch, and smell) as a reliable source of botanical knowledge. For him, local flora told as much about bodies as bodies told about flora. What is more, Unanue argued that regulating the use of plants could help to correct and direct human behavior. The explanation is simple: digestion annexed the various attributes of plants, and as a result, the character of a people responded to their alimentation. Managing a country wisely required selecting the plants to be produced and controlling the circulation of those consumed.

The Patchwork of Climate

The Creoles had discovered sociobotany: collective conduct is predetermined by the virtues of the plants we consume. They therefore felt no qualms about accepting the notion that a large part of what we are derives from what we eat. The Creoles knew nothing about evolution, but they had a mechanism that explained how the behavior of humans was closely linked to the characteristics of the plant life surrounding and nourishing them. We have seen how Creole criticisms of the applications of a particular type of botany (Linnaean) required a change in the scale at which science was conceptualized and practiced. A closer look at the classification of vegetation and the social function of flora led to the destabilization of Linnaean botany and also a deconstruction of nature as an easily handled notion. Creole botany and nature required that the characteristics and behaviors of local communities be taken into account. The final step in the Creole reform of botany was to reconceptualize climate. The Creole conception of "climate" was expansive and included the physical and moral identity of all the inhabitants (plant and animal) of the different American territories. For them, a proper notion of climate took into account characteristics of a people as diverse as its traditional farming and pharmaceutical practices to its techniques of classification and nomenclature. Climate, in short, posited fundamental links between temperament, temperature, and nurture. Caldas, the Creole from New Granada, once again comes to our rescue, for his definition of climate is pertinent here: "By climate, I understand not only the degree of heat or cold in each region, but also the electrical charge, the quantity of oxygen, the atmospheric pressure, the profusion of lakes and rivers, the distribution of the mountains, forests and pastures, the level of population, or the deserts, winds, rains, thunder, clouds, humidity, etc."[42]

The Creole notion of climate had the outward appearance of a scientific object, but no laboratory could contain it: an object can only fit between walls when it is constructed following a finite and quantifiable set of values. And such a set of values was exactly contrary to the Creole concept of climate at that time. In other words, climate, as mobilized by the Creoles, broke the rules of containment and became an object as much of politics as of science—an object to be smuggled across the border between nature and culture. Furthermore, this concept of climate promoted the notion of the nation as a natural construction. Nations, from a Creole point of view, were to be defined by their inhabitants' changing physical reactions to food, flora, and atmospheric phenomena. Governments, from this point of view, could be seen only as shepherds of living organisms.

Chapter 9

How Derivative Was Humboldt?

*Microcosmic Nature Narratives in Early Modern Spanish America
and the (Other) Origins of Humboldt's Ecological Sensibilities*

Jorge Cañizares-Esguerra

In this chapter I offer an example of how a more generous understanding of the Atlantic can yield new readings of the past, altering time-
honored narratives. Alexander von Humboldt (1769–1859) has long
been hailed as a founding father of the science of ecology and a genius.
He has been credited with single-handedly creating a new discipline that
relied on painstaking measurement to identify hitherto uncharted regularities in the mechanism of the planet as a whole. Humboldt, for example, demonstrated that individual species were part of larger plant communities and that these communities were distributed geographically
according to environmental variables such as elevation above sea level,
temperature, and soil composition (see figure 9.1). After a lengthy visit
to Spanish America (1799–1804), Humboldt went on to publish some
thirty volumes on subjects ranging from botany to the political economy
of Cuba and Mexico, consolidating his reputation as one of the leading
nineteenth-century scientists.[1]

Heroic narratives are today out of favor, and historians no longer portray Humboldt as a lone genius. Janet Browne, for example, has situated
Humboldt's ecological thinking within the larger history of the discipline of biogeography and shown that Humboldt drew on the ideas of
contemporary German scholars such as Johann Forster (1729–98),
Georg Forster (1754–94), and Karl Ludwig Willdenow (1765–1812). But
for all her efforts to historicize figures such as Humboldt and to show
his indebtedness to other Europeans, Browne remains wedded to the
notion that exotic places are worth mentioning solely as backdrops to
the exploits of European naturalists; she thus has no place in her study
for the discourses and ideas circulating in Spanish America that could
have influenced Humboldt.[2] In contrast, the British historian David
Brading has shown that Humboldt's works on the political economy of

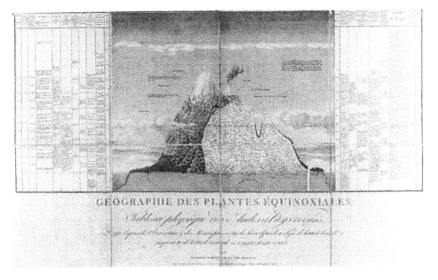

Fig. 9.1. Alexander von Humboldt's cross-sectional map of the Andes, in which correlations between plant communities, soil composition, and heights are shown. From Alexander von Humboldt, *Essai sur la géographie des plantes* (Paris, 1805).

Mexico and Cuba were possible only because he drew on the reflections of scholars in New Spain and on decades of data collection by colonial bureaucrats.[3] I have demonstrated that Humboldt's antiquarian scholarship on Mesoamerican societies grew out of his encounter with rich empirical and interpretive traditions in Mexico.[4] Here I propose a new narrative for the history of biodistribution that takes seriously the intellectual milieu Humboldt encountered in Spanish America and that takes into account the discourses about space and nature formulated in Spanish America during the seventeenth and eighteenth centuries.

The Euro-Creole Origins of Biodistribution

A generation ago the Catalan geographer Pablo Vila maintained that "geo-botany was born of the encounter between two sages," namely, Humboldt and the late eighteenth-century Colombian naturalist Francisco José de Caldas (1768–1816). Pointing to the remarkable similarities between Humboldt's research program and that of Caldas, Vila forcefully argued for the "Euro-Creole" origins of the new science.[5] Caldas was a self-taught naturalist and astronomer who, after having dazzled Humboldt, became one of the leading naturalists in José Celestino Mutis's botanical expedition to New Granada. Lionized by Humboldt for his fabulous collection of South American plants and natural history books on the "tropics," the powerful Mutis (1732–1808) hired Caldas, who spent the following three years (1802 to 1805) traveling throughout the Ecuadorian Andes identifying and classifying varieties of quinine plants for his patron. Impressed by the results, Mutis called Caldas back to Bogotá, the capital of the kingdom of New Granada, to direct a brand-new astronomical observatory. Once in the city, Caldas also edited a weekly, the *Semanario de la Nueva Granada*, and became increasingly involved in the local patriotic societies that sought to change society through enlightened reforms. During the wars of independence triggered by the political vacuum in the colonies caused by Napoleon's invasion of Spain in 1808, Caldas joined the patriot armies as both ideologue and military engineer. Captured by Spanish armies in 1816, he was shot in the back by a firing squad, signaling the end of an entire generation of patriot naturalists.[6]

In 1801, when he encountered Humboldt in Colombia, according to Vila, Caldas was already charting the geographical distribution of plants in the northern Andes. By the time Humboldt published his 1805 *Essai sur la géographie des plantes* in France, Caldas had already produced several biogeographical maps of the northern Andes (1802) (see figure 9.2), a memoir on the geographical distribution of plants near the equator (1803), and a study of distribution of quinine relative to height above sea level and temperature (1805).[7] These documents clearly indicate that Caldas was thinking about mapping biodistribution in terms identical to those later made public by Humboldt. Vila, therefore, insists on the "Euro-Creole" origins of theories of biodistribution.

For all Vila's insights, the evidence clearly shows that Caldas learned cross-sectional mapping of heights from Humboldt, not the other way around. The Prussian naturalist found a lonely and self-taught Caldas botanizing in southern Colombia while making a living as an itinerant merchant. Humboldt was impressed by the creativity of this Andean "genius," for Caldas had built instruments from scratch, kept extraordinarily accurate astronomical observations, and invented a mathematical formula to calculate altitude by noting the temperature of boiling water at different heights. Caldas seems at first not to have been overly impressed by the Prussian and was skeptical of Humboldt's reliability as an observer: "Can we hope to get anything useful and knowledgeable from a man who would traverse our kingdom with so much haste [four to five months]? Isn't he going to broadcast prejudices and false information to Europe as almost all travelers do?" Caldas, however, soon changed his tune and looked forward to benefiting from his encounter with Humboldt (including the promise of a trip to Europe that Humboldt later

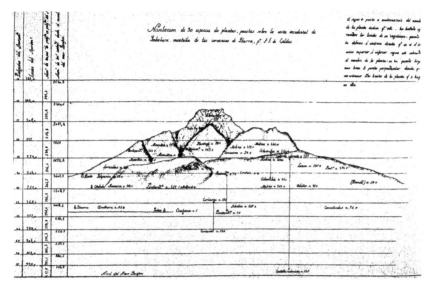

Fig. 9.2. Cross-sectional map of the Andes indicating correlations between altitude and the distribution of thirty plants. This is one of many maps drawn c. 1802 by Francisco José de Caldas that demonstrates striking similarities with those later published by Alexander von Humboldt. Caldas was inspired by a map Humboldt drew in October 1801. From Archivo del Real Jardín Botánico de Madrid (ARJBM), Division III, Signatura M-529. Published by permission of the ARJBM. Many thanks to Daniela Bleichmar for helping the author to obtain this image.

withdrew in haste when in Quito he met the handsome and aristocratic marquis Carlos Montúfar, whom Humboldt subsequently took to Europe): "I will seek to learn and suck knowledge from this sage to gain some small measure of enlightenment and overcome [my] barbarism." In fact, Caldas first obtained from friends a cross-sectional map of Andean heights that Humboldt had completed sometime in October 1801, and he was thus inspired to develop his own. Clearly Caldas was the junior member in this so-called "Euro-Creole" partnership.[8]

Vila's assertion of the Euro-Creole origins of biodistribution understandably bolstered patriotic pride in Colombia, but it left Eurocentric narratives in the history of science unchallenged.[9] For Vila's insights can be easily made to fit into diffusionist narratives of scientific discovery: Caldas emerges from our more careful analysis simply as the precocious disciple of the learned European traveler. In this chapter I seek to present an alternative narrative, one that focuses less on the origins of Humboldt's cross-sectional maps and more on the origins of his ideas about the Andes as a microcosmic space, a natural laboratory for testing theories of biodistribution. Although Humboldt might have arrived in Spanish America with a scientific agenda already framed by the writings of Karl Willdenow and the Forsters, he encountered a local intelligentsia obsessed with describing the rich ecological variations within their polities. Humboldt learned to read the Andes as a natural laboratory for the study of the geography of plant communities in part because local Spanish American scholars had for decades (if not centuries) been developing this idea.

The Microcosm and Paradise

Over the two hundred years before Humboldt's arrival, a tradition of natural history writing had arisen in Spanish America that considered the Andes a providentially designed space, a land seemingly endowed with all the climates of the world and thus potentially capable of housing or producing any natural product. This tradition resulted from the meeting of Indian and European conceptions of space. On arrival the Spaniards encountered civilizations in the Andes that from a European perspective exhibited curious patterns of settlement. Instead of relying primarily on markets to access "exotic" commodities, Andeans sought to control faraway resources by sending migrants to occupy distant ecological niches. Andean groups were fissiparous communities deployed in "vertical archipelagos." Spaniards soon learned to take advantage of these peculiar spatial arrangements for purposes of labor mobilization and commercial agriculture.[10]

The immensely rich diversity of ecological niches encountered by the Spaniards in the Andes prompted colonial scholars to associate the region with the biblical paradise. Paradise, it was thought, had once contained all the fauna and flora of the earth. In an effort to re-create this primitive space, naturalists in the Renaissance established botanical gardens. As John M. Prest has argued, the so-called discovery of America set off a vogue for collecting exotic plants in hopes of reviving paradise.[11]

In the early modern period, mountains were second only to botanical gardens as sites for envisioning paradise. As late as the eighteenth century, Carolus Linnaeus (1707–78) imagined paradise as a tall equatorial peak with a multitude of climates. The many microclimates of this mountain had once sustained all the fauna and flora of the world. As the oceans receded, however, species began to colonize distant geographical regions from the tropics to the Arctic as they sought environments that resembled the niches in paradise for which they had originally been designed. Linnaeus used the ancient construct of paradise as an equatorial mountain to explain biodistribution.[12]

Steep equatorial mountains with microclimates that reproduce those of the rest of the world, however, were not bygone primeval spaces. They could be found in America. Columbus was perhaps the first to think that the lands he had just discovered had been home to the garden of Eden. Like his contemporaries, Christopher Columbus thought that paradise was at the top of an extremely tall mountain, the nipple of a breast-shaped peak that reached beyond the sublunary sphere. To be perfect, paradise had to transcend the laws of physics, and in classical cosmology heavenly matter in the celestial sphere was not subject to change. Only above the spheres of earth, water, air, and fire could the generation and transmutation of the elements be avoided.[13]

Spaniards did not find in the New World peaks so tall as to be impervious to the laws of matter in the sublunary sphere. Nevertheless, they found in Andean mountains a way to explain why the Torrid Zone was in fact temperate even though the ancients had predicted it to be uninhabited owing to the scorching heat of the equatorial sun. Naturalists such as José de Acosta (1540–1600) held the Andes in awe as they discovered that climate was as much a function of elevation above sea level as of temperature. The equatorial mountain ranges Acosta encountered contained within relatively small vertical spaces all the climates of the terrestrial sphere. Spanish naturalists seem to have ascribed paradisiacal properties to the Andes (and the New World generally) as they sought to identify the meteorological mechanisms that kept the scorching tropics temperate.[14]

The first author to make this connection explicit was Antonio de León Pinelo (1590–1660). León Pinelo's *Paraiso en el Nuevo Mundo*, written between 1645 and 1650 but not published until the mid-twentieth century, sought to prove that paradise had once been located on the eastern slopes of the Andes (see figure 9.3). León Pinelo's work grew out of his dissatisfaction with all extant literature attempting to pin down the original position of the garden of Eden. Ancient learned consensus held that paradise had been situated somewhere in the Middle East or Asia. The new philological and geographical knowledge of the Renaissance gave novel twists to these age-old speculations. León Pinelo dismissed both new and old theories and argued that the correct reading of Genesis placed paradise in the Andes.

To prove this, León Pinelo engaged in high-flying philological speculation. He demonstrated that the Amazon, Magdalena, Orinoco, and Plate Rivers had the properties ascribed in Gen. 2:6–15 to the four rivers of paradise, namely the Gihon, Tigris/Heidekel, Euphrates/Perath, and Pishon. He showed that the reference in Gen. 3:24 to an angel with a flaming sword guarding the entrance to the garden was simply a metaphor for Andean volcanoes surrounding Eden. He also argued that the

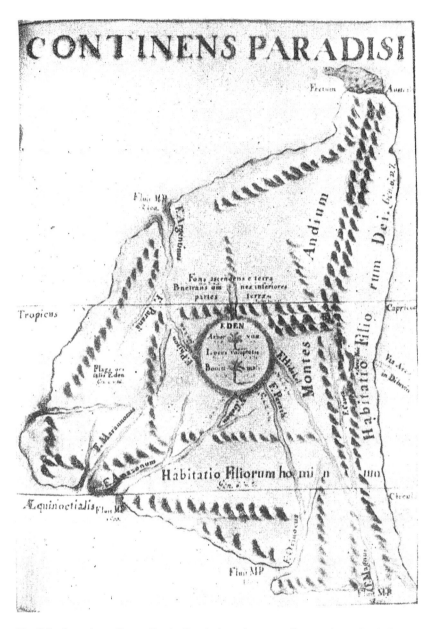

Fig. 9.3. Location of paradise in South America according to Antonio de León Pinelo. From Antonio de León Pinelo, *El paraíso en el Nuevo Mundo*, ed. Raul Porras Barrenechea, 2 vols. (Lima: Imprenta Torres Aguirre, 1943), vol. 1, 138.

tree of knowledge, whose fruit when tasted caused the Fall of Man, had most likely been the Peruvian granadilla (passion fruit or *Passiflora edulis*), for its flowers and leaves resembled the instruments of Christ's passion on the cross (nails, sponge, lance, wounds, bindings, and crown of thorns). The tree thus pointed to both the original sin and its redemption (see figure 9.4). More important for my argument, León Pinelo maintained that of all places on earth only the South American tropics near the Andes enjoyed the topographical and meteorological conditions that could have been home to a garden as temperate and as bountiful as Eden.

León Pinelo was skeptical that paradise could have been on top of a mountain, for life in the Andes proved that the thin air of very high altitudes made breathing difficult. Nevertheless, he maintained that of all places in the world only the Andes could have reached the middle region of the sphere of air, where corruption and the transformation of the elements were considerably retarded. In addition, the Andes helped him explain how a place on the equator, which should have been rendered uninhabitable by the scorching heat of the sun, was in fact the most temperate environment on earth. Andean heights offset the tropical position of Peru on the terrestrial sphere, yielding a perfect meteorological balance.

Once he showed that paradise on the equator was not an oxymoron, León Pinelo set out to demonstrate that the natural history of Peru was sufficiently rich to make his case. His catalog of local fauna and flora, however, was somewhat atypical because he offered only a list of wonders. Forced to offer reliable criteria to measure the organic capacity of the terrain, León Pinelo turned to a description of curiosities, believing that the more wonders brought forth from the land, the more likely it had once been home to paradise. It is not clear how León Pinelo drew the connection between natural wonders and the sacredness of the place, but early modern scholars thought that God best revealed his omnipotence through nature's play (artistry) rather than through nature's regularities.[15] Judging by the sheer size of León Pinelo's catalog of local wonders (it took up at least one-third of his treatise), Peru far surpassed any other competitor (see figure 9.5).

León Pinelo also suggested that the abundance of microclimates in the Andes was the cause for the wealth of wonders he had cataloged. The Andes allowed him to explain why Peru was so bountiful. León Pinelo identified three habitats in the Andes, each distinctively rich in its own way: the low-lying areas of the coast and the Amazons, the middle ground or *llanos*, and the high-altitude sierras. These multiple ecological niches rendered the area particularly productive, for as a crop withered in one niche it flourished in another. More remarkable was the fact that

**GRANADILLVS FRVTEX INDICVS
CHRISTI PASSIONIS IMAGO.**

Fig. 9.4. The passion flower with all the different instruments used in Christ's Passion. From John Parkinson, *Paradisi in sole paradisus terrestris* (London, 1629). Leoń Pinelo was clearly drawing on a well-established tradition of theological scholarship concerning this flower. This tradition begins in 1609 with the publication of works by Donato Rasciotti (*Copia del fiore et fruto che nasce nelle Indie Occidentali, qual di nuovo e stato presentato all Santità di NSP Paolo V* [1609]) and Simone Parlasca (*Ill fiore della Granadiglia, overe della passione di nostro signore Gieso Christo; spiegato, e lodato con discorsi, e varie rime* [Bologna: Bartolomeo Cocchi, 1609]).

Fig. 9.5. A monster in Peru. From Antonio de León Pinelo, *El paraíso en el Nuevo Mundo*, ed. Raul Porras Barrenechea, 2 vols. (Lima: Imprenta Torres Aguirre, 1943), vol. 2, 116.

its many microclimates made Peru hospitable to all crops and products. Whereas the plants of America were not easily acclimatized in Europe, all European crops yielded immense harvests in Peru.[16]

The Political Economy of Paradise

León Pinelo's natural history was chiefly concerned with cataloging wonders and curiosities, not with thinking broadly about ways in which the microcosmic attributes of Andean space could be used to generate wealth. His forceful, patriotic argument seemed disconnected from a discourse on the economy. It fell to eighteenth-century intellectuals to undertake this task. To capture this transition we may consider similar developments in Europe. Lisbet Koerner has shown that Linnaeus's taxonomy and natural history were intimately linked to cameralist (statist) discourses seeking to transform Sweden into a self-sufficient economy. Linnaeus sent students abroad to collect flora in the hope of weaning the polity from its dependency on imports. According to this utopian view, naturalists would provide, through careful acclimatization of plants in botanical gardens, all the raw materials needed for the kingdom of Sweden to become autarkic. Given Linnaeus's views on biodiversity, the effort consisted not merely in reproducing paradise but in making it economically viable.[17]

Spanish Americans who lived in the Andes did not have to send naturalists abroad to create wealth. They simply turned to the microcosm next door. Unlike Linnaeus, however, Spanish American intellectuals did not seek to make their national economies autarkic. They sought instead to transform their kingdoms into commercial emporiums by using the microcosmic ecological attributes of the Andes, namely, by supplying the consumers of the world with all they needed. A flurry of utopian debates on how to harness the untapped wealth of the Andes greeted Humboldt on his arrival in the kingdoms of New Granada and Peru. To understand these debates, we need first to understand the institutional and cultural context in which they took place.

Having been soundly defeated by the British in the Seven Years' War (1756–63), the Spanish Bourbons sought to introduce aggressive economic, administrative, and cultural reforms in every corner of their far-flung empire. Like their British-American cousins who felt entitled to their "English freedoms," Spanish American settlers had enjoyed unparalleled degrees of autonomy and self-rule until the post–Seven Years' War reforms. Spanish American societies were kingdoms, not colonies, autonomous polities in the loosely held composite monarchy that was the Iberian Catholic monarchy. These "kingdoms" (hierarchical polities organized on the principles of socioracial estates and corporate privileges) enjoyed numerous forms of local political representation (from city councils to cathedral chapters) that came under attack with the Bourbon reforms.

Determined to transform these kingdoms into colonies, the Spanish Bourbons turned to the new sciences. The Spanish Empire had long been losing territories along with status and prestige in the New World to other European powers. Some Spanish intellectuals maintained that the loss of territories began with losses in the struggle over naming, surveying, and remembering. The writing of histories of "discovery" and colonization and the launching of cartographic and botanical expeditions therefore became priorities for the state, and many such expeditions visited the New World. Naturalists sought to benefit the economy by identifying new products (dyes, spices, woods, gums, pharmaceuti-

cals) or alternatives to already profitable staples from Asia. Spanish botanical expeditions to the Andes, for example, put a premium on finding species of cloves and cinnamon to challenge British and Dutch
monopolies in the East Indies.[18] The logic behind sending botanical expeditions to the New World was best expressed in 1777 by the architect
of these policies, the physician Casimiro Gómez Ortega (1740–1818),
who promised José de Gálvez (1720–87), minister of the Indies, that
"twelve naturalists . . . spread over our possessions will produce as result
of their pilgrimages a profit incomparably greater than could an army of
100,000 strong fighting to add a few provinces to the Spanish empire."[19]

One of these expeditions was organized to survey and give names to
the resources of New Granada, a territory that had been recently transformed by the Spanish Bourbons into a viceroyalty (kingdom) with its
administrative center in Bogotá, in the hopes of bringing an end to the
immensely profitable British and Dutch illegal trade off the coast of Venezuela. José Celestino Mutis was the head of the expedition, and his
ideas typify the spirit of the enterprise; they also capture how quickly
notions of the Andes as a microcosm were grafted onto the original
Spanish expeditionary project.

Mutis arrived in Bogotá in 1761 as part of the viceroy's entourage. He
quickly set out to explore the land, and when he heard about the official
campaign to send botanical expeditions to the New World he requested
his efforts be acknowledged. In 1783 he found himself in charge of the
so-called "Botanical Expedition of the New Kingdom of Granada."[20]
Spanish merchants had long benefited from their monopoly over the
trade in quinine. In a century in which "fevers" were at the center of
medical thought in Europe, the febrifuge virtues of quinine made it extremely profitable. Quinine came from the bark of a tree found in a
small area of Loja on the eastern slopes of the Equatorial Andes. Mutis,
however, was determined to find quinine-producing cinchona trees on
the Andean slopes of Colombia as well. Mutis eventually found new species in Colombia, although they were different from the trees of Loja.
He simply assumed that similar areas (elevation above sea level, temperature, distance from the equator) should produce similar trees.[21]

Many of the efforts of the expedition under Mutis were fueled by the
assumption that similar environments engendered similar botanical species and that the Andes were a treasure-trove of microclimates. Thus, in
1785 Mutis alleged to have found in Colombia a substitute for Asian tea.
Mutis launched a campaign to convince Spanish authorities that his Colombian product was as good if not better than the tea Europeans had
been importing from China. Behind these efforts lay the idea that the
Colombian Andes were providentially designed with microclimates
capable of furnishing the world with any product. "Countless are the

natural productions with which Divine Providence has endowed this New Kingdom of Granada," Mutis argued in a letter sent to José Moñino y Redondo (1728–1808), count of Floridablanca, the Spanish state minister whose recent illness the new Colombian tea would likely cure. The striking organic potential of New Granada, Mutis maintained, was due to the fact that this kingdom "was like a center of the Americas in which similar or equivalent productions to those found in the immense space of the Old and New Worlds have been gathered."[22]

The members of the expedition led by Mutis did their utmost to spread the news of the fantastic economic potential of New Granada. For example, around 1790 the Creole lawyer Pedro Fermín de Vargas (1791–1830), a member of the first phase of Mutis's expedition (1783–91), portrayed New Granada as a land of unparalleled commercial potential. According to Vargas, this kingdom enjoyed a privileged geographical location where it was possible "to find almost all the climates of the globe." Colombia was a microcosm owing to the multitude of ecological niches created by the Andes and to the endless agricultural cycle of its equatorial climate. It was also a potential economic leader of the world. If an enlightened ruler were to build roads, and to protect and increase the population to accelerate the "circulatory rhythms" of the country, Colombia, according to Vargas, would be poised to supply the world with cinnamon, cloves, tea, betel pepper leaves (a stimulant that is chewed in Southeast Asia and that could have been replaced with coca leaves), and indigo of even better quality than equivalent Asian merchandise. The coastal plains of Cartagena and Santa Marta alone would provide the cotton needed by all factories of the world.[23]

Other members of the expedition used the pages of *Papel Periódico de la Ciudad de Santafé de Bogotá* (1791–97), a periodical created by fiat of the viceroy in order to stimulate a colonial public sphere and to spread the optimistic message that the key to the future prosperity of the kingdom lay in its microcosmic qualities. The future director of the Royal Botanical Garden at Madrid, Francisco Antonio Zea (1770–1822), led the charge for this message in 1790. Under the pseudonym Hebephilo, Zea called on the youth of Bogotá to become republican patriots, interested solely in the greater good of society. Such virtuous citizens armed with the tools of the new sciences would one day witness New Granada becoming a trade emporium, for it was a privileged land, "a favorite of Nature. For here Nature has shown herself in all her magnificence; here [she] has revealed even to the blind and the ignorant the bright pageant of her marvels."[24] A few days later a reader of *Papel Periódico* made Zea's views explicit by clarifying the reasons to be hopeful. "Nueva Granada," the reader argued, "is surrounded by the most beautiful and most diverse variety of climates, which are located at very little distance from one another." Moreover, according to this reader, the land could produce any type of natural commodity in the world, including balms, gums, medicinal plants, cotton, wheat, legumes, fruits, cattle hides, wool, birds, precious stones, reptiles, metals, coveted mercury for amalgamation, and even East Indian cinnamon and cloves. Only idleness, ignorance, and lack of republican virtue, the reader posited, could keep the new kingdom of Granada from fulfilling its unlimited economic potential.[25]

It is not surprising, therefore, that when Humboldt arrived in New Granada with rigorous new techniques to measure and chart biodistribution, Caldas embraced them almost overnight, producing studies and maps before Humboldt had the chance to publish his own. It was this prompt embrace that led Vila to suggest that biodistribution was an idea created by both Humboldt and Caldas. Yet this was not so. Caldas's charts and maps were simply spatial representations of much older ideas.

Like Mutis, Vargas, and Zea, Caldas was deeply committed to the notion that New Granada was a microcosm providentially designed to enjoy unlimited economic potential. "From the bosom of New Granada," Caldas insisted, "all the perfumes of Asia, African ivory, European industrial commodities, northern furs, whales from the South Sea, [in short] everything produced on the surface of our world [can be obtained]."[26] The microcosmic attributes of the Andes prompted Caldas to present New Granada as a natural laboratory to study correlations between behavior, race, and climate.[27] In addition to being a microcosm New Granada was geographically designed to be a trade emporium, a new Tyre or Alexandria. It was located at the center of the world and equipped with navigable rivers to carry staples from the interior to the coast as well as ports facing both the Atlantic and the Pacific. "Nueva Granada," Caldas maintained, "appears destined for greatness by its geographical position for universal commerce."[28] This type of logic led another member of Mutis's expedition, the naturalist Jorge Tadeo Lozano (1771–1816), compiler of a yet unpublished "Fauna of Cundinamarca," to predict in 1806 that his *patria* was poised to become in "a few centuries a vast empire that . . . will equal the most powerful in Europe."[29]

These ideas also surfaced in Peru, another place that witnessed botanical studies sponsored by the Spanish Crown. Hipólito Unanue (1755–1833), editor of a periodical in Lima, *El Mercurio Peruano* (1791–95), and a physician committed to reform, typifies scholars in Peru who gave León Pinelo's old ideas a new twist. Like Caldas, Unanue thought that Peru was destined to become a trade emporium. In addition to the microcosmic qualities of the Andes, Unanue focused on the physical features of the land, pointing to Peru's yet unfulfilled potential. "It seemed," Unanue argued, "that after having created the deserts of Africa, the fragrant and lush forests of Asia, and the temperate and cold climates of Europe, God made an effort to bring together in Peru all the productions he had dispersed in the other three continents. In this manner God has sought to create [in Peru] a temple for himself worthy of his immensity, [a temple] majestically surrounded by all the treasures hidden in this kingdom."[30] Peru was, in short, "the most magnificent work Nature has ever created upon the earth."[31] God had revealed a predilection for Peru in the subtleties of its physical structure. For instance, Peru had been chosen by God to keep the balance of the planet. The massive weight of the Andean mountains was responsible for tilting the earth's axis and thus for the very existence of Europe, which otherwise would have remained under water.[32] Like Mutis, Unanue speculated that certain local products were suitable substitutes for popular products whose monopoly was in the hands of Spain's European rivals. Coca, whose sharp, acrid particles stimulated circulation and digestion, could one day replace tea and coffee in the global economy.[33] Enjoying such unparalleled physical properties, Peru was poised to supply the world with all it needed.

Curiously, these ideas about the microcosm found audiences throughout the Spanish American lands, including the mountainous areas of Mexico and flat plateaus such as those of Buenos Aires. For example, Creoles in Buenos Aires, who had long called the pampas a "desert," useful only for wild-cattle grazing, also imagined the kingdom of La Plata as a microcosm. In their imagination La Plata became a land of multiple ecological niches poised, like ancient Tyre, to be "the center of all the commercial circulation of the world" and, like ancient Alexandria, "a port communicating the East and the West." In his 1799 inaugural address to the Nautical Academy, financed by the Consulado of Buenos Aires to train qualified sailors for a merchant navy, Pedro Antonio Cerviño (d. 1816) called attention to the Argentine capital's privileged central position in the world. "Our location [on the globe] is a most felicitous one," Cerviño maintained, "[because] North America, Europe, Asia, and the Pacific Ocean are all equidistant to us. This marvelous location assures us an immense commercial traffic. [We] will become the warehouse of the universe."[34] In 1801 Francisco Antonio Caballe, editor of the short-lived periodical of Buenos Aires, *Telégrafo Mercantil*, presented the viceroyalty of La Plata as a land capable of supplying the world with hides, tallow, wheat, cocoa, quinine, indigo, copper, henequen, "all sorts of resins and drugs, not to mention precious and abundant gold and silver . . . [as well as the equally precious] saltpeter, pearls, and seashells that can be found in the spacious Chaco." Caballe concluded, "without recourse to hyperbole," that "round the globe there is not any other land as rich, holding as many variety of products . . . and [thus] as suitable for establishing strong and powerful commercial institutions" as the viceroyalty of La Plata.[35] By 1802 it had become a truism among Creoles that their kingdom was "like a sea, [in which] we lose ourselves in the horizon . . . a land of wondrous mountains with the best wood in the universe," a land "located [right] at the center of the commercial world and deliciously situated at the margins of a mighty river"—in short, a land "with the greatest productive power in the globe."[36]

Mexico also found an avid audience for these kinds of microcosmic narratives. Juan Manuel de San Vicente, for example, argued in 1768 that Mexico, like Babylon, was "the world writ small (an *epitome*)"; its markets demonstrated the abundance of "this second Terrestrial paradise."[37] The physician Juan Manuel Venegas in 1788 offered cures for all sorts of diseases with prescriptions based on Mexican plants. For Venegas, New Spain was "the purse of Omnipotence; an Eden capable of providing Europe not only with precious metals, but also with many of the noblest vegetables, roots, woods, fruits, gums, and balms."[38] And José Mariano Moziño (1757–1820), one of many late eighteenth-century Creole naturalists intent on writing a *materia medica* based on local plants and Nahua herbal lore, was convinced that "every single medicinal substance [in the world], with the exception of some three or four, can be abundantly supplied by our land. [Mexico] produces, if not the same medicinal botanical species, others that are of equivalent or perhaps of superior efficacy."[39]

Conclusion

Natural history and botany played significant economic and ideological roles in the early modern world. In the first phase of colonization of the New World, Europeans single-mindedly pursued mineral riches. From Hernán Cortes to Walter Raleigh, conquistadors and explorers saw the New World both as an obstacle on the way to Asia and as an endless source of gold and silver. This was all to change by the seventeenth century. The emergence of a fledgling mass-consumption society in northern Europe set off a plantation boom on the Atlantic shores of the Americas, based on the ruthless exploitation of slave and indentured labor. The new wealth of the Americas suddenly turned "green." As the other contributors to this volume make clear, growing, harvesting, and distributing sugar, tobacco, coffee, indigo, rice, and quinine, to name only a few food staples and drugs, became sources of fabulous new wealth for both governments and merchants. The Iberian Catholic monarchy, however, proved slow to adjust to this new era: Under Philip III

(r. 1598–1621) and Philip IV (r. 1621–65), the Portuguese footholds in the Indian Ocean and the China Seas were picked off one by one by the Dutch, who transformed the loose multiethnic, maritime trading networks of Southeast Asia into monopolies to control the production and distribution of nutmeg, cloves, and pepper. The Catholic monarchy could not stem the northern European barrage in the New World either. It was only in the late eighteenth century that, under the command of a new Bourbon dynasty, the Spanish monarchy began to compete in the new global agricultural-botanical markets. By then it was simply too late. The possessions of Spain in the New World had already become independent kingdoms, with traditions of self-rule and historical identities resistant to easy colonial subordination. The well-intentioned plans of botanists to use the New World to grow cloves, cinnamon, and other spices to break the Dutch and British monopolies proved to be a mirage.

But if the great Bourbon botanical plans never materialized, the cultural transformation they brought about was profound. In societies that had long considered themselves kingdoms, the new botany became new cultural capital, allowing local intellectuals to develop providential idioms and discourses highlighting the untapped economic potentials of each polity. Projects designed to turn local societies into subordinate appendages of a new, revitalized modern empire unwittingly offered ideological tools that allowed those communities to think of themselves, literally, as middle kingdoms. Like their contemporary Qing counterparts in China, Creole intellectuals came to think of local polities as the center(s) of the world. The old microcosmic narratives initially deployed to prove that America had been the original location of paradise were suddenly redeployed to rethink the future of local kingdoms under duress.

The encounter with the Andes and the long-standing tradition of thinking about local kingdoms as microcosms could not have failed to impress Humboldt. His trip to Spanish America was not planned in advance: he had actually been trying to go to Egypt. He did not voyage to South America deliberately looking to prove the Forsters' and Willdenow's speculations on biodistribution true; his encounter with the Andes was serendipitous. Prompted by the ceaseless rhetoric about the microcosmic virtues of the Andes, Humboldt began to think of these mountains as a laboratory for testing theories of biodistribution. Historians have managed to write histories of biogeography without acknowledging that crucial components of Humboldt's ideas did not emerge in Europe. Humboldt arrived in a Spanish America humming with discourses of nature in which each *patria* was cast as a microcosm wondrously poised to become a trade emporium. Humboldt learned to read the Andes as a natural laboratory for studying the geography of plant communities only because local scholars had for years been toying with this idea. Bringing a Pan-Atlantic perspective to bear on seemingly eso- teric subjects such as the origins of the science of biogeography can yield strikingly rich harvests that challenge narratives of subjects that have managed to remain firmly Eurocentric in an age of transnational and global historiographies.

III.
Cash Crops: Making and Remaking Nature

Chapter 10

The Conquest of Spice and the Dutch Colonial Imaginary
Seen and Unseen in the Visual Culture of Trade

Julie Berger Hochstrasser

A *banketje* (banquet piece) painted by the Dutch artist Willem Claesz. Heda (1596/4–1680) in 1635 presents us with an elegant meal (see figure 10.1). Laid out on a tousle of shimmering white damask is an array of imports: French, German, or Spanish wine; a Mediterranean lemon; Baltic grain; Caribbean salt; Peruvian silver. Amid this copious spread one detail may at first glance seem inconsequential: on the small plate jutting out over the front edge of the table at far left, a little curl of paper (a used-up page of an old almanac) spills forth ground pepper. There is, however, much more here than meets the eye.

Throughout most of the seventeenth century pepper was the principal commodity in the luxury trade that made wild fortunes in the Netherlands for merchants and investors alike: Malabar pepper imported from the west coast of India; additional stock grown in Java, Sumatra, Malaya, and Borneo; and *China Peper*, bought from Indonesia by the Chinese.[1] Trade in this and the other fine spices—cloves, nutmeg, mace, cinnamon—was the driving force behind the greatest of all Dutch trading companies, the Dutch East India Company (Verenigde Oostindische Compagnie, or VOC).[2] Company directors insisted that "special attention must be paid to driving a peaceful trade throughout all Asia, which is what keeps the cooking going in the kitchens of the fatherland."[3]

The genesis of still life painting as an independent genre coincides strikingly with this key phase in the birth of consumer society, when the United Provinces of the Netherlands achieved a position of primacy in global trade. Period household inventories show works of art featuring still life compositions to have become increasingly widely owned over the course of the seventeenth century and to peak, along with the affluence of the republic, in the 1660s.[4] What is at stake in these representations of the fruits of Dutch commerce—at exactly this time?

Figure 10.1. Willem Claesz. Heda, *Pronk Still Life with Oysters*, 1635, oil on panel, 34 × 44″ (87 × 113 cm), Rijksmuseum, Amsterdam. This painting is a characteristic "monochrome *banketje*," a particular type of still life of the laid table with a unified tonal palette, painted especially in Haarlem in the early decades of the seventeenth century. Reproduced by permission of the Stichting Het Rijksmuseum, Amsterdam.

Such paintings are revealing primary documents in the visual culture of trade and, indeed, of colonial botany. Inquiry into the Dutch spice trade reveals much that is *not* articulated in the still life painters' mute assemblages of ready comestibles; but, as we explore the Dutch imaginary regarding this colonial trade, we shall find formidable historical significance in both what *is* and what is *not* represented. In fact, we shall discover also that this visual culture played a surprising role within the very workings of history, one all the more sobering because patterns of imaging and consuming established in the early modern period continue to reverberate today.

By the seventeenth century pepper had long been in circulation in Europe—mostly thanks to Iberian merchants—and its uses were already familiar. In the Netherlands pepper had appeared frequently in cookbooks as early as Gheeraert Vorselman's *Nyeuwen Coock Boeck* (New Cook

Book), published in 1560 in Middelburg. In the era before refrigeration, demand was understandable for a product that could smother the flavor of bad meat; its use was widespread enough that *peper* was also the name for a typical winter sauce for meat dishes. Pepper was prized as well for inherent properties that complemented the qualities of other foods in accordance with the medieval theory of the four humors, which still governed dietary theory during this period. In his influential *Schat der Gesondheyt*, Dordrecht physician Johan van Beverwyck (1594–1647) recommended hot, dry pepper to counterbalance cool or moist foods such as melon or oysters.[5]

Dutch painters' evocations of the precious commodity in their still lifes of laid tables accord well with van Beverwyck's recommendations. Heda's picture, for example, explicitly pairs pepper with oysters. Throughout the century pepper reappears in works by a variety of artists, in contexts invariably consistent with van Beverwyck's dictates.[6] Earlier Haarlem artists such as Floris van Dijck (1575–1651) and Floris van Schooten (1605–56) had painted some banquets containing pepper, while Heda and his contemporary Pieter Claesz (1597/8–1660) painted the most such depictions.[7] Willem Kalf (1619–93) served up the pepper paper in a fine porcelain dish in midcentury Amsterdam, while even Jan de Heem (1606–83/4), the most highly acclaimed still life artist of his time, remembered pepper in his copious banquets, now in a silver pepper shaker or *peperdoesje* [sic] (see figure 10.2).[8]

Conventional art historical interpretations have held that images such as these levied moral critiques against excessive consumption, but the adherence to dietary decorum they exhibit contradicts that notion, as does the sheer attraction of these virtuoso works. If the whole purpose was to warn viewers away from the items depicted, why lavish on them such luscious painterly attention? Granted, moralizing literary tropes employed just such semiotic inversion at the time.[9] But given the visual force of these exquisitely painted images, one senses that whatever else they may claim to do, the rich assemblages of imported luxuries effectively served to celebrate the achievements of Dutch trade.

Seventeenth-century commentaries offer clear indications of how remarkable Dutch commercial accomplishments were considered for their time. By 1669 Sir Josiah Child (1630–99), the aggressive merchant who would become chief shareholder in the English East India Company, marveled, "The prodigious increase of the Netherlands in their domestic and foreign trade, riches and multitude of shipping is the envy of the present and may be the wonder of future generations."[10] Denis Diderot (1713–84) would later write: "The Dutch are human ants; they spread over all the regions of the earth, gather up everything they find that is scarce, useful, or precious, and carry it back to their storehouses.

Figure 10.2. Jan Davidsz. de Heem, *Pronkstilleven*, 1665–70, oil on canvas, 54³/₄ × 45¹/₄″ (139.2 × 115.1 cm), Centraal Museum, Utrecht. This still life is typical of the more elaborate banquet pieces painted midcentury and later, which by the eighteenth century had earned the name *pronkstilleven* for their conspicuous display (Dutch *pronken* = "to show off"). The silver pepper shaker is at the very center of the composition here. Reproduced by permission of the Centraal Museum, Utrecht.

It is to Holland that the rest of Europe goes for everything it lacks. Holland is Europe's commercial hub. The Dutch have worked to such good purpose that, through their ingenuity, they have obtained all of life's necessities, in defiance of the elements."[11] The Dutch were equally effusive on this point. In an ode written for the VOC directors, Joost van den Vondel (1587–1679) proclaimed: "Wherever profit leads us, to every sea and shore, for love of gain the wide world's harbours we explore."[12] In van Beverwyck's manual the chapter on spices opens with a verse by the venerable Jacob Cats (1577–1660), who likewise seems unable to conceal his pride in the ability of the Dutch to overcome the geographical limitations of their small country:

All that one calls Holland is but a few towns,
And Holland's friends are but a few members,
   And of little scope; but nonetheless
   Everywhere various wonders are hidden here . . .
Everything the Heaven sends, or grows out of the earth,
That comes to us by sea imported into our harbors. . . .[13]

True to form as the country's leading moralist, Cats exhibits a more pious tone than Vondel's brash bravado. His passive voice almost magically banishes Dutch agency ("everything" simply "comes to us"), reattributing it to divine intervention:

God is like a sun who beams a thousand golden rays
Continuously upon our little garden;
   Whatever hung on trees, or stood in the field,
   Here that comes falling into the mouths of the people. . . .[14]

Cats cites the spice trade explicitly, reaffirming both its exotic content and its sheer amplitude:

The rich Indian plants of pepper, mace, nutmeg,
Are here thrown out of cellars like grain:
   Here one picks no cinnamon, no other noble herb,
   Yet we distribute it by whole shiploads.[15]

He concludes with a show of gratitude that virtually preordains the whole happy miracle of this bounty:

Consider this, Dutch folk, consider what a high blessing
Has been vested in you so wonderfully by the hand of God:
   Your fields are lacking in rich plants
   And yet what you don't have, you have after all.[16]

If Cats, the moralist, could wax this smug about the awesome undertakings of the Dutch spice trade, then surely painted depictions of its commodities could inspire similar pride in the citizens who viewed them. Wealthy merchants were among the major consumers of paintings at this time in the Netherlands, so this signification would not have been lost on them.

The Dutch spice trade surely was an awesome undertaking.[17] The first Dutch voyagers to make it around the Cape of Good Hope to Indonesia suffered sunk ships and heavy loss of life, yet the pepper brought back from Bantam more than paid for the venture. So in 1598 the "Company of Far Lands" sent off a second excursion, of eight ships. When the first four returned fifteen months later, an anonymous participant marveled, "So long as Holland has been Holland, such richly laden ships have never been seen."[18] Droves of smaller boats welcomed the vessels, trumpets blew, and flags waved as the great ships known as "East India travelers" fired gun salutes. A civic reception was given for the chief officers and merchants, while "the bells of Amsterdam rang peals of joy."[19] Altogether Jacob van Neck's (1564–1638) fleet brought back 600,000 pounds of pepper and 250,000 pounds of cloves, along with nutmeg and mace. The ultimate return on the combined initial investment was a staggering 400 percent.

It sounds too good to be true, and as things turned out, it was. The story carried with it a subplot of violence, exploitation, suffering, and death: the human and social costs of Dutch overseas trade were to prove appalling. First, for those in the employ of the VOC, the "great voyage" to the East Indies was fraught with danger. Dutch handlers reputedly outdid their competition by keeping their costs low through minimal provisioning and manpower, leaving their crews overworked and underfed. Between this, tropical disease, and shipwreck, a quarter of the men did not survive the voyage, while the unhealthy climate in Batavia took a further toll. Factoring out a small percentage to defection, odds for survival remained staggeringly poor: over the two-hundred-year life of the VOC, by various estimates, one in every two or three sailors who embarked never returned.[20]

Moreover, the trading companies only achieved control of this commerce through outright military force against rivals and rapid subjugation of indigenous peoples; Dutch patriotic fervor condoned a multitude of sins in this regard. In the case of the VOC, some of the most egregious were committed under the direction of Jan Pietersz. Coen (1587–1629) of Hoorn, recognized today as the real founder of the Dutch colonial empire in the Indies.[21] Appointed governor-general of the East India Company in 1618 at a critical time in its history, Coen quickly conquered the fort at Jacatra (now Jakarta) to found Batavia, which would serve as the long-awaited rendezvous for the Dutch spice trade in the East.[22] Highly effective but ruthless, Coen wrote to the company directors: "There is nothing in the world that gives one a better right, than power and force added to right."[23] His words are chilling in light of his actions, which spoke even more loudly.

To build the Dutch outpost of Batavia, he leveled the native port of Jacatra; but most appalling were his actions on the Moluccan Islands. The spice-growing districts of the Moluccas were concentrated on the island coasts, exposed to the reach of Dutch sea power; while Dutch ships were built as both freight carriers and fully outfitted warships, local rulers possessed no warships of any strength. Dependent on rice, cotton textiles, and other necessities imported from Java, Malaya, and India, inhabitants of the Spice Islands were thus particularly vulnerable, unable to fight back against the Dutch as could mightier kingdoms on the Asian mainland. Hence "the Heeren XVII enthusiastically supported, when they didn't actually initiate, aggressive action in the Moluccas, at times when they deprecated or forbade the waging of offensive warfare elsewhere."[24]

In 1621, in punishment for the Bandanese elites' continued ploys to circumvent their 1602 contract with the VOC for exclusive provision of mace and nutmeg, Coen arrived in Banda with a large fleet and 1,200 soldiers, and in the largest massacre of the company's history, they effectively obliterated Bandanese society. Fifteen thousand Bandanese were slaughtered, driven away, or enslaved in Batavia. Of the 789 shipped to Batavia, many died of disease or were executed for treason or conspiracy.[25] The remaining thousands on Banda were blockaded in the mountains until an attack found most of them starved to death; some of the survivors jumped to their deaths from the cliffs. Coen recounted coldly: "About 2,500 are dead either of hunger and misery or by the sword. So far we have not heard of more than 300 Bandanese who have escaped from the whole of Banda. It appears that the obstinacy of these people was so great that they had rather die all together in misery than give themselves up to our men."[26]

While nutmeg first originated on the island of Banda alone, pepper-growing regions spanned the southwest coast of India and the island of Sumatra. Principal ports were Jambi and Palembang on Sumatra's east coast, where the VOC had small buying posts, and Bantam on the northwest coast of Java, which the Dutch monopolized in 1684.[27] Still, the expanses that needed policing to suppress local trade were far too vast, and the coastlines of the many islands too complex, ever to achieve a complete pepper monopoly.[28] Ruthless tactics notwithstanding, the struggle for monopolies on the spice crops dragged on for years. Boxer cites its end in 1684, with Malacca, Macassar, and Bantam all under VOC control and the local population laid to waste: "Indonesian shipping and trade in the Spice Islands had been virtually annihilated, with dire consequences for the economy and living conditions of their inhabitants."[29]

The inhuman policies of Coen and his successors did not go uncriticized in Holland. It may have been Lorenzo Reael (1583–1637), later to serve as governor-general, who wrote angrily: "What honourable men will break up their homes here to take employment as executioners and gaolers of a herd of slaves, and to range themselves amongst those free men who by their maltreatment and massacre of the Indians have made the Dutch notorious throughout the Indies as the cruellest nation of the whole world?"[30]

Disturbing, however, is the fact that colonial botany is discussed in seventeenth-century publications with scarcely any mention of the atrocities of the VOC's colonialist regime. Van Beverwyck is coolly scientific in his descriptions of pepper, citing its distant origins and outlining the various types, likening leaves and growth patterns to those of more familiar plants. His accounts of the origins of other spices are similarly detached, sparing his Dutch readers the (literally) gory details: "Cloves grow only in the Moluccan Islands, on trees like laurel-trees. . . . Nutmeg and mace grow not far from the Moluccan Islands, in Banda, also in the islands of Javu [Java] and Sunda on trees not unlike our pear-trees."[31]

Travelogues are still more revealing in this regard; texts and illustrations alike yield more information than just the botanical and geographical details they were devised to record. They too are valuable documents in the history of colonial botany, offering various insights into the Dutch colonial imaginary. Consider the *Itinerario* of Jan Huyghen van Linschoten (1563–1611), an account of his travels between 1579 and 1592, first published in 1596 and rapidly republished in many languages. Having begun his career in the service of the Portuguese, van Linschoten was instrumental in inspiring Dutch merchants to follow in the footsteps of their Iberian rivals in the spice trade, by exporting pirated navigational information to the Netherlands, as well as through tempting descriptions of the botanical bounty to be won (on van Linschoten, see also Claudia Swan's chapter in this volume).[32]

Van Linschoten's 1596 *Itinerario* contains, for example, an illustration of local flora such as "Oriental Pepper" twining around an *Arequeira* palm near "Indian Fig Trees," all clearly labeled for our edification (see figure 10.3). The vision of lush abundance echoes van Linschoten's textual suggestions of apparently limitless supply. Compare the vast forest stretching into the hills at left with his verbal description of Ceylon: "the best cinnamon of the whole Orient grows here in whole woods and wildernesses, whence it is exported and distributed throughout the whole

Figure 10.3. From Jan Huyghen van Linschoten, *Itinerario* (Amsterdam: Cornelis Claesz, 1596), engraving. Van Linschoten's report on his voyages and experiences with the Portuguese was an important early source of information for Dutch investors; his tempting descriptions of the rich harvests of pepper and other spices inspired the eventual formation of the Dutch East India Company. Reproduced courtesy of University Library of Amsterdam, Rare Book Department.

world."[33] He is equally wide-eyed about the supply of cloves in the Moluccas: "These Islands have nothing but Cloves: but in such great abundance, that they seem to fill the world."[34]

The figures in this print comfortably fit this mood of virgin paradise. Front and center, a putto-like child nibbles on what is labeled a "fig," which he holds in one hand, and a "fig leaf"—again clearly labeled—is trailing from the other.[35] Behind him a couple strolls along the riverbank, one nude, the other draped; other natives capsize a canoe, display their vaunted skill at climbing a tall palm, and crank a spit over an outdoor fire. While van Linschoten's text reports poverty and weakness among the Canarijn people depicted here, this scene—idyllic, devoid of European presence—recalls nothing so much as a tropical garden of Eden.[36]

In contrast, the several massive publications by Johan Nieuhof (1618–72) recounting his travels in the service of the West and East India Companies bear witness to the changes that took place over the course of the seventeenth century. So comprehensive that they were still consulted as authoritative anthropological sources long into the eighteenth century, Nieuhof's voluminous tomes treat in extensive detail not only colonial botany but also the appearance, customs, and beliefs of indigenous peoples; and yet their cataclysmic histories during this period are reduced to justifications of VOC actions in pursuit of the elusive spice monopoly.

Nieuhof's 1665 account of a journey in the service of the VOC incorporates both verbal and visual documentation of the various botanical commodities that concerned the East India Company on this expedition.[37] The plate illustrating a *Nagel Boom* (Clove Tree) is much more focused than Linschoten's: here, indigenous figures are intent on the activities of harvest, corroborating the development—in the view of Europeans—of the commodity value of these natural wonders (see figure 10.4).

In his discussion of pepper, Nieuhof, like van Beverwyck, resorts to analogies with plants more familiar to Dutch readers, such as currants or grapevines, but then, as befits an eyewitness account, describes in more specific and accurate detail the types of pepper and its cultivation.[38] Yet this pseudo-objectivity does not conceal the inherently superior attitude Nieuhof betrays toward indigenous peoples. His explanation of the process of shelling black peppercorns to make white pepper lets slip some unabashed prejudice: "after it has been steeped in Salt-water . . . the outward skin . . . swells, so that the white Pepper-corns within may be taken out with little or no trouble. . . . And if the lazy Indians were not so averse to labour, there might be much more white than black Pepper."[39] Nieuhof's casual aside about "lazy Indians" offers but a hint of the vast cultural hubris that had made way for Coen's ruthless policies.

More baldly, Nieuhof's 1682 account of his travels through the West and East Indies, nearly a century after Linschoten's publication, does report repeatedly on indigenous peoples who were killed or whose villages were burned in retribution for resistance, revolt, or violation of the Dutch spice monopoly—though Nieuhof consistently justifies the VOC's actions, and celebrates VOC victory.[40] His illustrations too register a decisive addition: a print similar to Linschoten's, again clearly labeling the pepper (at lower left), depicts another lush tropical scene inhabited by indigenous people—only this time Dutchmen make their entrance (see figure 10.5). Again the towering vegetation dwarfs all human presence—a visual corollary to its botanical importance to scientist and merchant alike. Equally striking is the interaction between Dutch and indig-

Figure 10.4. *Nagel Boom* [Clove Tree], engraving from Johan Nieuhof, *Beschryving van't Gesandschap der Nederlandsche Oost-Indische Compagnie aen Den Grooten Tartarischen Cham nu Keyser van China* [Embassy of the East India Company to the Great Tartar Khan, or Emperor of China] (Amsterdam: Jacob van Meurs, 1670). The writer and draughtsman Johan Nieuhof was sent along by the Dutch East India Company to record his impressions on the company's first trade venture to Peking, resulting in this publication. Reproduced courtesy of the Koninklijke Bibliotheek, The Hague.

enous people pictured. At right they join in a smoking circle, but at left is a quite different scenario: a second Dutchman stands stiffly, cutting a prettified figure and looking unbearably stuffy in his full European attire, while behind him a serving boy holds a parasol poised at the ready to shade him from the sun. One can understand why—how uncomfortable that elaborate outfit would be in the tropical heat!—and yet one cannot mistake his imperious pose and his utter disregard for the servant at his back.

Similar presumption is still more painfully obvious in the widespread imagery of the allegorical personification of the "Maid of Amsterdam," who presides over scenes of trade to represent Amsterdam as its world center. She appeared as early as 1606 on the lid of the Amsterdam city clavier, and in 1611 Claes Jansz. Visscher (1587–1652) presented her in a print, in which she is receiving a procession of individuals offering commodities from various regions of the world (see figure 10.6).[41] By

Figure 10.5. Engraving from Johan Nieuhof, *Gedenkwaerdige Zee en Lantreize door de Voornaemste Landschappen van West en Oostindien* [Noteworthy Travels by Sea and Land through the Principal Territories of the West and East Indies] (Amsterdam: widow of Jacob van Meurs, 1682). Nieuhof eventually served both the East and West India Companies, which by this point conducted active trade throughout the world. His voluminous texts and images were long valued as authoritative anthropological sources; yet this depiction betrays an imperious attitude toward indigenous people that Nieuhof's text corroborates. Reproduced courtesy of the Koninklijke Bibliotheek, The Hague.

Figure 10.6. Claes Jansz. Visscher, *Profile of Amsterdam from the IJ (Allegory on the Prosperity of Amsterdam)*, 1611, etching and engraving, 10 × 44″ (25.5 × 112 cm). This image from early in the seventeenth century casts Amsterdam allegorically as a maid to whom all the fruits of trade are offered up. Reproduced by permission of the Stichting Het Rijksmuseum, Amsterdam.

midcentury we find the motif of Amsterdam as the doyenne of trade writ large, in grand sculptural form, in the great Town Hall of Amsterdam, completed in 1653; here, the visual culture of trade was cast programmatically in stone. While imagery of Amsterdam's economic and political power permeated the iconographic program of the decor inside and out, the sculptures carved by Artus Quellinus (1609–68) from the designs of Jacob van Campen (1595–1657) for the pediments of the building, completed in 1665, bring global subjugation into focus. Atlas balances the globe of the heavens above the back pediment, while inside its pyramidal frame the Maid of Amsterdam dominates, with allegorical representations of the four continents then known bowing down or bearing tribute to her; the medieval cog-ship from Amsterdam's coat of arms is traced behind her in bas-relief, and the world lies (literally) at her feet (see figure 10.7).[42]

Such female allegorical figures representing Amsterdam (or, alternatively, Holland or all of Europe) became extremely common, reappearing in frontispieces of books, such as the 1663 "Historical Description of Amsterdam," in which the peoples of the world again genuflect before her. The ship in the background contextualizes the whole business in commercial terms (see figure 10.8).[43] The same symbolism that early in the century proclaimed optimistic pride of accomplishment stands at ever greater odds, as the century progresses, with the darkening tarnish of Dutch exploits.

If from such revealing images we might at least infer the more disturbing presumptions of this trade history, its darker side is conspicuously absent from the tables of Dutch still lifes. What, after all, does the violent subjugation of native peoples in the far-flung regions of the globe from

Figure 10.7. Artus Quellinus and assistants, back pediment, Amsterdam Town Hall (now Royal Palace), marble bas-relief, 1650–65. Here the Maid of Amsterdam returns, with all the continents of the world bearing tribute, in the stone sculpture adorning the rear pediment of the newly constructed Town Hall. Reproduced by permission of the Gemeentearchief Amsterdam.

which the Dutch derived their great wealth in trade have to do with these paintings? None of that ugly history shows up alongside the pepper in Dutch banquet pieces, served up on silver platters, as it were, worlds apart from such unpleasant conditions. Cats's poetic sleight of hand echoes here—these painted commodities too leave agency neatly out of the equation—and perhaps that was precisely the point. As the history grew more egregious, the still lifes seem to become more adamantly opulent: Jan de Heem's ebullient cornucopia (see figure 10.2), which features pepper served in a costly silver pepper shaker, is characteristic of these richer images at midcentury and later—so sumptuous as to come to be referred to as *pronkstilleven*.

Ironically, then, with regard to the prevailing moralizing interpretation of still lifes of the laid table, there was a great deal these painters *could* have moralized about that they did *not*. Did this contribute to the allure of still life for the wealthy patrons of the period? This very ambiguity, which permitted Dutch viewers to marvel at the visual feasts painters had laid before them, replete with overtones of pride in the power of

Figure 10.8. Engraving of Amsterdam receiving the tribute of four continents, frontispiece to *Historische Beschryvinghe van Amsterdam* [Historical Description of Amsterdam] (Amsterdam: Jacob van Meurs, 1663). The now-familiar imagery of the Maid of Amsterdam reigning over all the world recurs often in print as well. Reproduced courtesy of the Koninklijke Bibliotheek, The Hague.

the Dutch to assemble such a cosmopolitan array of riches, without any of the more troubling particulars—did this help make still lifes of the laid table so popular at just this historical moment? Of course, it remains difficult to ascertain how much viewers and painters of such pictures could have known of the grim particulars at the time. We cannot blame the painters or equate their motives anachronistically with those of unscrupulous corporate advertisers. On the contrary, their concerns seem to focus more on sheer pictorial virtuosity, as the quality and power of their renderings attest. But the fact remains that here, what was *not* depicted in these images is every bit as revealing for us, as students of visual culture and of the visual culture of colonial botany, as what *was*.[44] By extricating these commodities from the incriminating facts of their acquisition, these images relieved their viewers too of the distressing details of colonial domination.[45]

Other ironies trouble this history as well. First, all told, the Dutch attempt at securing a spice monopoly proved so costly that, although one source credits pepper with a third of the total VOC income at spice auctions, real profits fluctuated from large to little or none.[46] Complicated bookkeeping methods prevented the VOC directors from accurately calculating their expenses throughout the life of the company so that, unbeknownst to them, the daunting costs of enforcing the monopoly "with fleets, forces and garrisons" actually consumed their profits in the long run.[47] Still more ironically, the elision of historical evidence in period accounts (both written and pictorial) reveals more about the eventual fate of the Dutch economy than everything that was asserted more deliberately at the time: it was precisely what they overlooked that caused the downfall of the Dutch Golden Age. Historian Jan Luiten van Zanden has recognized that the exploitation of colonial populations and portable slave stock fed the growth of the Dutch economy. But, he argues, the nonsustainability of this labor supply, coupled with the long-term decimation of the Dutch population through sea travel, contributed directly to the decline of the Dutch economy: "the payment of labour was insufficient to reproduce labour power," resulting in a series of changes in the structure of the labor market that help to explain both the transformation of Holland's economy between 1580 and 1650 and its stagnation after 1670.[48]

Thus the social costs of exploitation both at home and abroad were conveniently overlooked by society just as they were in elegant compositions of the laid table. Eventually, though, they took their toll. In one respect the Dutch did indeed recognize the economic and social realities of the costs of pepper: that is why it *was* so costly. Karl Marx's dictum vividly applies: "What is the value of a commodity? The objective form of the social labour expended in its production. And how do we measure the quantity of this value? By the quantity of the labour contained in it."[49] Figured by this equation, the costs of pepper were wildly extravagant, even unbearable, as in fact they proved to be when one factors in untold costs in human lives, both Dutch and other. As van Zanden contends, by failing to acknowledge these social costs, the Dutch republic precipitated the end of its own wrongly won prosperity.

Sadly, these inequities are not yet behind us. While copious subsequent scholarship on the spice trade has better illuminated the historical record, a recent flurry of popular histories further attests to its contemporary relevance: today the tiny island of Banda, once the world's only source for cloves, is impoverished and nearly barren, its spice profits denied to residents through artificial price suppression by an international cartel—based in Rotterdam.[50]

We are heirs to this legacy in a much broader respect as well. Deceptively facile consumption of goods derived from distant markets runs rampant in the hyperdeveloped consumer society of late capitalism; the patterns of consumption established during the early modern period have proliferated to a scale and extent that strain the social and ecological capacities of the globe to sustain them.[51] By now this is decidedly no accident: glitzy advertising is ruthlessly deliberate in its commercial agenda in a way that Dutch still life painting was not. Yet seductive Madison Avenue advertising photography does bring uncannily to mind the selfsame sheen that Roland Barthes saw in seventeenth-century Dutch still life painting: "Still life painters like . . . Heda always render matter's most superficial quality: sheen. Oysters, lemon pulp, heavy goblets full of dark wine . . . what can be the justification of such an assemblage if not to lubricate man's gaze amid his domain, to facilitate his daily business among objects whose riddle is dissolved and which are no longer anything but easy surfaces?"[52] Just such "easy surfaces" surround us today. Even with the dramatically improved communications of the twenty-first century, staying informed about the true social costs of First World appetites requires sustained effort.[53]

However, if the Dutch led the way during their Golden Age in consuming beyond their means, they have also led the way in recent years in identifying the implications of this imbalance and calling for the problem to be addressed. In 1992, Friends of the Earth Amsterdam published their investigation of allowable annual levels of globally equalized and sustainable consumption, concluding that in almost every area, the average person in the Netherlands is consuming far beyond his or her means and is thereby depriving people in poorer countries of the ability to meet their basic needs.[54] In their turn, in 1994 the Washington, D.C. chapter applied the Dutch estimates to the United States with equally alarming findings.[55] Urban planner William Rees similarly observes that the consumption of an average person living in a high-income country actually requires far more land than is sustainable within its borders. He cites the Netherlands as his first example, due to the unusual density of its population, estimating that the country currently consumes an output equivalent to some fourteen times the productive land within its borders, "in part by drawing down their own natural resource stocks and in part through international trade that allows them to expropriate the resources of lower-income countries."[56]

The reader will now readily recognize that the crisis of global imbalance facing the world today has roots reaching back to the seventeenth century. One is haunted by the recollections of Cats and his fellow Dutchmen, who prided themselves on their ability to roam the world, gathering whatever they needed (or desired)—whatever they did not have at home—from far beyond the borders of their own country. Their strength (or so it was then perceived) has become our weakness (for so it is finally recognized). What does and does not figure in the visual imaginary of Dutch colonial botany carries a more urgent lesson than we might ever have thought. For the question still haunts us: who consumes what, and at whose expense?

Chapter 11
Of Nutmegs and Botanists
The Colonial Cultivation of Botanical Identity

E. C. Spary

The introduction and naturalization of plants and animals for human profit and benefit have been linked to the rise of early modern European empires, with all their problematic implications for local peoples and environments. For this reason, the global travels of plants between different cultures and natures have attracted much historical and ecological interest. Particularly during the eighteenth century, European scientific travelers and settlers played more important roles than ever before in the distribution of new species or the redistribution of known ones.[1] "Big histories" of colonial botany, however, pass swiftly over the highly contested, complex procedure by means of which the living plant in its "other" location became interchangeable with the metropolitan substance of commerce, cuisine, or medicine.[2] In this chapter I present a microstudy of a highly publicized case of exotic plant introduction: Pierre Poivre's successful effort to naturalize the nutmeg, a commodity monopolized by Dutch traders. For over two decades, beginning in the 1750s, Poivre (1719–86), a missionary naturalist turned colonial administrator, attempted to introduce nutmeg cultivation to French colonies. The quest for the botanical identity of this species was intimately tied to the difficult task of achieving scientific authority. Inattention to this problem, I argue, undermines simplistic narratives of the relationship between botanical expertise, colonization, and empire.

As Steven Shapin and others have shown, the credibility of a knowledge claim depends on the credibility of the scientific practitioner who advances it. The status of the scientific expert is both fragile and ephemeral, an outcome of negotiation at several levels: social, political, epistemological. Yet "big pictures" of colonialism are winners' histories: they ascribe undisputed, univocal authority to scientific experts, a status which is only retrospectively attained.[3] Representing a given natural production as historically unitary is, similarly, achieved only at the cost of silence concerning the work done by scientific practitioners to define and circumscribe it, and to attach it firmly to selected previous meanings, transactions, or spaces while leaving room to fashion new ones. Between its point of origin and its point of consumption, a species, the unit of botanical analysis, was subjected to many operations and transformations that simultaneously endangered and constructed its identity.

In part, identity was stabilized through the mastery of classification. Natural history supplied preexisting frameworks of description and differentiation: by slotting a given natural body into taxonomic categories, one could fashion its scientific identity. The creation of identities for individual plants allowed naturalists to undertake a variety of operations with the species while maintaining authority over it as a single, historically continuous entity. For a plant such as nutmeg to be the subject of scientific experimentation and agricultural exploitation, it had to be assimilated to existing classificatory and descriptive schemes, yet clearly distinguished from similar species.[4] Such work was still necessary even after nutmeg was already (according to the "big picture" histories) officially "known" to European botanists. Only these social and botanical maneuvers ensured that metropolitan scientific and medical practitioners could stand as credible guarantors of the authenticity of a particular species. It is a testament to the difficulty we *should* ascribe to the identification process that criteria such as odor and appearance were not enough to settle the issue in the eighteenth century, even for consumers more familiar with nutmeg (then a widely used seasoning and medicament) than we are.

The settlement of disputes over authenticity retrospectively stabilized both botanical expertise and the criteria of authentication. These disputes, however, reflected botanists' place within the parallel patronage hierarchies that generated the colonialist and imperialist policies of absolutist France. Many served as administrators and spies, on a mission to capture natural resources for the state. These experts in the management of exotic natural productions thus mediated between state colonial ambitions and metropolitan botanical networks. They simultaneously engaged in the micropolitics of botanical identity and the macropolitics of eighteenth-century colonial and social life.

The process of settling political disputes fueled the manufacture of scientific authority over the management of nature. Botanical experts retrospectively worked to make nature univocal in written accounts of controversies over botanical identity, which ascribed disinterested access to truth to the winning side and political interestedness to the other.[5] The diverse forms of delegated authority in colonial settings and the tenuousness of metropolitan control over colonial agency, coupled with the lack of agreed standards both for botanical identification and for judg-

ing botanical expertise, ensured that under certain conditions the reso-
lution of botanical controversies over an unfamiliar species could be virtually impossible. Rather than following a macrohistorical approach, I
will stress the difficulty encountered by contemporaries, even in a single,
narrowly defined cultural space, in attempting to create and maintain
the identity of useful plants. I will show how social, political, botanical,
and classificatory relations were coproduced in the process of settling
the identity of individual species. In the course of doing so, I shall venture a larger historiographical claim: namely, that the history of colonial
botany be written as process, negotiation, and exchange, rather than as
event(s).

Spicing Up History

French agronomic projects of the later eighteenth century were conceived on a grand scale. The colonies were to become the sites of a commerce in exotic natural luxuries that would, some claimed, replace the
costly foreign trade in such goods with an internal colonial production
supplying metropolitan consumption. Supporters of plant introduction
schemes envisaged the extension of the plantation system, by which
sugar was mass-produced in the colonies, to other natural productions,
particularly those monopolized by France's European rivals. The Seven
Years' War was a crucial turning point in the success of such enterprises.
French empire building was curtailed by the confiscation of the North
American colonies following the Peace of Paris in 1763. On the Indian
subcontinent and in the West Indies, only commercial installations survived; the French were banned from building military fortifications or
making treaties with local rulers. The defeat was taken by ministers to
demonstrate the weakness of France's navy; for many, it underlined the
importance of competing on a different terrain, that of agriculture.
Over subsequent decades ministers and scientific administrators treated
the improvement of colonial cultivation as an essential precondition of
national military and commercial strength. The history of France's acquisition and acclimatization of nutmeg will demonstrate the extent to
which botanical knowledge depended on the relations between metropolitan patronage, scientific institutions, and colonial agriculture.

Today, as in his lifetime, Pierre Poivre is principally known for his
involvement in breaking the Dutch hold over the global trade in fine
spices. Although sporadic attempts were made to support its introduction to French colonies from the 1680s onward, nutmeg, along with
cloves, entered French colonial cultivation only in the 1770s. This followed a prolonged period during which the Dutch policed these crops
with naval precision at the centers of their spice trade, the islands of

Amboina and Banda, and exterminated plants growing wild elsewhere.[6] From the 1720s onward the French spice trade was monopolized by the Compagnie des Indes (East India Company), which assumed governmental and administrative responsibility for several colonies. Among the administrators employed by the company was Poivre, the one-armed son of a Lyons silk merchant, whose training as a missionary had equipped him with extensive knowledge of the Far East. Poivre promised his superiors that it was possible to staunch the flow of French wealth to foreign purses by expropriating the original plants from which the spices came. Beginning in the 1750s he made or organized several voyages to Indian Ocean islands, seeking nutmeg and clove plants for the leading French Indian Ocean colony, the Isle de France (Mauritius). After a lengthy stay in the Philippines, Poivre returned to the Isle de France for a brief visit in 1754. He brought with him just five nutmeg plants, obtained by prolonged negotiation with Spanish governors of islands near the Moluccas.

The possession of the nutmeg plants did not, however, confer automatic credibility on Poivre's enterprises. Credibility would hinge on botanical expertise—specifically, the establishment of rules for differentiating "true" species and varieties of food and spice plants from others that were similar but lacking in the appropriate qualities. This is evident in the interpretation of the most famous of the spice piracy voyages, which took place soon after Poivre's appointment as royal intendant of the Isle de France in 1767. The first, in 1769–70, was captained by one d'Etcheverry, a lieutenant in the French royal navy, who claimed in his report to have obtained "twenty thousand nutmegs, as seeds or plants, and three hundred clove trees" for the intendant. On board was a botanical expert, Mathieu-Simon Provost, who two years later accompanied the chevalier de Coëtivy, commander of the Crown vessel *Nécessaire*, in a top-secret expedition back to the first source, the outlying islands of Gebe and Patani in the Moluccas, nominally under Dutch rule. On both occasions all the resources of diplomacy were deployed in negotiating with local inhabitants and Dutch malcontents, and thrilling escapes from threatening Dutch vessels were recounted in the expedition reports. The voyages took an explicitly imperialist form: d'Etcheverry planted the French standard on Gebe, symbolically laying claim to this territory, which the Dutch had been unable to subdue. The nutmeg and the clove thus symbolized the inseparability of European consumption and European imperialism.[7]

Nutmeg and clove also encapsulated Poivre's scientific status. Poivre's claims to possess the "true" nutmeg and clove became the central feature in heroic narratives of his life. One such account, on which subsequent histories were based, was a 1786 eulogy presenting him as a philanthropic and agronomic sage. Its author was Pierre-Samuel Dupont de

Nemours (1739–1817), better known as a political economist and the namesake of the pharmaceuticals company. Dupont presented Poivre as a fellow physiocrat, an ancestor of Dupont's economic views, emboldened by a vision of peaceful national strength and philanthropic improvement. His edition of Poivre's works juxtaposes hagiography with extracts from d'Etcheverry's journal describing the daring escape from the Dutch, and with legitimating texts asserting the authenticity of the plants Poivre brought to the Isle de France. The tale ends with the success of the spice missions of the early 1770s.

In Dupont's account, the claim to have acquired the "true" nutmeg and clove was connected with the claim to present the "true" version of events leading up to the d'Etcheverry voyage and the "true" interpretation of Pierre Poivre, the man. Accepting it meant buying into a specific understanding of how nature could work in a French colonial setting, a particular—and disputed—politics of nature. This model of Poivre's activities and expertise has persisted in recent biographies. Manuscript materials suggest a very different story, however, in which the identity of the plants in question is not clearcut, and there is no single moment of triumph for the spice project. In fact, Poivre made several attempts first to obtain plants, next to cultivate and propagate them, and finally to obtain acceptance of their authenticity.[8]

Poivre's nemesis was Jean-Baptiste-Christophe Fusée Aublet (1720–78). After studying and practicing pharmacy in Montpellier, Spain, and Paris, Aublet was appointed in 1751 as the Compagnie des Indes's apothecary on the Isle de France. After nine years on the island during which he devoted himself to natural history and agronomy, Aublet resigned when he was accused of having interfered with Poivre's cultivatory enterprises. Poivre had an adventurous background on a par with that of Aublet. While training as a religious missionary he had spent many years in China and Cochinchina (Indochina) but entered colonial administration before taking orders. Both Poivre and Aublet went on to assume other posts in colonial administration. In 1762 the Compagnie des Indes sent Aublet to botanize on Cayenne, the island colony off the coast of French Guiana, and soon afterward, the commander of the French Antillean stronghold of Saint Domingue (Haiti) hired him to found a new settlement on that island at Môle Saint-Nicolas. Meanwhile Poivre returned to the Isle de France in 1767 as its intendant.

At every point in their travels these men—one diverted from the familiar late seventeenth-century "career path" of the traveling medical practitioner, the other from that of the missionary naturalist—pursued policies of agronomic improvement and botanical inventorizing. In administrative terms, both shared similar aims: to form a botanical garden on the Isle de France for the propagation of useful plants for distribution to the inhabitants. Aublet's first garden, in the Pamplemousses area of the island, was, he claimed, vandalized by Poivre's supporters; later he formed one at the governor's estate of Le Réduit, where he claimed to have "collected everything I could get by way of rare, useful or curious plants, from various parts of the world."[9] Simultaneously Poivre was creating a botanical garden at the intendant's estate, Monplaisir. Both men came from Lyons; both had support among the large de Jussieu family, also of Lyons origin, whose members dominated Parisian botany through the eighteenth century.[10] Their self-presentations as apolitical, disinterested improvers, and the accusations of wanton botanical destructiveness that they hurled at one another were almost identical. Apparently engaged on very similar projects, they acted out a conflict between rival powers in the French Compagnie des Indes and Crown administration, as well as a dispute over the nature of scientific expertise, in the form of a disagreement about botanical identity. This combination of individual freedom of action constrained by tacit indebtedness to patrons was typical of absolutist scientific practice. The rambling, delegated nature of political authority in old-regime France encouraged competition among administrators answerable to different metropolitan patrons, even if they were all ultimately answerable to the Crown.

The Controversy

In 1754 Aublet had taken a dim view of Poivre's treasured haul, even though others on the Isle de France welcomed it: "I neither could nor would acknowledge that tree and those berries to be the true nutmeg of commerce."[11] This skepticism opened a Pandora's box: for two decades the validity of Poivre's claims to possess true nutmeg would be disputed. As metropolitan ministers and colonial governors came and went, the respective political and epistemological credit of Poivre and Aublet waxed and waned. At the end of 1755 René Magon, a friend of Aublet's main metropolitan patron, Paul-Henri-Thiry, baron d'Holbach (1723–89), became governor of the Isle de France. He succeeded two Poivre supporters, members of the same Compagnie de Indes family. This patronage, combined with Aublet's Parisian botanical connections, gave Aublet the power to cast official doubt on the legitimacy of Poivre's glorious enterprise, on his knowledge claims, and on his learned expertise. This was the status quo while the company continued to retain responsibility for the island. In a memoir to the island council written in 1755, after his second, unsuccessful spice-hunting trip, Poivre blamed Aublet for the death of the five precious plants during his absence, but his accusations did not tarnish Aublet's good name. The disappearance of his patronage within the Compagnie des Indes ultimately forced Poivre to abandon the island and return to France. Although Aublet left the island too, he continued to work in colonial administration.

While Dupont's account focuses on Poivre's good character and is silent concerning his patronage affiliations, contemporary manuscripts and reports make evident the highly political character of Poivre's activities. Poivre did not confront only Aublet; he also clashed with many other powerful figures in the Isle de France's administration. He ascribed the failure of his early attempts to bring spice to the island to opposition from among certain familial factions at the head of the Compagnie des Indes, notably Pierre Duvelaër, a director of the company. Rather than accept claims that his botanical judgment was at fault, Poivre characterized skepticism about the nutmeg as a conspiracy against the naturalization project that infiltrated the entire company, via the close family ties within that body. Without naming names, he attacked Duvelaër as a traitor in Dutch pay (a true patriot would be a Poivre supporter), and in a memoir of August 1758 he defamed Aublet's character as a "man without manners, ability or experience," notwithstanding Aublet's Parisian botanical credentials and connections. Elsewhere he accused the botanist of having poured hot water on the germinating nutmegs and of being in cahoots with the Dutch.[12]

The conspiracy theory usefully portrayed Aublet as a lone bad nut, or at most the political creature of a particular self-serving interest group within the Compagnie des Indes. Poivre's conspiracy theory could achieve this end only by strategic silences: about his own political allegiances or about the astronomical costs of equipping spice-hunting expeditions, which Poivre sought, with some success, to have funded by the French Crown or the Compagnie des Indes. In fact, both Poivre and Aublet, like most contemporary French scientific travelers, depended on particular patrons in the ministries, the Compagnie des Indes, or the colonial administration. In their role as administrators, botanists played an active part in colonial politics, and alongside their scientific observations such travelers were also under (sometimes secret) orders to report on the condition of colonial government or to propose locations for new settlements and military fortification. Natural historical observation must therefore be viewed as a form of colonial government, in which cataloging existing resources and acquiring new ones served the ends of European imperialism.

The asymmetry of Poivre's account is typical of patronage narratives in that it conceals the fact that both protagonists utilized similar political strategies but belonged to different patronage groups. By 1765, when the French Crown took governing responsibilities for the colonies back from the Compagnie des Indes, which was facing financial ruin, Poivre enjoyed the patronage of Étienne-François, duc de Choiseul (1719–85), the French minister of foreign affairs. Consequently he was in a much stronger position to defend his nutmegs on his return to the Isle de France in 1767.[13] Throughout the period under discussion, however, many chains of command linked the colonies to the metropolis; each metropolitan patron in the ministries or the company could field his own colonial protégé, and conversely, particular metropolitan political configurations permitted greater or lesser freedom of agency to colonial administrators such as Poivre and Aublet.

Historical accounts have simplified and polarized these complex political worlds and the scientific enterprises that were born of them. Many biographers, following Poivre, portray Aublet as opposed to the spice project on grounds of jealousy and lack of patriotic sentiment. Aublet's self-exculpation in his *Histoire des plantes de la Guiane Françoise* of 1775, however, presented quite different reasons for his denial of Poivre's credibility. Questioning Poivre's priority claims, he asserted that the nutmeg had first arrived on the Isle de France in 1753, when one Aubry, a ship's captain, distributed to three cultivators plants that later died. In 1755 Poivre had called on the island's governing council to "receive and have examined by capable persons, the seeds and plants which he had brought back from his voyage," a meeting which Aublet was ordered to attend: "We were all, I believe, animated by the desire to fulfill our duty and by our attachment to the Company's interests; but I was the only one for whom the knowledge and research of plants were the principal occupation." Nonetheless Aublet's botanical expertise was questioned. His arguments against authenticity on botanical grounds were constructed by comparing the morphology of the Poivre specimens with a description of the nutmeg by the Dutch-based colonial naturalist Georg Everhard Rumphius (1626–93), in his *Herbarium Amboinense* (Amsterdam, 1741–55) (see figure 11.1). Aublet's invocation of botanical knowledge was unavailing: "the details or characters of plants pass, among people who are not Botanists, for overly minute and inessential research," he complained. Against his will, Aublet was commanded to undertake the cultivation of the last remaining specimens, which were in poor condition. In October 1755 he presented one dead plant and a seedling to the council, which dismissed him. In his retrospective defense, Aublet angrily refuted suggestions of deliberate sabotage: "Would I not have shared the glory of having enriched our Colonies with [the spices]? And what man, above all what Traveler, does not have the desire to be cited by posterity as having been useful to his Fatherland and to men in general?"[14]

To defend himself against his critics, Aublet challenged Poivre's botanical account of the identity of the nutmeg plants brought back from the Philippines in 1754. He counteridentified the seedling he had been

Figure 11.1. The Amboina nutmeg. From Georg Everhard Rumphius, *Herbarium Amboinense*, 6 vols. (Amsterdam, 1741–55), vol. 2, pl. IV, opposite p. 18. Reproduced by permission of the Syndics of Cambridge University Library.

given: it was either the seed of an areca palm or "a Nutmeg similar to those which are sold at Keyda (Kedah, Malaysia)," the so-called bastard nutmegs with little aromatic virtue. Had Poivre, the great explorer, made a detour to this well-known trading post to obtain false nutmegs and passed off a different species for the real thing? More seriously still, Poivre could be viewed as taking unpardonable political and economic liberties. Like others who bombarded the ministers with unrealizable projects and never made good on the money they were advanced, Poivre's "project to have the fine spices cultivated in our Colonies" was a project "with which the Ministers have been fatigued for thirty years, & which has already amounted to a considerable cost, both in advances and in rewards."[15] The economic argument Poivre made for introducing the spices was flawed too because the Dutch spice income derived from artificial manipulation of the market by destroying large crops and wiping out plants growing outside Banda, so as to keep prices high. In short, almost every detail of Poivre's self-accounting could be doubted: his motives (as an agronomic swindler), his expertise (wasting resources on the wrong plants), the authenticity of the voyage (the suspected trip to Keyda), his predictions of a triumphal economic future (market collapse if the Dutch monopoly were broken).

In one version of the nutmeg story Poivre emerges as sage, philanthropist, patriot, and economist. In another he appears as squanderer of public resources and spreader of vile rumors. His heroic and scientific identities were thus inseparable from the identity of the plant. Conversely, by endangering Aublet's reputation, Poivre ensured that Aublet's botanical knowledge-making strategies failed to secure the status of universal truth. In such encounters both men resorted to tactics ranging from secret memoirs presented to colonial and metropolitan authorities, to rumor-mongering and, eventually, printed denunciations, which often took an autobiographical form.

Poivre's Politics

The success of Poivre's autobiographical maneuverings depended on the validation, particularly by other scientific practitioners, of the objects brought back. Where this could not be obtained, he endeavored to redefine the context of proof so as to exclude (on the grounds of bad politics and bad character) those opponents who could not be silenced. Following the Aublet fiasco, on subsequent occasions when Poivre made a bid to claim that "the" nutmeg was in his hands, he began by defining the "environment" in which authentication could survive. This often meant clashing noisily with fellow administrators jockeying for metropolitan political support. In Poivre's 1766 instructions for his administrative position as intendant on the Isle de France, the minister for the colonies added that Poivre would "immortalize his name if he could bring the colony into competition with the Moluccan Isles for [spice] production." With such clear-cut metropolitan support, Poivre was emboldened to act unilaterally. To the dismay of Benoît Dumas, the military governor, Poivre immediately provoked a political situation on the island by encouraging the syndics' assembly, the third royal authority on the island besides the intendant and the governor, to oppose the "tyranny" of Dumas's orders. This opposition took an explicitly political form, with representatives sent to the metropolis to present *rémontrances* against the governor. *Rémontrances* were a type of antimonarchical protest characteristic of seventeenth-century *frondeur* politics and perpetuated in the *parlements*, a noble judiciary system that often opposed the French Crown.[16] In borrowing forms of *parlementaire* resistance, as well as ones that appeared close to those adopted by the Revolutionary Assemblée Nationale some twenty years later, the syndics' assembly could well be said to be creating, as Dumas put it, a "republic in the bosom of the monarchy." Dumas reimposed order by a military coup, marching into the assembly in February 1768 at the head of the island's troops and proclaiming that it had exceeded its authority. The balance of power on the island thus depended on alliances between different Crown representatives, encouraged or repressed by ministerial patronage. In this case the governor was the official head of Crown authority on the island, but Poivre's metropolitan support meant that, while he was only admonished by the minister, Dumas was recalled and the island assembly temporarily dissolved.[17]

Although ministerial intervention terminated Poivre's "republican" ambitions, the assembly remained a valuable locus of support for Poivre's spice enterprise and continued to create situations of political unrest for future governors such as François-Julien du Dresnay, chevalier des Roches. As in the case of the Aublet/Poivre clash, interactions with the governors, even over the nutmeg question, should be viewed in light of the ongoing rivalry among colonial officials for metropolitan patronage. By 1770, with other sources of island authority weakened, it was possible to engineer a context of discovery that would silence dissent, on the island at least. Poivre choreographed the return of Provost and d'Etcheverry in detail, based on the strategy used against Aublet in 1755. On disembarking, Provost submitted a declaration concerning his voyage and its results to the two administrators, Poivre and des Roches, and then requested a memorandum attesting to the authenticity of the plants and seeds brought back. Des Roches, however, initially refused to sign this document. Instead, the memorandum was read to the administrative council by the world-renowned naturalist Philibert Commerson

(1727–73), who was on the island thanks to a ministerial directive solicited by Poivre. The governor annotated his copy of this document with comments that demonstrate his concern about the political form of the assembly: "I responded clearly that I was persuaded the Minister would disapprove of such an operation. In consequence I did not authorize it and I did not care to attend." Despite his broad support for the spice project, des Roches judged that those present at the meeting were not sufficiently expert to determine whether the plants they were being shown were nutmegs or not. Doubt still hung over Poivre's nutmegs, and accusations and defenses of expertise proliferated. Poivre portrayed des Roches as an inexpert administrator and poor sailor, while des Roches insisted that he had "been everywhere, seen everything, thought of everything." Although he eventually conceded, des Roches's initial resistance was clearly a response to Poivre's attempt to authenticate the specimens by political means: "I opposed any act of authentication which might excite the resentment or jealousy of our rivals and competitors if we have succeeded, and which, in the opposite case, would give everyone a good laugh."[18]

A fortnight later a ceremony took place at Poivre's house, in the presence of des Roches, Commerson, and other leading military, naval, and medical figures. After the memorandum had been read, the company proceeded to Poivre's garden, where its members were shown eighty flourishing nutmeg plants, five clove plants, and two germinated seedlings in a box containing one hundred others, and five chests containing some ten thousand nutmegs, mostly germinated. These were formally identified by Commerson, citing Rumphius, as true nutmegs and cloves. A second chest contained "bastard nutmegs" with longer fruit, larger leaves, and only a faint spiciness. The task of legitimation was not, however, left to the judgment of the colonial experts. The results of the French spice piracy were also evaluated back in Paris, by botanists at the Académie Royale des Sciences, the Jardin du Roi, and the Trianon garden. Reporting on the success of the Coëtivy expedition in 1772, two leading botanists, Bernard de Jussieu (1699–1777) and Michel Adanson (1727–1806), approved the scanty specimens they had received as coming from the true spice plants, and they even went so far as to crown the travelers with the title of "French Argonauts," although "those who made the famous conquest of the Golden Fleece, certainly did not have so useful an object in view, nor perhaps such great perils to fear."[19]

The naturalization of a useful plant was, however, not a historical event but rather a collage of failed attempts, possible successes, future potential, and the social labor that underpinned the credibility of all of these. At which point could the nutmeg be said to have been naturalized on the Isle de France? One might be charitable to Poivre and agree that he distributed "true" nutmeg plants to his agricultural supporters in 1754. And yet, even if the plants were there in body, Poivre could neither marshal univocal support for their authenticity nor guarantee their durability. Only the mastery of both social assent and cultivatory expertise counted as "true" acclimatization. Poivre made several other attempts to obtain nutmegs in germinating condition from the 1750s to the 1770s, but historical accounts generally take the moment at which d'Etcheverry sailed away from Gebe with his cargo as the pivotal one.[20] Although it certainly had the makings of a superb public relations exercise, closer attention reveals that powerful contemporaries viewed this first voyage as inadequate for acclimatization purposes. Indeed, a second voyage was launched on the heels of the first, to obtain a greater number of plants and seedlings. The issue divided the very body to which Poivre and his protégés appealed for scientific legitimation. In 1779 an astronomer of the prestigious Paris Académie Royale des Sciences, Guillaume-Joseph-Hyacinthe-Jean-Baptiste Le Gentil de la Galaisière (1725–92), published an account of his travels in the South Seas. Le Gentil, who had been on the Isle de France when d'Etcheverry returned, discredited the results of this first voyage: "The nut that one saw at the Academy, in 1773, was judged by the Commissioners to be the true nutmeg of commerce; it is certain that that of 1770, which I saw on the spot, was not, at least four-fifths of it." These contested four-fifths of Provost's haul were, Le Gentil claimed, "for the most part, bastard nuts, that is to say, large, oblong, such as I had seen at Manilla . . . , and which cause it to be said that the Spanish have the nutmeg in [the Philippines]; but this species has little scent by comparison with the other: now, four to five times as many of that oblong species were brought back to the Isle de France than of the other which is small and round."[21] Le Gentil had been on an official government-funded voyage with the purpose of observing the passage of Venus, and his book was published by the Imprimerie Royale under the academy's aegis. As such his refutation of Poivre's enterprise possessed semiofficial status.

Poivre's devoted protégé Jean-Nicolas Céré (1737–1810), who eventually took over the running of the Monplaisir garden, carried the battle to authenticate the spice plants into the Paris periodicals, challenging Poivre's opponents, including the new minister for foreign affairs, Emmanuel-Armand Vignerot du Plessis, duc d'Aiguillon (1720–88).[22] Aublet's *Histoire des plantes*, with its damning versions of Poivre's motives and expertise, appeared at this juncture, in 1775, when Poivre had already retired to his estate in France. Aublet's Parthian shot drew together his arguments about botanical expertise, his attacks on Poivre's authenticity, and his irrefutable denials of the logic of identity:

everything leads us to believe that one has not yet been able to obtain the true Nutmeg in germinating condition, nor plants of the commercial Nutmeg tree; that the Nutmeg tree and Nutmegs which the Nutmeg Argonauts have procured for themselves, are wild Nutmegs with which the Isles of the Indian Archipelago are filled, & that many of these Nutmegs which come to us on Asian vessels, are the Nutmeg common to the Philippines, Manillas, Moluccas, to Keyda on the Malabar coast, and this one is sold fresh in the Bazaars or Markets of India. . . . The Dutch have not destroyed the species which are of no utility in commerce, by their lack of quality; & it does not require much courage or trouble to procure those species.[23]

The Poivre publicity apparatus swiftly swung into action to refute Aublet. Even once cloves and nutmeg were growing at Monplaisir, Poivre made sure to recruit highly visible botanical supporters who were qualified to demonstrate the identity of his products with the true nutmeg "of commerce." In his *Voyage à la Nouvelle Guinée* of 1776, Pierre Sonnerat (1745–1814), Poivre's relation by marriage and Commerson's protégé, graded the different varieties of nutmeg according to goodness. Against Aublet, Sonnerat asserted that there were three different species: the male nutmeg, most esteemed in Europe; the female nutmeg, less highly regarded in Europe but still aromatic; and the false or bastard nutmeg, with no aroma. This last could not, as Aublet had implied, have been unfamiliar to Provost on the d'Etcheverry voyage for he had already brought some specimens to the Isle de France in 1768 as a curiosity. The "truth" of the nutmeg was now a matter of relative expertise: Aublet's familiarity with "the" commercial nutmeg could be deemed a product of inadequate knowledge about the number of nutmeg varieties in the Indies.[24]

That the path of spice credibility had been substantially cleared by 1776 is demonstrated by the relative facility with which cloves, having been acquired only in 1770, were introduced to French colonies. The first colonial cloves were presented to the king in 1777. Their rapid multiplication and ease of cultivation ensured that they were the first fruits of the nutmeg war. Not until December 1778 was the "first French nutmeg" ceremonially harvested by Céré before the administrators of the Isle de France, an event publicized in the French periodical press. According to Céré, the crop was visited by more than five thousand people; through his propagating activities he was able to send shipments of clove and nutmeg plants to other French colonies, notably the Seychelles and Cayenne in French Guiana. Soon after young clove plants sent from the Isle de France to Cayenne had fruited for the first time, the abbé Guillaume-Thomas Raynal (1713–96) received a specimen from a Cayenne correspondent; he passed it on to an agronomist, the abbé Henri-Alexandre Tessier (1741–1837), who published an enthusiastic article on the spice project and visited Poivre, "that estimable Citizen," on his estate (see figure 11.2). Tessier's article swallowed Poivre's legitimating strategies hook, line, and sinker, giving as evidence for the clove's authenticity the dates and content of the signed memorandum from the Isle de France and panegyrizing Commerson's "indefatigable zeal for Botany, in which he was singularly well versed." As further proof, Tessier compared his own botanical description of the specimen with that of Rumphius and commissioned an illustration. The future of the nutmeg and clove project was fought out in the public sphere; specially organized visual displays and textual strategies served to strengthen the link between nutmeg or clove authenticity and the expertise of Poivre's faction.[25]

Identity and Authenticity

The identity of the "true nutmeg of commerce" in this dispute is not historically accessible. There was no single point on which the identity of the nutmeg could be constructed so as to generate criteria of sameness to which all key participants assented. In 1755 the Isle de France's council rejected arguments based on botanical morphology, while Aublet rejected those based on Poivre's good faith. Neither scientific appeals to natural reality nor social appeals to patriotism and personal reliability could produce settlement in one direction or another. Neither Aublet's denunciation of Poivre's first nutmeg nor the authentications of his later ones by Commerson and de Jussieu were based on prior personal observation of a plant bearing a commercial nutmeg. Both Aublet and Commerson used a single textual resource as their principal means of evaluation, Rumphius's *Herbarium Amboinense*. No French botanist had seen a nutmeg plant, so Poivre's specimen could not be compared on the basis of experience; because it could not be compared, it could not be identified unless hierarchies of expertise were established beforehand. In this sense the nutmeg fulfills the criteria of Bruno Latour's account of black-boxing, in which a natural object can only be permitted to become natural once the social setting in which it will take on meaning has been controlled.[26] The necessity of resorting to such complex strategies of proof demonstrates just how hard it was, in high-profile botanical enterprises, to convince contemporaries of the authenticity and properties of introduced species. Asymmetrically, the success of the 1770–72 expeditions became a demonstration of the "natural" truth of Poivre's versions of the spices, while earlier failures were ascribed to social factors: political opposition, deliberate sabotage, or inadequate botanical expertise. The manufacture of one of these introductions as *the* introduction, and the others as irrelevant, demonstrates the conversion

Diequemare del

Juillet 1er

Figure 11.2. The "first French cloves." From Henri-Alexandre Tessier, "Mémoire sur l'importation du Géroflier des Moluques aux Isles de France, de Bourbon & de Sechelles, & de ces Isles à Cayenne," *Observations sur la Physique, sur l'Histoire Naturelle et sur les Arts* 14, no. 2 (1779): pl. I, opposite p. 54. Reproduced by permission of the Syndics of Cambridge University Library.

of historical process into historical event. In the course of that transformation, the social labor involved in these repeated introductions and the threat posed by each to the fragile identity of the nutmeg and the botanist is effaced.

This chapter also has an important historiographical point to make about the history of colonialism and botany. Aublet and Poivre began with similar projects and backgrounds, and both drew their claims to botanical expertise from their association with Parisian patrons such as the de Jussieu family. Neither Aublet's medical nor Poivre's missionary training was out of keeping with the backgrounds of other naturalists of the period. To term either a "botanist" unproblematically is to neglect the difficulty of sustaining such a role and the associated expertise in the absence of formal modes of training and accreditation. Retrospectively, Aublet retained botanical authority, while Poivre lost it. Even so, Poivre won the nutmeg identification dispute, thanks to his skills in coordinating political, social, and botanical support in the colony and the metropolis. "The botanist" as a figure of epistemological expertise was thus a retrospectively fashioned identity, reflecting the outcome of local disputes. This was particularly true in colonial situations since travel disrupted individuals' patronage relations and thus caused the sources of their scientific expertise to be destabilized. What should we learn from the case of the nutmeg for histories of colonialism and science? It seems that the "big histories" of empire and botany depend for their analytical categories on a retrospective reliance on the stability of actors' and objects' identities—a stability which is, however, dependent on the prior settlement of disputes like the nutmeg controversy. Hence, such accounts are of necessity teleological. Only by considering such controversies in the making can we see how categories such as "empire," "the botanist," and "the nutmeg" are simultaneously constructed by localized negotiations between participants.[27]

Chapter 12

Out of Africa
Colonial Rice History in the Black Atlantic

Judith Carney

> *Wade in de water, Wade in de water, children.*
> *Wade in de water, God's a goin' to trouble de water.*
> *See dat band all dress'd in red,*
> *God's a goin' to trouble de water.*
> *It looks like de band dat Moses led.*
> *God's a goin' a trouble de water.*

The African American spiritual "Wade in the Water" recalls the passage to freedom that the parting of the Red Sea gave Moses and the enslaved Israelites. It also provides a powerful metaphor for examining rice cultivation and its origins in the Americas. Rice is the only grain that demands copious amounts of water. Its caretakers wade through fields of water for its cultivation. Water is also essential for preparing the cereal for consumption. The children of Africa did not wade, as in the spiritual, but were carried in shackles across the troubled waters of the Atlantic slave trade. Enslaved West Africans brought an indigenous knowledge system that would establish rice as a subsistence and plantation crop over a broad region from South Carolina to tropical South America. With them, rice arrived in the Americas in the holds of slave ships, crossing over the ocean grave of the Middle Passage as provisions for its survivors. The cultivation, processing, and preparation of rice reveal a profound knowledge system brought to the Americas by those enslaved from West African rice-growing societies. The importance of rice cuisine to the African diaspora serves even to this day as recipes of memory and cultural identity throughout the black Atlantic.

The African diaspora was one of plants as well as people. Rice figured significantly among the African plants and agricultural systems that shaped environment, food preferences, economies, and cultural iden-

tity throughout the era of plantation slavery. During the eighteenth century rice produced by enslaved labor made colonial South Carolina the
wealthiest plantation economy in North America. The foundation for its
economic prominence rested with West Africans skilled in growing the
crop in diverse environments. This chapter illuminates the African origins of rice history in the western Atlantic, emphasizing the technologies
and knowledge systems brought to the Americas by enslaved West Africans, especially those long associated with African women's work.

Of Rice and Slaves

Asia has long been associated with the origin of rice, but the grain was
also independently domesticated in West Africa. Two key properties of
African rice (*Oryza glaberrima*) illuminate why its history is less well
known than that of the Asian species (*Oryza sativa*). *Glaberrima*'s yields
are lower than *sativa*'s and when milled by machines the grains of African rice break apart. These crucial factors privileged Asian rice in global
economic history and favored the diffusion of *sativa* elsewhere. Higher-
yielding Asian rice, however, failed to dislodge *glaberrima* from its position as the preferred species in West Africa, where sativa was introduced
during the period of Atlantic slavery. Demand for rice grew by leaps and
bounds in eighteenth-century Europe. The grain was used for brewing
beer, and making paper, and it was increasingly favored among middle-
class Catholics to accompany fish on meatless Fridays and during Lent.[1]
Slaves accompanied the first Europeans arriving in South Carolina from
Barbados in 1670. Rice cultivation was well under way in the colony by
the 1690s, with the cereal's transition to a plantation crop completed by
the early eighteenth century.

   Prior to the Civil War, rice was grown within forty miles of the Atlantic
coast along the floodplains of sixteen tidal rivers, from the North Carolina–South Carolina border to the Saint Mary's River, which demarcated
Florida from Georgia. On the eve of the Civil War nearly one hundred
thousand slaves cultivated some seventy thousand acres of tidal floodplain swamps on about five hundred rice plantations.[2]

   Until the 1970s the historiography of the colonial rice economy routinely attributed its origins to the ingenuity of white European planters.
Accounts of colonial rice history praised the early planters for discovering how to grow an unfamiliar tropical crop in the swamps found along
the coastal corridor.[3] This long-standing view changed with the publication of *Black Majority* by the historian Peter Wood in 1974. He argued
that the English and French Huguenot planters, who settled the colony,
had no experience growing rice. This was not the case with slaves who
originated in West Africa's indigenous rice region—a vast area that ex-

tends along the coast from Senegal to the Ivory Coast and inland for a thousand miles to Lake Chad. Wood contended that they alone possessed the requisite knowledge, experience, and skills for developing rice cultivation.[4]

Wood's research resulted in a revised view of the African role in shaping wetland landscapes planted to rice. However, questions remained over whether planters recruited slaves from West Africa's rice region to help them develop a crop whose potential they independently recognized or whether African-born slaves initiated rice cultivation in Carolina swamps through their efforts to grow a favored dietary staple for subsistence. The political-ecological analysis presented here seeks to resolve this issue. It reconsiders the way that historians have conceptualized rice. Instead of treating the cereal solely as a commodity consumed and traded, rice is examined as the product of an indigenous knowledge system whose expression in different environments across geographic space was mediated by cultural traditions and power relations. Shifting the perspective on rice from commodity to knowledge system focuses attention on the environments cultivated to the cereal during the era of transatlantic slavery while facilitating recovery of the crop's cultural origins in the American Atlantic.

Archival and historical materials clearly establish the presence of rice in West Africa prior to European maritime voyages as well as in the environments planted in South Carolina and Brazil during the early colonial period. The case for African agency in establishing rice culture in the Americas is made first for South Carolina and then extended to Brazil.

Historical Accounts of Rice Systems in the African Atlantic

In West Africa rice is cultivated in both upland and wetland environments. While more than twenty microenvironments are planted to rice, there are three principal systems of production: rain-fed, inland swamps, and tidal floodplains.[5] The first system relies strictly upon precipitation, while the second two access supplemental sources of water. Inland swamps receive additional moisture from subterranean reserves (for example, high groundwater tables or underground streams), while floodplain rice is irrigated by tides. Tidal rice developed in two distinctive environments: along freshwater rivers and on the floodplains of coastal estuaries within reach of marine tides. Each ecological zone typically forms part of a broader system of production, which enables farmers to utilize different environments along a landscape gradient. The objective is to spread out labor demands while minimizing crop loss.

Historical accounts affirm the antiquity of these systems in West Africa while revealing the long-standing social and ecological aspects of African rice cultivation. Portuguese caravels venturing south of the Senegal River in 1446 encountered floodplains of rivers and estuaries "sown with rice."[6] A decade later Alvise da Cadamosto reached the Gambia River, where he noted the significance of rice in the regional diet. Over the remainder of the fifteenth century, Portuguese commentaries proliferate on the geographical extent of West African rice culture.[7]

Throughout the era of transatlantic slavery, European mariners called the region between Senegal and Liberia the "Grain or Rice Coast," after its specialized production of grains such as millet, sorghum, and rice. Portuguese caravels depended on African surplus grain production to restock provisions, thereby establishing a pattern followed by slave vessels from other European countries. South of the Senegal River ships arrived in a region abundant in cereals; east of Liberia grain cultivation gave way to root crops such as yams. While reference to the Upper Guinea coast invokes images of the Atlantic slave trade, the term "Grain or Rice Coast" does not. Yet, African acumen in agriculture and plant domestication nurtured the dense populations of the Upper Guinea coast that Europeans subsequently enslaved, while routinely delivering surpluses to the slave ships that plied the Atlantic Passage.

In the sixteenth century the Cape Verde Islands emerged as a crucial trading entrepôt for the transatlantic expansion of Portuguese commerce. Located just five hundred miles west of Senegal, the islands provided ships with water, wood for fuel, and salt, as well as slave-produced cotton cloths. Outbound ships on Atlantic voyages headed there in the fall and winter with the prevailing northeast winds and then followed the southward flow of the Canary Current before continuing on to Brazil, West Africa, or India. Previously uninhabited, Cape Verde soon became an Atlantic extension of West African culture. Slaves grew African plant domesticates for subsistence; by the early 1500s rice was grown on Santiago, the island most favorable for agriculture.[8] Food was occasionally sold to Portuguese caravels, although the islands also depended on surpluses produced along the Upper Guinea coast. Rice appears on the cargo lists of ships bound for Portuguese America in 1514; another record from 1530 records the transport of rice seed and cane cuttings from Santiago to the nascent Brazilian sugar-plantation colony Bahia.[9]

With the arrival of ships from other European nations to the Upper Guinea coast and the growing momentum of transatlantic slavery, references to rice purchases increased during the late sixteenth century.[10] Relying on information supplied by Dutch merchants operating along the Windward coast (the region between Sierra Leone and Liberia), the seventeenth-century Amsterdam geographer Olfert Dapper (1639–89) described rice cultivation along a landscape gradient where production transitioned from freshwater river floodplains to inland swamps and rain-fed uplands: "They sow the first rice on low ground, the second a little higher and the third . . . on the high ground, each a month after the previous one, in order not to have all the rice ripe at the same time; this would bring them into difficulty with regard to cutting the rice, since it is cut ear by ear or stalk by stalk—a very wearisome task. . . . The first or early rice, sown in low and damp areas . . . the second, sown on somewhat higher ground . . . the third, sown on the high ground.[11] Dapper's remarks reveal a concerted European interest in West African rice culture.

No system, however, drew as much commentary as the tidal system that developed on the floodplains of coastal estuaries between Gambia and Sierra Leone.[12] Within reach of marine tides, this system is often termed "mangrove rice," after the vegetation that is cleared for cultivation. Mangrove rice embodies the sophisticated principles and achievements of indigenous African rice culture. Oceanic tides keep the soil saturated prior to cultivation, while rainfall is used to desalinate and irrigate the rice perimeter. Farmers turned a constant menace—the risk of marine tides overflowing a rice field and leaving undesirable salt deposits—into a source of innovation. They devised an elaborate system of water management by enclosing the mangrove swamp with embankments to impede tidal flow during the cultivation season. Water control and irrigation were achieved through a system of canals, sluices, and dikes.

In 1594 the Luso-African trader André Alvares de Almada (d. 1594) drew attention to the elaborate "system of dikes" that enabled mangrove rice farmers to "[harness] the tides to their own advantage" more than a century before a similar floodplain system would shape the rice-plantation economy of colonial South Carolina.[13] His description was one of the first to detail African expertise in growing tidally irrigated rice. De Almada's words reveal that West Africans had developed, and were practicing, a wet-rice culture as fully evolved as any found in Asia over the same period.

As the Atlantic commerce in human beings grew, so did European interest in peoples specialized in mangrove rice cultivation. Samuel Gamble, who slaved off the Guinea coast between 1793 and 1794, sketched and described the Baga system that yielded the surpluses that he purchased to provision the slave ship across the Middle Passage: "The Bagos are very expert in Cultivating rice and in quite a Different manner to any of the Nations on the Windward Coast . . . they have a reservoir that they can let in what water they please, [on the] other side . . . is a drain out so they can let off what they please."[14]

Attention to peoples practicing rice culture is also evident in the comments of Francis Moore, a factor for an English trading company along the Gambia River during the 1730s. Moore described the seasonal shift in land use between Mandinka farmers and Fulani pastoralists in freshwater floodplain systems, noting that rice land became cattle pasture in the dry season.[15]

European accounts also reveal the important role of women in the cultivation, processing, and economy of rice. On food purchases by Dutch traders near the Liberian border with Sierra Leone, Samuel Brun noted in 1624 that "for the rice they wanted only glass corals [beads] for their wives, because rice is the ware of women."[16] During the same decade Richard Jobson detailed the role of females in milling rice along the Gambia River: "I am sure there is no woman can be under more servitude, with such great staves wee call Coole-Staves [pestles], beate and cleanse both Rice, all manner of other graine they eate, which is only womens worke, and very painfull."[17] Writing about Sierra Leone in 1678, Jean Barbot added: "The land abounds in millet or white maize [sorghum] and in rice which they have as their main food. The women pound the rice in slightly hollowed tree-trunks."[18] Francis Moore noted in the 1730s that rice was solely a woman's crop in Gambia. "For every Town almost having 2 common Fields of cleared Ground, one for their Corn [millet and sorghum], and the other for the Rice . . . The Men work the Corn Ground and Women and Girls the Rice Ground."[19] He drew attention to an important distinction between floodplain rice systems. On freshwater floodplains rice was a woman's crop. However, mangrove rice depended on the participation of both males and females. Men built the embankments and canals with long-handled shovels, while women performed hoeing and weeding.[20]

These accounts reveal the existence of a fully elaborated rice culture in West Africa. Rice attracted European attention for the surpluses Africans produced, the cereal's availability for purchase, and the peoples skilled in its cultivation. The Portuguese did not introduce rice culture to the region from their voyages to Asia, as historians would later claim.[21] Rain-fed, inland swamp, and tidal rice cultivation long antedated their arrival.

Antiquity of African Rice Development

African rice (*Oryza glaberrima*) was domesticated along the inland delta of the middle Niger River in Mali. But broader scientific knowledge of West Africa as an independent center of rice origins dates only to the twentieth century. Early Portuguese accounts revealing the extent and significance of African rice cultivation had faded from memory with the conclusion of four centuries of Atlantic slavery and the imposition of colonialism in the late nineteenth century in Africa. Scholars attributed

the presence of rice culture in West Africa to the Portuguese, who introduced the cereal from Asia.[22] These views were reassessed when French
botanists, working in the region of *glaberrima* domestication, argued that
the grain's unusual characteristics suggested a separate species. Nineteenth-century botanical collections taken from the Upper Guinea coast
also revealed specimens sharing similar features and the same red color.
By the second half of the twentieth century scientists agreed that rice
was independently domesticated in West Africa. Research now places
the domestication of African rice between three thousand and forty-five
hundred years ago.[23]

Rice History in South Carolina

Nowhere in the Americas did rice play such an important economic role
as in South Carolina. Within a decade of the colony's settlement in 1670,
slaves were growing rice for subsistence.[24] By the 1690s the grain was
being cultivated for export. Annual exports from South Carolina exceeded sixty-six million pounds on the eve of the American Revolution,
making rice the first globally traded cereal.

About a hundred African slaves accompanied the arrival of Europeans
to the Carolina colony. The number brought directly from Africa rapidly
increased in a short period of time. By 1708 the black population exceeded that of whites.[25] While slaves cultivated the colony's rice and indigo exports, they also grew the subsistence crops consumed by blacks
and whites alike. In a pattern similar to the black settlement of the Cape
Verde Islands, slaves from West Africa's rice region began planting preferred food staples. Rice became a subsistence crop in the early Carolina
settlement period.

There were multiple introductions of rice seed to the colony between
1685 and the early 1690s, both deliberate and casual. Among the earliest
types was the "one called Red Rice in Contradistinction to the White,
from the Redness of the inner Husk or Rind [bran] of this Sort, tho'
they both clean and become white alike."[26] In the early twentieth century, when the southern historian A. S. Salley researched the types of
rice initially grown, French botanical research on the African red rice
was not broadly known.[27] However, Salley did not address the commonplace knowledge of Thomas Jefferson and his contemporaries—that rice
was grown along the West African coast during the era of Atlantic slavery. In attributing the earliest seed introductions to Asia and ship captains arriving from the Orient, Salley remains silent on what commerce
brought such ships to South Carolina in the seventeenth century, what
African ports of call were visited en route, and the type of cargo they
carried to the colony.[28] In his view, rice culture was an outcome of the

age of sail, the result of deliberate exchanges of seed between learned and well-traveled gentlemen.

One record from the colony's early settlement period, however, reveals yet another way that seed rice arrived in Charleston, as leftover provisions from a slave ship: "a *Portuguese* vessel arrived, with slaves from the east, with a considerable quantity of rice, being the ship's provision: this rice the *Carolinians* gladly took in exchange for a supply of their own produce. . . . This unexpected cargo was distributed, which gave new spirit to the undertaking, but was not sufficient to supply the demand of all those that would have procured it to plant."[29] Entering the colony as provender from slave ships, this rice originated, along with its human cargo, in Africa. For reasons detailed below, the rice was likely African *glaberrima*, the source of the red type planted in the early settlement period. For leftover grains to serve as seed rice, as the quotation indicates, the cereal must have crossed the Middle Passage in the husk, that is, unmilled.

Along the rice-growing Guinea coast slave-ship captains regularly bought rice to provision slave ships. Thomas Phillips purchased some five tons of rice in 1693 for his supplies, while the cereal figured in the provisions procured by James B. Barbot in 1699. In 1750 John Newton purchased nearly eight tons of rice to feed the 200 slaves he carried across the Middle Passage. Another captain, John Matthews, estimated that 700 to 1,000 tons of rice would feed the 3,000 to 3,500 slaves he bought in Sierra Leone. For the 250 slaves he carried to Jamaica in 1793, Samuel Gamble purchased more than eighteen tons of rice.[30] Gamble's journal records a demand for both milled and unmilled rice, but purchases of unhusked rice exceeded those of the white milled product, as it was cheaper and served as inexpensive provision for slave ships. Unmilled rice was also less susceptible to moisture spoilage than was its milled counterpart.[31] Gamble's journal mentions that the unhusked rice was red, a detail that suggests it was African *glaberrima*.[32]

"Seed rice" refers to rice grains that have not been milled. Rice in the husk therefore could double as germ plasm for planting, if any remained from a slave voyage (as occurred with the Portuguese slaver mentioned above), and there was interest in growing the cereal. Those forlorn passengers of the Middle Passage, already familiar with planting and processing the cereal, became the agents for pioneering its cultivation in the Americas. Through their conscious efforts to grow a dietary staple, they revealed the cereal's commercial potential to those who held them in bondage.

Additional insight into the development of rice culture in the western Atlantic is gained by examining the cereal's milling. Preparing rice for human consumption requires removal of the indigestible husks or hulls.

Until the advent of mechanical devices in the late eighteenth century, rice processing depended on hand-milling the grains in an upright hollowed-out cylinder carved from a tree trunk. It involved striking the grains with a handheld pestle to remove the hulls and minimize grain breakage. Milling, like food preparation more generally, was traditionally the work of African women.[33] Slave ships replicated existing cultural practices by relying on enslaved females to process and cook the food for all the slaves on board.

One journal entry from the slave ship *Mary* (outbound from Senegal), dated Monday, 19 June 1796, noted enslaved females at work in food preparation: "Men [crew] Emp[loye]d tending Slaves and Sundry Necessaries Jobs about the Ship. . . . The Women Cleaning Rice and Grinding Corn for corn cakes."[34] "Cleaning" rice refers to its milling. The reference suggests that the ship purchased unmilled rice that required processing on board. More detail is provided in a doctor's report from 1795 or 1796 of a visit aboard a slave ship recently arrived from West Africa's rice region. Observing what the slaves were eating, Dr. George Pinckard wrote:

their food is chiefly rice which they prepare by plain and simple boiling. . . . We saw several of them employed in beating the red husks off the rice which was done by pounding the grain in wooden mortars, with wooden pestles sufficiently long to allow them to stand upright while beating in mortars placed at their feet. . . . They beat the pestle in time to the song and seemed happy; yet nothing of industry marked their toil, for the pounding was performed by indolently raising the pestle and then leaving it fall by its own weight.[35]

Providing another instance of rice as the ship's provision, Pinckard's commentary also reveals the use of mortars and pestles aboard slave ships. Their presence suggests that the rice was purchased in the husk, that slavers relied on their enslaved captives to mill the cereal, and that it was processed with a mortar and pestle—the only way to mill the African species.[36] While gender is not explicit in Pinckard's account of rice processing, the role of enslaved women in beating and pounding rice aboard slave ships becomes clearer in Henry Smeathman's account from the 1770s: "Alas! What a scene of misery and distress is a full slaved ship in the rains. The clanking of chains, the groans of the sick and the stench of the whole is scarce supportable . . . two or three slaves thrown overboard every other day dying of fever, flux, measles, worms all together. All the day the chains rattling or the sound of the armourer riveting [*sic*] some poor devil just arrived in galling heavy irons. The women slaves in one part beating rice in mortars to cleanse it for cooking."[37]

On slave ships wooden fences or barricades separated women and children from male slaves.[38] Females on a ship were usually placed aft in

Figure 12.1. Pretextat Oursel, "Transport des nègres dans les colonies."
Colored lithograph. Photo by Michel Dupuis, Ville de Saint-Malo. Reprinted by
permission of the Musée d'Histoire, Saint-Malo, France.

the quarterdeck near the lodgings of the ship's officers.[39] To minimize
the potential for sabotage, slavers relocated the ship's galley from the
hold to the deck, typically placing the cast-iron cooking hearth in the
area reserved for enslaved females. A door in the barricade facilitated
the passage of the crew and food to male slaves at mealtimes when they
were allowed on deck.

Figure 12.1, an early nineteenth-century lithograph, shows the barricade that segregated the sexes, the enclosed cooking area (with chim-
ney) on the female side of the deck, as well as a cauldron of food being
readied by crew members for passage through the fence to male captives. It is quite possible that the wooden device depicted center-left of
the lithograph, placed underneath the pot of food from which enslaved
females are eating, is an African mortar. The device does not form part
of the ship's rigging, while its hollowed-out cavity would provide a convenient pedestal for holding the round-bottomed cooking pots used in the
eighteenth century.

Figure 12.2 provides more substantive evidence for mortar-and-pestle
processing by females aboard slave ships. This oil painting (ca. 1785) of
the Danish slave ship *Fredensborg II* shows two female slaves at work on
the right side of ship (the quarterdeck near the mizzenmast). Each one
holds a lifted pestle, for use in the grain-cereal-containing mortar below.

Figure 12.2. Painting of the Danish slave ship *Fredensborg II*, c. 1785. Reprinted by permission of The Danish Maritime Museum, Kronborg, Denmark.

The slave ship has already embarked on the transatlantic journey, with most of the sails set for the passage.[40] The painting provides visual confirmation of the fragmentary written record that enslaved females processed cereals aboard slave ships, while it also underscores the gendered organization of work and association of mortar-and-pestle milling with African women.

Wherever blacks were forcibly settled in the Americas, the African mortars and pestles were used to prepare food. The device played a crucial role in the evolution of the Carolina rice economy since prior to the development of mechanical mills in the mid-eighteenth century it was the only means to process the cereal without breakage (see figure 12.3). A milling technology brought to the Americas by enslaved females thus enabled the Carolina rice economy to realize global prominence.

Although Asian rice eventually captured plantation fields, other types were grown for subsistence. Well into the antebellum period slaves were reported planting "Guinea rice" in their dooryard gardens, along with other African plant domesticates such as "Guinea corn" (sorghum), okra, black-eyed peas, and watermelons.[41] Choice of the toponym "Guinea" for both rice and sorghum suggests their African provenance. The deliberate cultivation of African rice also undoubtedly reflected taste preferences, as one slave-ship captain expressed in 1828: "African rice has more taste and solidity than the Carolina rice, although it is not so white."[42] In cultivating African seeds and food technologies, the en-

Figure 12.3. Gambian woman with mortar and pestle. Photo courtesy of David P. Gamble.

slaved transformed dooryard gardens and provision fields into botanical gardens of the dispossessed. Their efforts paralleled the crop experimentation and seed exchanges carried out by Euro-American scientific societies in the same era. In this manner, Africans and their descendants profoundly shaped the agricultural systems and foodways of the Americas.

African Technology Transfer in Carolina Rice Culture

Reexamination of secondary sources and archival records reveals in greater detail the African antecedents to Carolina rice production. The forms of cultivation as well as processing and cooking methods are startlingly similar and relied on similar implements and devices. The three principal African rice systems were established in the Carolina colony by the 1730s.

The initial cultivation of rice as a subsistence crop had much to do with its complementarity with cattle rearing. Replicating the African pattern, the rain-fed system developed in rotation with cattle raising.[43] Many slaves entering South Carolina undoubtedly possessed knowledge of both rice farming and cattle rearing since herding was widespread along the Upper Guinea coast and South Carolina has been identified as a possible point of origin for the North American ranching tradition.[44] Rain-fed rice was planted in the colony until the early 1700s, while beef entered the transoceanic export trade as salted meat.

The dramatic increase of slaves from three thousand in 1703 to twenty-nine thousand in 1739 provided the labor to clear swamps and shift a burgeoning rice economy to the higher-yielding inland swamps.[45] Inland-swamp cultivation involved impounding water from rainfall for soil saturation and manipulating subterranean springs, high water tables, and creeks for supplemental water. Plot berms and sluices enabled controlled flooding and drainage, which drowned weeds and minimized the labor spent removing them.[46] An identical rationale and system characterized African rice cultivation in inland swamps.

One of the earliest references to tidal rice production appeared in a notice printed in Charleston, South Carolina, in 1738 advertising sale of river-front land.[47] Rice cultivation on the floodplains of freshwater rivers provided another boost to the region's economic prominence, as the soils were annually rejuvenated with alluvium.[48] While creation of tidal plantations required enormous inputs of labor, the elaborate infrastructure of embankments, berms, sluices, and canals comprised a highly productive system. A slave was consequently able to manage five acres instead of the two typical with inland swamp cultivation.[49]

Tidal rice plantations resembled their West African counterpart, the mangrove rice system. As in the African freshwater floodplain system, rice was sown directly on Carolina tidal plantations, rather than transplanted.[50] The significance of African expertise in tidal production was evident by the mid-eighteenth century, when slaves from West Africa's indigenous rice area became the preferred workforce.

Planters knew which African peoples specialized in the cereal's cultivation. Newspaper advertisements from the eighteenth century reveal this awareness in notices announcing shipments and sales of slaves skilled in rice culture. In the period preceding the American Revolution one Charleston newspaper advertised the sale of 250 slaves "from the Windward and Rice Coast, valued for their knowledge of rice culture" (see figure 12.4). On 11 July 1785 another notice announced the arrival of a Danish ship with "a choice cargo of windward and gold coast negroes, who have been accustomed to the planting of rice."[51] Carolina planters in particular stated a preference for slaves from Gambia and

Figure 12.4. Advertisement from the period prior to the American Revolution announcing the arrival in Charleston, South Carolina, of Africans from Sierra Leone specialized in rice cultivation. Courtesy of the Library of Congress.

the Windward coast (Sierra Leone)—two key rice-producing areas of West Africa where slave markets were well developed. In the crucial third quarter of the eighteenth century, when the colony was shifting to tidal production, Africans from rice-growing areas formed the majority of all slaves imported into Charleston.[52]

African antecedents to Carolina rice culture are also evident in winnowing, which complements mortar-and-pestle milling. Winnowing involves placing the mortar's partially milled contents, a mixture of grains and empty hulls, in shallow oval baskets. The grains and hulls are rotated and repeatedly tossed in the air, which leaves the heavier husked grains inside the basket. Carolina rice plantations relied on the same method. Even the weaving style of the Carolina winnowing baskets reveals an African genesis as they were coiled, the same weaving method used for winnowing baskets in Senegal and Gambia. The colony's native peoples used only plaited and twilled weaving patterns in basket design.[53]

Methods for cooking rice indicate other linkages to Africa. The Carolina kitchen favored grain separation, following African preferences. The preparation of rice with beans also followed Africans into the diaspora with the appearance of dishes such as Carolina's Hoppin' John, cooked with black-eyed peas.[54] African women transferred the cooking method known as parboiling; parboiled rice is another name for converted rice. That this method is the same as that used to produce the brand name Uncle Ben's Converted Rice sheds light on the image represented on the label.

From field to kitchen, an indigenous African knowledge system took root in South Carolina at the end of the seventeenth century. It reflected the acumen, skills, and techniques developed over millennia by African ethnic groups specialized in rice culture. The development of rice farming in deep-water swamps required far more than the mere availability of seeds for planting. It demanded the presence of people who knew how to transform diverse swamp environments into productive rice landscapes. These were the enslaved from the Upper Guinea coast, who established a subsistence preference in the Carolina wetlands. Their efforts and expertise ensured the survival of a crop that still remains central to cultural identity in the black Atlantic.

Rice in Colonial Brazil

More than a century before its cultivation in South Carolina, rice was introduced to Brazil. Written accounts attribute the cereal's introduction to Portuguese initiatives, even though rice was not planted in Portugal during Brazil's colonial period.[55] The Portuguese did not serve as the source of expertise for floodplain cultivation or mortar-and-pestle milling.[56] The first documented reference to rice cultivation is provided by Gabriel Soares de Sousa. Writing in the period between 1570 and 1587, Soares noted slaves growing the crop for subsistence and its cultivation with rainfall and in inland swamps.[57] By the second decade of the seventeenth century, rice had become a dietary staple throughout northeast Brazil.[58]

Soares attributed the impetus for its cultivation to the arrival of seed rice from the Cape Verde Islands in the 1530s. Archival materials reveal that red rice—possibly African *glaberrima*—figured among the earliest types grown in Brazil. The cultivation of red rice continued in the eastern Amazon until 1772, when the colonial government imposed jail sentences on those continuing to plant it. This punitive measure accompanied metropolitan efforts to establish a rice-plantation economy with Carolina seed and to prevent contamination of the white-grained export rice with the traditional red type.[59]

From the mid-eighteenth century Portugal aimed to achieve self-sufficiency in the production of rice by developing a plantation economy in the eastern Amazon modeled after Carolina. Emphasis was on floodplain cultivation and husking by the water-driven mechanical mills then being developed. Rice plantations flourished in Maranhão, along the Itapicuru, Mearim, and Pindaré Rivers near the capital, São Luís. The metropole relied on the direct import of African slaves familiar with rice cultivation. Between 1755 and 1777 nearly twenty thousand slaves from the rice-growing region of Cacheu and Bissau arrived in Maranhão.[60]

While the rice plantation era ended with Brazil's declaration of independence from Portugal in 1822, its long-term consequence was to add another historical layer to Brazil's history, one that celebrated Portuguese initiative and stimulus in rice culture. However, in the former plantation area the rural descendants of Maranhão's slaves contest this narrative, claiming instead that a female ancestor introduced rice by smuggling the grains in her hair aboard a slave ship.[61]

Conclusion

Rice cultivation accompanied the forced settlement of African slaves to the western Atlantic throughout the early modern period. In another prominent wetland area of the Americas—near Tabasco along Mexico's Gulf Coast—a Spanish land grantee noted as early as 1579 the cultivation of rice in provision gardens by Africans enslaved for tobacco production.[62] Fugitive slaves also planted the grain in the maroon communities they established in northeastern South America.[63] Rice culture accompanied slavery throughout the Americas—for subsistence in plantation provision fields, export as a plantation crop from South Carolina and Maranhão, and as a food staple in maroon communities. Rice remains the signature cereal of the African diaspora—in the Hoppin' John of the Carolina kitchen, Louisiana gumbo (with African okra), *moros y cristianos* (rice and beans) of Cuba, and *gallo pinto* of Nicaragua.

The history of rice in the Americas reveals an important West African legacy. The grain's arrival in the American Atlantic and role in plantation societies represent a remarkable form of technology transfer from West Africa under conditions of forced labor that are difficult to imagine. We are just beginning to explore the plants and knowledge systems that facilitated survival and cultural identity across the troubled areas of the black Atlantic.

IV.
Technologies of
Accumulation

Chapter 13
Collecting *Naturalia* in the Shadow of Early Modern Dutch Trade

Claudia Swan

As socially and culturally salient entities, objects change in defiance of their material stability. The category to which a thing belongs, the emotion and judgment it prompts, and the narrative it recalls, are all historically refigured.

—*Nicholas Thomas*, Entangled Objects

The famously disproportionate relationship between the tiny geographical area of the northern Netherlands and its vast wealth in the seventeenth century is a function of the Dutch role in global trade, especially of the kind monopolized by the East and West India Companies (founded, respectively, in 1602 and 1621). Among Dutch imports, plants and plant products figured prominently: pepper, cloves, cinnamon, nutmeg and mace, coffee, tea, and sugar were the most lucrative. Many other plants too were absorbed into the Dutch market and assimilated to European natural history in the course of the seventeenth and eighteenth centuries. While some of these exotic varieties were already familiar to Europeans by way of overland spice trade dating to the Middle Ages, others were introduced to western European markets in the early modern period. It was not exoticism, however, that inspired the extraordinary reach of the Dutch overseas, but profit. Pepper (*Piper nigrum*), for example, from Sumatra and Malabar, comprised one-third of the imported goods sold at auction in the Netherlands in the seventeenth century; on average, the Verenigde Oostindische Compagnie (Dutch East India Company, or VOC) shipped six million pounds of pepper to Europe per year. Militarily empowered and driven by economic ambitions as much as, if not more than, by patriotic fervor, the Dutch competed and fought with the English, the Spanish, and the Portuguese frequently, violently, and often at the cost of native lives to secure and maintain control over the spice trade. Local rulers were the occasionally unruly subjects of a multinational global safari, whose primary purpose was to recuperate investment in the outfitting and completion of voyages east and west and to turn a hefty profit in the process.[1]

While economic and political gains were the principal motivations for foreign trade, encounters with far-flung regions had a profound effect on Dutch culture and ways of seeing, thinking, and knowing. It would be misleading to state that the arts and sciences played a generative role in Dutch global pursuits, but it is certainly true that they were—differently at different times—affected by the commercial and colonial enterprise overseas. This chapter studies one aspect of early Dutch engagement in the East and West Indies: the assimilation of plants, plant products, and information about them at the turn of the seventeenth century. It asks by what means botanical specimens were acquired and circulated and how they came to be valued. Naturalists and medical professionals play signal roles in this account—their botanical work was often performed in the context of and process of accumulating collections of *naturalia*. How were foreign specimens amassed, organized, described, and represented and how did such processes make sense of the newly imported goods?

This chapter opens with a survey of different kinds of collections assembled in the northern Netherlands in the seventeenth century's first decades: one belonged to the Enkhuizen doctor Bernardus Paludanus (1550–1633); one is the garden of Leiden University; and a third belonged to the pharmacist Christiaen Porret (1554–1627). Later the discussion turns to another, related mode of collecting, exemplified by published accounts of foreign goods. In 1605 Carolus Clusius (1526–1609), director of the Leiden garden, published *Exoticorum libri decem*, the first study issued in the Netherlands of plants (and animals and other items) imported from points east and west. The *Exoticorum*, I will argue, amounts to a natural history collection in print. Producing the *Exoticorum* was a social affair: Clusius depended on a circle of contacts and informants for information and goods. Collecting *naturalia* was a social and epistemological endeavor, especially in the context of (pre-) colonial botany.[2]

Collecting Nature and the Nature of Collecting

Strictly speaking, the first Dutch account of Asian flora and fauna is a travel narrative by the voyager-merchant known as "the Dutch Magellan," Jan Huyghen van Linschoten (c. 1562–1611).[3] Van Linschoten's *Itinerario* (1596) played a formative role in the rise of the Dutch "seaborne empire."[4] By providing practical information—maps, navigational and meteorological tips—and general inspiration, it helped spur a burst of trade activity that resulted in the establishment of the VOC and broke the Portuguese monopoly in the east. The appeal of van Linschoten's *Itinerario* lay in the access it provided to the regions described, as much to the armchair traveler as to actual seafarers. The book, based on van Linschoten's experiences in the service of the Portuguese on Goa and in the Azores as "Factor for the King's Pepper," offers its readers the equivalents of state secrets. While the *Itinerario* contains vital navigational and cartographic information that had been fiercely protected by the Portuguese, it also participated in an established genre of travel writing. The *Itinerario* adheres to the model of European travel narratives, or quasi-anthropological accounts of foreign and unfamiliar locales, people, and customs, that was in wide circulation in the early modern era. Early in the text van Linschoten writes that he was determined to spend time on Goa, on the west coast of India, in order "to investigate the manners, nature, and form of the lands, people, fruits, wares, products, and other things."[5]

A merchant by trade, van Linschoten was not particularly interested in botany. The portions of his *Itinerario* that touch on plants are either entirely market-oriented (where he discusses the trade in pepper), brief (his poetic evocation of trees that blossom in the night, for example), or added to the text by the Dutch doctor Paludanus, his friend. When he returned to the Netherlands from his travels, van Linschoten settled in Enkhuizen, the northern port town, where he cultivated a working friendship with Paludanus that lasted until van Linschoten's death in 1611. It has been suggested that Paludanus "produced" the *Itinerario*; in any case, he provided quantities of botanical information in the form of annotations throughout the text. Even here the account is derivative; the annotations rehash contents of an earlier publication on the plants of Goa, the groundbreaking *Coloquios dos simples e drogas he cousas mediçinais da India* by the Portuguese doctor Garcia da Orta (1500–1569).[6] Van Linschoten's contribution to botany in the context of early Dutch voyages east is hardly monumental or even original. Nonetheless, several aspects of his endeavor are relevant here.

A few general comments about botany of the Indies in the late sixteenth and early seventeenth centuries are in order, to clarify the nature of van Linschoten's enterprise and its impact. First, the natural history of the Indies was often described in the course of more broadly painted accounts of the new worlds under discovery, and in as many cases the descriptions were derivative.[7] Second, the dissemination of botanical information depended on the more general market for information and goods from the regions recently encountered. Third, in the Netherlands around the turn of the seventeenth century, medical professionals played a pivotal role in the production of colonial botany. And fourth, the production of exotic natural history—the study of the flora and fauna of the newly discovered regions of the world—was already linked, within the medical realm as elsewhere, to the practice of collection.[8]

Van Linschoten's legacy survives through his publications, but he is also remembered for the many foreign goods he gave to his friend Paludanus, perhaps the most renowned collector in the Netherlands around the turn of the seventeenth century. Van Linschoten is known to have offered Paludanus two birds of paradise, a male and a female; an armadillo; a tortoise-shell comb; Chinese paper, pens, and ink; and various other exotica. Paludanus's collection, the fame of which was widespread throughout Europe at the time, encompassed many thousands of objects: ethnographic items such as clothing and armor; dried plants and seeds and resins; and, as one German visitor recalled, "all manner of beautiful and remarkable curiosities and foreign things from China, India, America, Africa, Asia, Peru, Egypt, the Mollucas, Spain, the Canary islands, Turkey, Greece, etc."[9] Before he settled in the northern port town of Enkhuizen as city doctor, Paludanus had traveled widely in eastern Europe, the Middle East, Egypt, Italy, and German territories. During his travels he acquired a medical education (he received his doctorate in philosophy and medicine in Padova), hands-on experience of some of the most celebrated European collections of the time, and collectibles.

Early modern Dutch natural history, medicine, and collecting are deeply intertwined. Lorraine Daston and Katharine Park have recently asserted that "the emergence of collecting as an activity not just of patricians and princes, as in the High and later Middle Ages, but of scholars and medical men as well" was "closely connected with [the] new surge of interest in natural wonders."[10] The relationship between the medical profession and *Wunder-* or *Kunstkammern* (cabinets of curiosity) in the early modern period was built of a common interest in the natural world—in natural philosophy and natural history and in the facts of matter, the particularities of nature. One contemporary, who visited the Enkhuizen collection in 1594, recorded that Paludanus "Showed me his collection, which had such varied and numerous items that I scarcely believed they existed in nature. Nature herself seems to have moved into his house, entire and unmutilated, and there is nothing written down in books that he cannot present to your eyes. That is why the great man Joseph Scaliger gave all his rarities (which were both numerous and spectacular) to Paludanus, saying 'Here are your things, which I have possessed unjustly.'"[11] The brilliant Dutch jurist Hugo de Groot (Hugo Grotius, 1583–1645) was particularly inspired by Paludanus's extensive possessions, which he described as "the treasury of the globe, collection of the whole, ark of the universe, sacred sanctuary of nature, and temple of the world" (Thesaurus Orbis, Totius compendium/Arca universi, sacra Naturae penus, Templumque Mundi).[12] A perfect microcosm, Paludanus's collection offered its visitors the experience of the totality of nature. He was master of the goods he had accumulated, which, as a microcosmic whole, signified a form of knowledge. The literature on early modern collecting tends to deemphasize the professional, medical interests of collectors.[13] Generally speaking, Paludanus's profession is perfunctorily passed over in studies of his collection. As Harold J. Cook has demonstrated, however, "medicine and natural history constituted . . . the 'big science' of the early modern period, soaking up enormous sums of money and energy contributed by countless people."[14] Medical and natural history collections were among the primary arenas of the "big science" of the day.

Paludanus and the other collectors under discussion here extended the practice of medical collection northward to the shores of the Netherlands; likewise, the range of foreign *naturalia* that their collections contained stretched to accommodate Asian, African, and American items, in step with the advances of Dutch trade. Medical collection was a sort of professional sine qua non and at the same time depended on the vagaries of travel and trade for its supply. It tended, in general, to focus on the natural world. Paludanus's collection, while quasi-encyclopedic in the range of items it contained, stopped short of works of art, for example.

In several instances the production of early modern colonial botany was closely linked with the cultivation of a collection. I say *cultivation* since the gardens in which living specimens were kept also pertain. The microcosmic, encyclopedic, inclusive collections in which exotic plants came to be viewed, studied, named, queried, classified, and propagated range from rooms such as Paludanus must have overseen, filled to capacity with *naturalia* of every conceivable shape and sort; to gardens private and public, such as the Leiden University *hortus*, which opened under the direction of Carolus Clusius in 1593; to books whose structure mimes the order of actual collections of *naturalia*. All of these "collections" were assembled in early modern Holland by doctors and other medical professionals. In many cases, additionally, there is a fluid extension from one form of collection to another. This is, for example, well known of Clusius, who served as imperial gardener to Emperor Maximilian I in Vienna, moved to Leiden to direct the new university garden, and published a collection of specimens, the *Exoticorum* of 1605. It is generally known that Paludanus tended a stupendous collection, and seldom pointed out that his professional identity was medical, and that he also cultivated a garden; but it is even less frequently recalled that the position Clusius came to occupy at Leiden was actually first offered to Paludanus.[15] Gardens became yet another repository of the objects of medical study and, by slight extension, the *naturalia* fit for *Wunderkammern*.

Universities throughout Europe established gardens during this period to further medical study and the more general study of what was then called "natural philosophy." The plots of land cultivated in Pisa (est. 1544) by Luca Ghini, in Bologna (est. 1568) by Ulisse Aldrovandi (1522–1605), and in Leiden by Pieter Pauw (1564–1617) and Carolus Clusius, for example, in their early years served a dual function of fostering study of the makings of pharmaceutical remedies and of accommodating rare and exotic specimens recently transmitted to Europe.[16] The late sixteenth-century garden was a space in flux: while it continued to be used for humanist recreation and for the cultivation of known varieties for medicinal (pharmaceutical) purposes, it grew to accommodate the rapidly developing accumulation and study of unknown, foreign, and rare plants. In the larger context of reformed medical study and the development of a market for rare varieties of flowers, the professional or institutional garden (as opposed to the private garden) was no longer simply a repository for time-tested plants that served as the ingredients for medicinal concoctions. Seeds, bulbs, and roots were exchanged avidly, planted and transplanted, and studied fiercely for their potential properties—whether pharmaceutical in the case of foreign specimens or financial in the case of exotic, prized varieties (or both).

While their roots lie in medical study of the plant world, early modern academic gardens bore the marks and accommodated the goods of colonial and trade endeavors. We know, for example, from personal letters, from Clusius's publications, and from a print of the Leiden garden issued in 1610 that two stalks of bamboo that were donated to the university remained highlights of the garden's offerings to its public (see figure 13.1). They are featured as framing elements at right and left and are labeled, as are the foreign curiosities delineated in the cartouche below the plan of the garden. These kinds of goods—bamboo, dried blowfish, crocodiles, tortoise shells, coral—had uncertain medical value but were nonetheless fervently assimilated to local collections of *materia medica*. In Leiden they were housed in the long gallery at the southern edge of the garden, the inventory of which is in essence the catalog of a *Wunderkammer*. Originally built to shelter students and visitors to the garden from rain and to provide protection for plants during the winter, by the second decade of the century, by which time its floor had been paved, the Leiden ambulacrum became a destination in and of itself. In 1614 it was described as follows: "This gallery [is] decorated and hung with many and various maps and geographical depictions [*Land-tafelen*], as with some foreign animals and plants, brought here from both of the

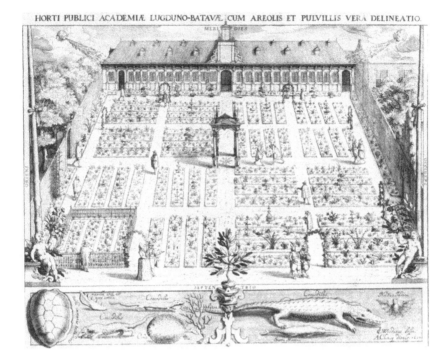

Figure 13.1. One of the earliest prints of the Leiden University garden, in which a variety of exotic specimens are displayed in the lower register. William Swanenburgh after Jan Corneliszn Woudanus, *Leiden Garden*, engraving, 1610, 330 × 400 mm., Holl. 29 Swanenburg(h), no. 32. Reproduced by kind permission of the National Herbarium of the Netherlands, Leiden.

Indies and other places." The earliest inventory of the contents of the Leiden gallery, which refers to the contents as curiosities (*"rariteyten"*),[17] records "foreign animals and plants," some of which may have arrived in the Netherlands on the first Dutch ships to return from the East Indies. They include the bamboo stalks (*Arundo Indica*), a webbed plant (*planta retiformis*), boxes of resins and extracts, and various fruits or nuts; animals are more numerous and range from crocodiles, penguins, and blowfish to parts of animals such as the foot of a cassowary bird, a walrus penis, various parts of a bear, and the "beak of a strange bird." Ethnographic items are also listed, such as pygmy vestments, two Indian hammocks, an Indian skirt, and an Indian ink pot (as in either the West or the East Indies). Also present were sheets of Chinese paper, some of which bore notations on plants. Among other things, items deemed exotic by contemporary natural history were the foreign, nonnative items no natural historian could procure without engaging market and trade relations. The precise sources for many of the objects displayed in the Leiden garden and ambulacrum c. 1600 and the social aspect of this process of accumulation are touched on later, but it is useful to recall at this point that the renown of gardens such as Leiden's rested in part on the presence of such items as the twin stalks of bamboo, which were donated to the garden by early travelers. Though technically speaking botanical, these goods were clearly appreciated for their exoticism, their foreignness, rather than for their medical applications.

That early Dutch collecting of foreign—exotic—plants and other goods was the province of medical professionals is borne out in various hitherto understudied instances. Consider, for example, the collection of the French pharmacist Christiaen Porret, who settled in Leiden as a young professional and remained there until his death in 1627. Little is known of Porret beyond his good connections: he was brought to Leiden by the great Antwerp publisher Christoffel Plantin (1520–89), who opened a branch of his printing empire in the university city in 1582 and in whose Leiden house Porret initially lived.[18] His connections to such prominent figures as Clusius provide insight into Porret's interests and standing. Porret is mentioned regularly in Clusius's publications (both his *Rariorum plantarum historia* of 1601 and his *Exoticorum* of 1605) as a source of goods and information. The goods and information he purveyed—to colleagues such as Clusius and, presumably, to his clients as well—he cultivated and stored in his home, which housed an amazing and little-known collection, and in his garden. In 1621, in an extended paean to his own home and garden in Zeeland, the minister Petrus Hondius (1578–1621) made a lengthy digression to praise Porret's talents and garden. Hondius introduced Porret as "Renowned Pharmacist, Simplicist, and Herbalist" (*Vermaerden Apothecaris, Simplicist ende Herbarist*) and praised his generosity and diligence. Hondius's lengthy dedication was written when Porret was sixty-seven years old. "Your old age," he wrote, "prevents you more and more from walking two and three times a day up and back to your garden outside the city."[19] That Porret's garden was important to him as a resource could hardly be more clearly spelled out.

Housed in his home on the Maersmansteeg in the center of Leiden was Porret's other collection—the other great *Wunderkammer* in the Netherlands after Paludanus's in Enkhuizen. In 1628, within a year of his death, Porret's collection was sold at auction. The small printed catalog is titled "Exceptional Items or Rarities/and/Unusual Sensualities (*Sinnelickheden*)/From Indian and other foreign locales conches/shells/terrestrial and maritime creatures/minerals/and also strange/animals; and some artfully made/handicrafts and paintings/Which Christiaen

Porrett [*sic*], late Pharmacist/collected in his *Cunstcamer*."²⁰ While the fact that this collection was assembled and maintained by a Dutch pharmacist has doubly condemned it to historical oblivion, it is a crucial and fascinating record of how *naturalia* were collected by medical professionals in the early era of Dutch global trade.

The contents of Porret's impressive collection ranged widely in kind and substance. The initial entries in the catalog name, for example, "two serpentine containers, used as cups or mugs" (no. 1); "two crystal glasses with white striping" (no. 2); "a platter of serpentine stone" (no. 3); "an ivory sphere or globe, with various balls that turn inside each other, on a pedestal or foot of ebony" (no. 5); "a spiral staircase made of ivory" (no. 15); Persian cloth in the shape of a turban (no. 16); ink from China (nos. 17, 26); and "two mother-of-pearl fishing rods from the Straits of Magellan" (no. 27). Reading the catalog is a staggering experience. Some of the *naturalia* listed are: "a sea plant like cauliflower" (no. 32)—likely coral; hundreds of shells in all sizes and shapes and colors, including at least one "mother of pearl shell, carved and painted" (no. 168); several nuts—a "covered sectioned nut from the Indies" (no. 42) and a "covered head, from a fruit from the Indies," either of which may have been a coconut; an emu's egg (no. 46); agates, quartz, sapphires, and other stones; ebony and ivory, bones and horns, and at least one fossil; a large piece of white coral, painted red and gilded (no. 69); a couple of beaks of birds from the Indies; a "Bird's nest in a red drawer, with five or six little birds very beautifully constructed of feathers in all colors" (no. 133); a blowfish; a large crocodile; a small crocodile; and drawers and drawers filled with resins, stones, minerals, and fruits. The contents of Porret's collection included not only *naturalia* but also *artificialia*—man-made or artfully natural items— ethnographic objects, and *scientifica*, although this is not the place to itemize more of the extraordinary contents of his collection or to call more than passing attention to the amazing range of crafted articles. In a manner as integral to early modern collections as it is foreign to modern taxonomic logic, the catalog moves from "seven crafted flowers from the Indies in a box" (no. 281) and "a box of various Indian fruits" (no. 282) to "A box with eleven black flutes/small and large/serving to attract different animals/or to imitate their voices" (no. 283).

Exceptional, curious, foreign, plants, animals, minerals, and art as well, all collected by a pharmacist and not in his shop but in his *Cunstcamer*: however amazing this seems, within the European context of the early modern collections assembled at courts from Prague to Vienna to Brussels, in ducal residences in between, and on a smaller scale privately, the combination in Porret's collection of natural items, works of art and handicraft, ethnographic specimens, and even optical devices is entirely congruent with more general developments. From a strictly functional point of view, however, Porret's collection makes little sense: it is not possible to explain how a pharmacist could have used its contents. While we know, for example, that pharmacists continued to supply such relative delicacies as sugar and hard-to-come-by supplies such as ink and distilled liquors and wine until well into the seventeenth century, and while it is true that unicorn horn (Porret owned one) and bezoar stone were still considered efficacious remedies, the uses of Turkish and Hungarian shoes, for example, or ivory lathe work or carvings of Chinese deities, such as he owned, are elusive.

That the history of collection and the birth of museums are linked closely with medical practice and, in particular, with pharmacists' interest in *naturalia* is clear. As Paula Findlen has shown, the continuities between pharmacists' collections and other early modern collections are manifold. They all equally pertain to the new model of the museum, which Findlen has described as "a site of encyclopedic dreams and human sociability . . . [and] a setting in which to examine nature."[21] Collections of *naturalia* pose interesting challenges, and a collection that pertains substantively to the model of a princely *Wunderkammer* but was assembled by a pharmacist doubly so.[22] Consider, for example, that the title page of a contemporary Italian pharmacist's collection tells us that it contains: "many natural and moral things worthy of philosophical consideration, and no less pertinent for setting out and explaining Medical things; [the collection] is not without great exotic things, as well as artificial things."[23] Looking back to Porret's collection and bearing Paludanus's and the Leiden gardens in mind, we observe the same relationship among philosophical consideration, medical explication, an interest in the exotic, and appreciation for what is artful. This is especially striking in the case of the goods explicitly identified as foreign.

Paludanus's collection contains numerous items identified as from the Indies—ranging from the items mentioned above to Indian cloths; "a tin spoon from the Indies" (no. 716); a "box with Indian beans" (no. 713); two Indian pikes (no. 679); Indian arrows, bows, and a shield (nos. 672–74); Moluccan chests (nos. 668, 669); an Indian melon (no. 655); and an ax, feathers, hats, baskets, swords, and such clearly botanical items as "a garland of Indian fruits" (no. 606). What is striking is how integrated these items are in the overall system of the collection. Their wonder is of course amplified by their foreignness, which translates into exoticism for all intents and purposes. One item, "A herbarium or plant book/printed in China" (no. 493), stands out as an example of foreign botany. But this is an exception to the rule of such collections, in which what was accumulated was not foreign knowledge or patterns of thought or practice, but objects—stubborn material objects that did, in Nicholas Thomas's[24] terms, change, that were historically refigured, but only by virtue of being categorized according to that European system of objects and of attention to the natural world also known as collecting. By encompassing such exotic goods, Porret's and Paludanus's collections evinced wonder, much as princely collections did, by virtue of the range and scope of textures of particulars.

Social Exchanges

Generally speaking, foreign goods were assimilated to European natural history wholesale in the early modern era; they came to occupy ingrained paradigms of study and of use. Botanical goods were imported; knowledge, by and large, was not. Existing regimes of classification offered templates for the absorption of new material and physical information in the later sixteenth and much of the seventeenth centuries. In many cases plants and plant goods were introduced as commodities rather than as medicinal remedies; and even when doctors introduced them, by way of publication or exchange, the terms in which they presented these foreign goods were saturated with social and commercial concerns.

The combination of travel and botanical study was not new to the seventeenth century, the heyday of global trade. In the course of the previous century, for example, naturalists returned to western Europe from foreign locales with specimens in hand. In the 1570s the Augsburg physician Leonhart Rauwolf (1535–96) made a three-year journey through the Levant (stopping in Tripoli, Damascus, Aleppo, Baghdad, Jerusalem, and elsewhere); on his return to Germany he published an account of his journey, a partial, botanical record of which also survives in the herbarium of dried specimens he assembled.[25] Rauwolf's brother-in-law, the Augsburg merchant Melchior Manlich, who conducted trade in the Near East, subsidized the trip, in the hopes of discovering new, salable goods and medicaments.[26] The case of Leonhart Fuchs (1501–66) is an inverse example of the motivating effect of mercantile interest on the production of natural history in the context of global trade. Fuchs set out to locate in Germany all of the plants cited by Dioscorides and Galen, and to describe them in his *De historia stirpium* (1542), partially in order to circumvent dependence on costly foreign imports from the East. While da Orta's efforts were more emphatically medical—his *Coloquios* is recognized as the first account of tropical medicine published for a European audience—it is nonetheless evident that early modern (sixteenth- and seventeenth-century) interest in exotic botany was already shot through with economic and social motivations. In a letter he wrote to Carolus Clusius—translator of da Orta's *Coloquios* into Latin and author of several key books on the plant world—ten years after returning from the Levant, Rauwolf asks Clusius to keep his eyes peeled for a patron or benefactor interested in purchasing his herbarium. He possesses, Rauwolf writes:

A number of other plants as well, which I brought from the east; some are very rare. They have been carefully pasted on to paper, so they maintain their colors, and are gathered in a volume such that they can be readily studied by whomever. These plants, which I obtained at the cost of great efforts, much hardship, and danger, I would willingly offer to a liberal and generous prince who would take pleasure in knowing them. Should you encounter such a person and should you have the occasion to discuss this in my name, I would be particularly obliged.[27]

This passage, the core of Rauwolf's letter to Clusius, is a representative instance of how early modern botany—and in particular, botanical study in foreign parts—garnered economic and social sponsorship. Rauwolf asks the well-positioned Clusius to find a backer for his efforts. As his herbarium is already ten years old at this point, he requests patronage ex post facto. The herbarium of oriental plants was as much a repository of botanical information—he says that it can be "readily studied by whomever"—as a commodity in advance of a marketplace.

During his tenure at Leiden University, Clusius was deeply involved in the exchange of such goods and of information about them. His *Exoticorum* is his most significant publication on foreign natural history, preceded by his translated editions of da Orta's colloquies in 1567, Nicolas Monardes's (c. 1512–88) three volumes on plants of the West Indies in the 1570s and 1580s, and Cristóbal Acosta's (d. 1592) follow-up to da Orta's description of East Indian botany in 1582.[28] All three of these treatises were translated by Clusius into Latin and revised, in order to make them more broadly accessible; they are also included in the *Exoticorum*. Clusius had long been interested in local or regional botany; he wrote volumes on Austrian and Spanish flora, for example. Aside from publishing catalogs of plants local and foreign, Clusius ran the Leiden garden; and he maintained a wide circle of correspondents all the while, many of whom supplied goods for either or both endeavors.[29] What is fascinating about Clusius's botanical work is its evident dedication to hunting and gathering, as it were. He is eager to acquire individual specimens, grateful to friends and acquaintances for providing actual plants and plant goods and descriptions of them, and quick to credit them and their collections. All of this is clear from the text of his *Exoticorum*, as well as from his other publications and his correspondence. The division

of his volume on exotic or foreign and unknown items into brief chapters is typical of early modern natural history. Clusius seems to engage
the same form of attention, encapsulated in the brief, descriptive entry,
in his discussion of specimens in letters: for example, in a letter dated
April 1598 to Paludanus, Clusius remarks on a box of dried fruits from
the Indies that Paludanus had sent to him to inspect and describe. As in
the *Exoticorum*—these fruits are the subjects of book I, chapter 20—
Clusius is interested in the provenance, the morphology, and previous
publication history. He tells Paludanus in the 1598 letter that he, Clusius, had never before seen three or four of the items sent to him, and
he asks for further information about where one of them ("the large,
black fruit that is also filled with seeds, as becomes evident when you
shake it") came from and what it is called.[30] He seems doggedly committed to descriptive—rather than prescriptive—values of plants. Original
or adapted uses are of no interest to him.

By such eager accumulation of specimens Clusius worked to fill the
Leiden garden with rare and interesting plants and even drafted instructions for sailors and ships' surgeons and apothecaries traveling to the
East. This means of acquiring exotic goods had already been exploited
by Pauw, a year earlier, in a set of instructions that are now lost. From
the internal VOC correspondence, however, it is clear that Pauw's principal aim in encouraging botanizing overseas was to acquire more simples for the collection at Leiden.[31] The VOC documents mention the
necessity of furnishing the garden with native and foreign simples and,
therefore, with "various Indian herbs, seeds, flowers, resins, roots, and
other such things," and furnishing the mineral collection with "spices,
medicaments, and mineral things that come from the Indies."[32] It is
clear from letters and other sources that items brought back on the earliest expeditions to the Indies (even prior to the foundation of the VOC
in 1602) were quickly and avidly consumed by local audiences. It is interesting that while Pauw's requisition must have stressed the academic
uses of the exotic goods he hoped to obtain, a document of the same
kind that Clusius submitted in 1601, two years later, is more materialis-
tic. Clusius's instructions to the traveling surgeons and pharmacists are
exacting. Branches bearing leaves, fruits, and flowers are all to be
brought back pressed between paper; and Clusius does not hesitate to
specify precisely which plants he wants: nutmeg, both male and female;
black pepper; white pepper; betel; cubebs (tailpepper); and cotton such
as grows in the vicinity of Bantam. He goes on to request branches of all
other sorts of trees that are foreign and asks that sketches be made of
how the trees grow, whether they are large or small, deciduous or not,
the names of the trees, and how they are used. Clusius explains, in a
succinct statement of his scientific motivation: "One must know all of

these things, in order to describe well"; and he concludes, "In sum, he who is attentive will find enough to bring back."[33] Interestingly, the plants that Clusius itemizes were among the most marketable items returned from the East Indies by the merchant ventures. These exotic specimens were eagerly subjected, once secured, to strategies of description and market pressures alike. Where the line between trade value and scientific value lay is unclear. The gifts made to the Leiden University collections c. 1600—bamboo, several fruits, and plants were given to Professor Pauw, rector of the university in 1601–2, by VOC merchants—emblematize the close partnership between mercantile and scientific interests. And in this form of social exchange, status accrued in both directions: to the giver of exotic goods and to the recipient/owner.

In the era under discussion the medicinal use of exotic goods does not yet seem to be their primary quality, in the eyes of natural historians. This is an era of description rather than prescription. Tropical medicine per se would not become an issue for the Dutch until later in the century, when the numbers of travelers to the East increased dramatically and, along with them, the need for remedies (see Harold J. Cook's chapter in this volume). The goods imported c. 1600 were quickly assimilated to collections—and thereby subjected to specific forms of attention.[34] In the gardens and in private collections, as in Clusius's publications, botanical specimens were acquired and accounted for as wonders. Their acquisition expanded the field of natural knowledge and experience, and in several cases they worked to individuals' social benefit. The objects changed, but the system of thought and the practices of medicine to which they came to belong did not. "The category to which [the bamboo given by the VOC to Pauw in 1602] belongs, the emotion and judgment it prompts, and the narrative it recalls," in Thomas's terms, were not—yet—the stories of colonial botany per se and tropical medicine. They still belonged to the arena of natural particulars, and of wonder.

Chapter 14

Accounting for the Natural World
Double-Entry Bookkeeping in the Field

Anke te Heesen

When Robinson Crusoe—seaman, merchant, and sole survivor of a ship-wreck—washed ashore on a desert island, he had with him little more than the wet clothes on his back. A pious and pragmatic Englishman and tradesman, Crusoe quickly took heart and sought to master his situation. In *The Life and Strange Surprising Adventures of Robinson Crusoe* (1719) Daniel Defoe (1660–1731) tells the story of man's installation in the world as the inverse of the creation story: the fact that Crusoe had already tasted of the tree of knowledge allowed him to rebuild the civilization he had left behind, in nature. After a year on the island, during which the Englishman procured essentials and settled down, Crusoe began to keep a diary. "I now began," he writes,

to consider seriously my condition, and the circumstance I was reduced to; and I drew up the state of my affairs in writing . . . and as my reason began now to master my despondency, I began to comfort myself as well as I could, and to set the good against the evil, that I might have something to distinguish my case from worse; and I stated it very impartially, like debtor and creditor, the comforts I enjoyed, against the miseries I suffered, thus. . . .[1]

Here the text contains a table with two columns, one labeled "Evil" and the other "Good." He closes his tabulations with the words "and let this stand as a direction from the experience of the most miserable of all conditions in this world, that we may always find in it something to comfort ourselves from, and to set in the description of good and evil, on the credit side of the account."[2] To order his thoughts and to clarify his situation before God as a believer, Crusoe engaged in a form of spiritual bookkeeping. In a sense, Crusoe's account enacts a secularized Last Judgment, in which he gives account to God but above all to himself and his reason. Only this weighing of debit and credit restores in him some measure of equilibrium, Defoe writes.[3]

One year after the publication of Daniel Defoe's account, the German scholar and physician Daniel Gottlieb Messerschmidt (1684–1735) set off from Saint Petersburg for Siberia on behalf of the Russian czar Peter the Great. The objective of Messerschmidt's seven-year Siberian expedition (1720–27) was to explore a theretofore little-known region of the Russian Empire. While the German physician's story is hardly as exemplary as that of Robinson Crusoe, he might have welcomed the comparison of his seven-year adventure with the Englishman's, although Messerschmidt's differed substantially on account of the climatic extremes and the fact that he was traveling on behalf of a secular power. He too kept a record of his journey; and it too contained an account of his experience. In Messerschmidt's case, more than Crusoe's, the journal revealed to him his own limitations; he became overwhelmed by the sheer quantity of natural objects he encountered and the corresponding notes he compiled, and he struggled with the sense of the futility of his work. What tormented him above all was his—as he saw it—inefficient use of time. His need to vindicate his efforts became particularly acute four years into his voyage when, on 25 November 1724 during a winter pause, he undertook a comprehensive temporal reckoning that takes up several pages of the journals he kept. Messerschmidt anticipated a day of reckoning—presided over by the bureaucratic administration in Saint Petersburg; he writes that he suddenly recalled, as he set to work, instructions sent to him with guidelines for accounting for his time and efforts.[4] The reckoning that Messerschmidt prepared providing evidence of the entire four-year expedition closes with a *Tabula*, or table, in which he sums up all of his working hours, calculated in a tremendously complex way, to account for his activities down to the minute. He concludes his account by remarking that he allowed himself—given the scarcity of time at his disposal—to be satisfied with his achievements, even comparing himself with the "learned men of Europe."

Both Messerschmidt and Robinson turned to the compilation of tables to order their thoughts and keep their spirits up under challenging circumstances. The penchant for accounting that they shared points up a common feature of early eighteenth-century novels and natural histories: the weighing and recording of debits and credits was a technique commonly used to narrate and order biography and natural history alike. Just as keeping accounts enabled Robinson to take his fate into his own hands, it strengthened Messerschmidt's resolve to continue his expedition. Taking Messerschmidt's expedition as a Robinsonian form of natural history, this chapter offers an analysis of such accounting, and on the mercantile patterns of thought and behavior that supported them as well as the mental and material techniques of recording and notation. What techniques were taken into the field of natural history in the early eighteenth century? Messerschmidt's accounts exemplify practices that were already common among world travelers, and that therefore shaped colonial botany from the beginning. In such recording practices, the connection between botany and commerce—even between botany and double-entry bookkeeping—is obvious and concrete.[5] Messerschmidt tackled both his own states of mind and the natural objects he encountered by means of various mercantile and scholarly bookkeeping techniques.[6]

Orders

Throughout the course of his seven-year expedition, and under even the most adverse climatic conditions, Messerschmidt recorded the Siberian landscape and its inhabitants in written accounts and by amassing specimens.[7] His chief activities consisted of ordering his notes, collating them with specific areas of research, and tracing links between them and the plants, animals, eggs, roots, and seeds he collected. To link the order of things with the order of paper, he moved through a series of steps, which he never varied, always completed, and regarded as his principal duty.

Messerschmidt's bookkeeping practice consisted of three principal steps. First, in the daytime he recorded his primary observations with the aid of a small (portable) ivory writing tablet (*Schreibtäfelein*).[8] In this step he also employed so-called *Schedis*, little pieces of paper he collected in his pocket. Second, in the evenings, using his notes and memory, he transcribed what seemed worth saving in his journal, which he kept throughout the expedition. He also kept more casual or "rough" (*unrein*) diaries, to be elaborated in greater detail after eight days, "so that I might not become confused over time."[9] His accounts contain extremely precise lists of collected plants, birds and their live weight, words he had picked up, and the medicines he had distributed to the needy. While carefully composed, his journal contains all sorts of references to expenses, material purchases, difficulties he encountered, and natural history descriptions that would serve him later in composing accounts to the court. Third, this enormous amount of information was ultimately redacted and recorded in a rigid sequence of books. In the case of Messerschmidt's botanical observations, for example, we find three final books. He called the small octavo volume in which he recorded plants alphabetically according to Tournefort's taxonomy the *Index Botanicus Siberia*.[10] Each page was inscribed with a letter of the alphabet, and remaining pages were left blank. Throughout this volume he recorded the plants he encountered, noting their names according to the literature he had with him, and citing specific references therein by page numbers. In addition he maintained *Catalogi plantarum officinalium*, which comprised more precise notes on *materia medica*. Finally, he collected his botanical results in the *Sibiria Perlustrata*, in which he provided a preview, organized by subject, of the scientific results of the journey.[11] His various collections of notes could thus be accessed in two ways: chronologically in the case of his daily bookkeeping, and according to themes and the alphabet in the more formal journals. How then to collate these accounts with the corresponding objects he collected?

In the case of specimens—objects—Messerschmidt also took a three-step approach. He began by placing all objects he had collected on a daily basis in a traveling box.[12] The traveling box was divided into several levels and drawers: dried plants were placed at the bottom, beneath the *partes avium*, and insects, seeds, stones, and such were placed above them. Later, he transferred the specimens to more specific containers: pressed plants were put in a temporary herbarium, and the seeds of plants intended for botanical study and for *materia medica* were placed in a "seed box," which he would later send to Saint Petersburg. He devoted particular attention to his seed collection since he planned to cultivate the plants he had observed during the short Siberian summer in the botanical garden in Saint Petersburg. Ultimately, he put the seeds in order, in a *seminarium*: "And thus [I] had the little bags of seeds . . . taken out of their box, arranged them on the ledges about the room according to the *classibus Institut[ionum] Tournefortii* and following this, beginning with the first class opened them, poured them out, sifted them, ventilated and cleaned them, laying aside a bit of each sort in paper, in order to transport them into the *seminarium*."[13] Messerschmidt was not content simply to collect and preserve the seeds for later use in the botanical garden of the new capital city; as in his written notes, here too he began to develop a more sophisticated mode of organization and presentation. (A *seminarium* was a collection container designed to order various seeds, with a view to completeness.) For a natural historian traveling through the Russian hinterlands, Messerschmidt maintained a remarkably systematic practice of organization during his travels. His technique for accounting for both his notes and the objects consisted of three steps referred to as *observatione*, *annotatione*, and *relatione elaborate*.[14] Observations were made in *curru et via*, along the way, annotations might be made in slightly more fixed locations, such as in a tent, but relations should rightly be carried out in a clean and tranquil place. Only at this final stage could one truly undertake the task of collating information and objects, a task "for which one must have all of the *serinia* [*sic*], boxes and cases open about one, in order to be able to revise everything everywhere."[15]

This synthetic account of the Siberian natural world constituted the key link between the ordering of his notes and the ordering of the natural objects. It represented the pinnacle of Messerschmidt's research and observations, the ideal around which his daily schedule was organized. He did not always succeed in achieving this ideal completely, however. Messerschmidt despaired at the many interruptions he withstood, the pointless discussions with village leaders he was forced to hold, and the stupidity of his servants. All these factors stood, in his view, in diametric opposition to his desire for order and regularity in his scientific routine. Moreover, he felt that his scholarly techniques—note taking, collection, ordering, and dissection—were too laborious and unwieldy for travel and the concomitant assimilation of the new.[16] Thus he complained that he was constantly forced to work *per intervalla* and did not have the leisure to perform all necessary tasks.[17] At the same time, he was so occupied by these activities that he wondered helplessly what to do when he did not have access to his books.[18] Messerschmidt's despair arose from transporting a highly sophisticated and codified practice into an uncharted and unpredictable environment. In order to retain these techniques, which kept him a "whole human being" even as he lived in a wet, filthy tent, he took stock. This stocktaking mediated between the two extremes of his ideal working conditions and the actual conditions of his journey, and it comprised a space for the performance of his "civilized" duties. In the midst of Siberia the German scholar in Russian service dedicated himself wholly to bookkeeping.

Keeping Accounts

Bookkeeping—in the general sense of maintaining a record of something for accounting purposes—was not a new technique in the eighteenth century. It had long been used in commerce, in the private household, and in the personal domain. While Defoe's hero keeps a diary that alternates between descriptive introspection and the depiction of external events, Messerschmidt's journal began as a work commissioned by the Medical Chancery and is closer in intent to a logbook than a personal record. In his journals Messerschmidt engaged in bookkeeping in the literal sense, by keeping a chronologically and thematically organized account that recorded and archived his stock and its movements on the basis of comprehensive records and receipts.

According to Johann Heinrich Zedler's (1706–63) *Universal Lexicon* of 1733, the bookkeeper "records, in a precise and orderly manner, that which is daily traded and turned over, received and delivered, taken in and paid out."[19] It was common knowledge that bookkeeping required industry, exactitude, stamina, and prudence. Moreover, bookkeeping revealed the relationship between the owner and his household.[20] When Messerschmidt wrote that he had "put something in order in my economy and museum,"[21] he referred to the entirety of his travel household, with all of its furnishings and management. To be sure, numerous manuals and *Apodemiken* (treatises on the art of travel) existed at the time, and they indicated what one should observe while traveling and which objects were worthy of collection or were particularly rare.[22] But popular handbooks alone cannot explain Messerschmidt's system for organizing objects and words. For an explanation of his accounts we need to look to a field that offered far more advanced techniques for systematic transcription than did natural history. This field, to which Messerschmidt had access, was mercantile accounting.

The practice of mercantile accounting comprised various methods and with them many bookkeeping techniques. The most prominent of them had asserted itself with the rise of the Italian trading cities of Genoa, Florence, and Venice in the thirteenth century: in so-called "Italian," or double-entry, bookkeeping every exchange of goods or money between the warehouse and the outside world was recorded as profit and loss, credit and debit.[23] This method was ideally suited to extensive transactions involving multiple traders and agents, great distances, and different currencies and for accounts settled over longer spans of time—all features of the early Renaissance economy. Double-entry bookkeeping allowed trading companies to maintain oversight despite multifarious and protracted transactions and—more important—allowed them to reconstruct at any given time even the most protracted transactions down to the details of the numerous receipts, orders, correspondence, interim bills, and so forth. To put it rather simplistically, double-entry bookkeeping was a principle for mastering the circulation of goods and their reinvestment. This principle asserted itself more and more in the course of the sixteenth century and was perfected not least by Dutch merchants.[24] The seventeenth and early eighteenth centuries saw the production of a growing number of introductions to the method of double-entry bookkeeping and manuals of instruction. These manuals were already widespread in the German territories at the time of Messerschmidt's expedition but were systematized in the second half of the eighteenth century.

In 1786 the economist and cameralist Johann Beckmann (1739–1811) reported that the general practice of "cameralism," or accounting for state treasuries, "had been elaborated with uncommon industry and much acuity" but it had the disadvantage "that not every entry is sufficiently explained to the reader, who thus cannot properly understand what manner of sums these actually are that he finds put to account there."[25] This problem, he added, could be remedied by the "Italian manner of bookkeeping." The entry on *Buchhaltung* in Zedler's 1733 lexicon still refers to the simultaneous use of more than fourteen separate books and accounting records. It is thus not surprising that double-entry bookkeeping, with its four principal account books, was praised as an invention by Beckmann and named as an educational tool by Johann Wolfgang von Goethe in 1795.[26] The decisive feature—for the organization of both large-scale trade and one's own life—was the possibility of drawing up a balance and recording accountability at any time. But what instruments were required? Double-entry bookkeeping is based on relations between debitor and creditor, between whom each commercial transaction is recorded, together with the names of the person who delivers and the person who receives the goods. "In a word, what I buy, receive or take into safekeeping becomes a *deber* [*sic*], along with the person to whom I pay something; in contrast, everything that I sell, deliver, or give out of my safekeeping becomes a *credit*, along with the person who pays me."[27] In order to manage the dual records of incoming and outgoing goods and money, double-entry bookkeeping required three account books, which were described by the English merchant William Webster (c. 1684–1744) in 1719: the first was the "waste book," the second the "journal," and the third the "ledger" or principal account book. The waste book was used to record commercial activities as they occurred, that is, in chronological order. In the journal the entries in the waste book were recorded more systematically, with specification of the debtor, the creditor, and the sum of money involved, as well as the object of trade (the goods) and the circumstances of the transaction. The ledger or principal account book was the largest and most comprehensive book. The first two served preparatory purposes, while in the third all persons, transactions, objects, and sums were listed in order, so that a balance sheet could be drawn up from the whole.[28] An index was also required in order to provide an overview of this last account book and to locate the individual entries (see figure 14.1).

The index lists proper names, places, individual journeys, and goods transacted, with each alongside corresponding page numbers. According to Webster, the waste book provided a chronology of events, and the ledger offered a final written form organized according to subject and accessible via the index. Bearing these specifications in mind and thinking again of Messerschmidt's bookkeeping methods, it is striking that Messerschmidt listed his income and expenses in his reports to the imperial court, and that his balance of pluses and minuses even extended to his calculation of time and thus to his own spiritual welfare. In both the commercial and the natural historical economies, information was registered systematically in a series of three books, each of which represented a different stage of revision and registration: information was transcribed and transferred from the form of cursory notes and chronological daily records to, ultimately, a large account book that contained all of the requisite information. The three-step process of calculation

A	B	C
Fol.	Fol.	Fol.
A. B. Executors 5	Brandy ——— 3	Cash ———
	Black Cloth ——— 7	Charges Merch. — 7
	Bills receivable — 9	Carefull (Peter) — 8
	—— payable —— 8	Chince (Indian) 10
	Broad Cloth in Company — 10	
	Balance ——— 15	

D	E	F
Fol.	Fol.	Fol.
Damner (Steph.) & Accompt Cur. 12	Eafy (Sam.) —— 4	Freeman (Tho.) —
—— Company 12	Ellis (Edward) — 6	Fletcher (Jacob) 14
Diffell (Ja.) Curr. 12		
—— Comp. 13		

G	H	I
Fol.	Fol.	Fol.
Goods for Acct. of Van Tromp 9	Harret (Edward) 4	Janes (Ben.) —— 4
	Holland ——— 6	Johnfon (Simon) — 5
	Houfe-Expences — 7	Johnfon (Edw.) and Comp. 5

K	L	M
Fol.	Fol.	Fol.
King (Sir Peter) 5	Long (James) —— 5	Moflin ———
Kind's Acct. Curt. 11	Lottery Tickets — 5	Mi..ton (Tho.) 9
—— Comp. —— 11	Long (Sir Nichol.) 14	Merchant (Tho.) 14
	Logwood ——— 14	

N	O	P
Fol.	Fol.	Fol.
..ble (Simon) — 5	Owners of the Ship Bonadv. 10	Port Wine ——— 5
..fo (John) — 8		Profit and Lofs — 1

Q	R	S
	Fol.	Fol.
	Refufal of Barg. — 7	Stock ——— 1
		Sherry ——— 3
		Ship James ——— 3
		Smith (Andrew) — 4
		Smart (John) —— 4
		Sugar ——— 6
		Sutor (John) —— 9
		Ship Bonadventure 11
		Sugar in Comp. 13

T	V	W
Fol.	Fol.	Fol.
Tobacco ——— 3	Voy. to Barbadoes 6	Williams (David) 6
Tamarins ——— 8	Voy. from Barbad. 8	
Tobacco & Comp. 11	V. Tromp's Curt. 10	
	—— Acct. Time 10	
	Voy. to Marfeilles in Comp. 13	

X	Y	Z

Figure 14.1. Example of an index from William Webster, *An Essay on Book-Keeping, According to the True Italian Method of Debitor and Creditor, by Double Entry* (London, 1719).

and keeping accounts is a feature of Messerschmidt's records and double-entry bookkeeping alike. Both instances involve a codified system of notation, in which observations or completed transactions recorded in the principal (final) account could be traced back, via the initial record books, to their respective circumstances, receipts, and connections. Messerschmidt was well aware that he alone was compiling notes and emphasized all the more that his reports were the results of an unvarying accounting practice. Incidentally, the sequential ordering of information at work in Messerschmidt's journals represents a practice that came to be even more warmly embraced by scholars than by natural historians. The retrieval system offered by accounting methods permitted scholars to quantify that which they had already processed, allowing them to trace information back to its source at any time.[29] The physicist Georg Christoph Lichtenberg (1742–99) also drew a direct connection between the art of double-entry bookkeeping and scholarly note taking. In 1775 he noted in his *Sudelbuch*:

Tradesmen have their waste books (*Sudelbuch, Klitterbuch*, I believe it would be called in German) in which they inscribe from day to day that which they sell and buy, everything [noted] together without any order; from here, information is transferred to the journal, where everything is recorded more systematically. Finally, it comes into the ledger *at double entrance* [English in the original], according to the Italian manner of bookkeeping. A separate account is kept herein for each man, as debtor and as creditor. This deserves emulation by scholars. First, a book in which I write everything as I see it or as my thoughts dictate to me, after which this can be transferred to another book, in which the material is more separated and ordered; the ledger could then contain the connection and the ensuing elucidation of the matter in an orderly manner of expression.[30]

Lichtenberg identifies what scholars were supposed to learn from double-entry bookkeeping, by emphasizing not the system of debtor and creditor but rather the keeping of sequentially refined accounts that aim to preserve what merited being remembered. Just as Messerschmidt divided his note taking into three stages, so too Lichtenberg explicitly referred to the three books used in double-entry bookkeeping.[31] Lichtenberg's passage points, additionally, to an accounting technique familiar to Defoe as well as to Messerschmidt, which Vincent Placcius (1642–99) called *gelahrtes Buchhalten* (learned bookkeeping).

Learned bookkeeping can be traced back to antiquity, though it was principally and most famously the expanding literature of the sixteenth and seventeenth centuries that the humanist culture of excerpts sought to tame and assimilate through orderly note taking. The polyhistoric tradition of excerpting was aimed at making the knowledge contained in books one had read accessible. Elaborate webs of reference were produced by drawing out units of information from books and recording them in one's own excerpt book under specific headings. Either this was done over time, so that an index had to be prepared at the end of such a book, or one organized the pages to be written on from the beginning according to subject headings, adding excerpts at the appropriate places.[32] This technique allowed one to process what one read and to keep it available for recall. It amounted to an *Aufschreibesystem*, or notation system.[33] The commonplace or mnemonic books that were its products—and vehicles—were also known as "collectanea," which the humanists used, "following antique rhetorical tradition, for *inventio*," that is, for gathering ideas for further works. Both the accounting and the rhetorical methods consist of the three steps of collecting, reviewing, and organizing material.[34] The *methodus excerpendi* of which they formed the key actions was outlined in numerous manuals that identified how, and to what purpose, one should excerpt. Such manuals were derived from instructions published by Conrad Gesner (1516–65), Daniel Georg Morhof (1639–91), Martin Fogelius (1634–75), and Vincent Placcius. One of the best known and most influential of such manuals was by John Locke (1632–1704), who promoted a special form of notebook, the commonplace book.[35] His *New Method of Making Common-Place-Books* was written in France in 1686 and finally published in English in 1706.[36] "Memory," Locke notes in the introduction, "is the treasurey or Store-house" and, as such, requires ordering. "It would be just for all the World as serviceable as a great deal of Houshold-Stuff, when if we wanted any particular Thing we could not tell where to find it."[37] This ordering process begins while one is reading. "The Places we design to extract from are to be marked upon a piece of Paper, that we may do it [make notations] after we have read the Book out."[38] One is instructed to read the text through a second time in order to decide what is worth recording. The systematic ordering of what is recorded is always already implicit; in Locke's method as in others, the index plays a key role. "I take a White Paper Book of what Size I think fit, I divide the two first pages which face one another, by parallel lines" and create an index (see figure 14.2).[39] Locke's instructions continue. In order to compile an index one must first determine keywords and place them in alphabetical order. One should also maintain a separate list of the keywords. In order to check information that had been registered, one consulted the most likely keyword, checked the index to see whether it was listed, and consulted the relevant page of the notebook. To add an entry, the same method applied. In this way Locke trained generations of English gentlemen, scholars, and boys and, indeed, propagated a sort of discipline of learned reading.[40] Here, too, two-way access to the material was of key importance. The commonplace book could trace a quotation back to its original context, while at the same time the contents of the book were a treasury on which memory or the author could draw. Like double-entry bookkeeping, Locke's mode of recording is characterized less by pluses and minuses than by the two-way street along which information was registered and accessed.

Messerschmidt's journals, the merchant's records, and a gentleman's notebook share a consistent method of processing information that moves from initial, rough notes to a polished state and then, finally, to the indexical registration of all units of information. Writing was not the only discipline that demanded this kind of order. It applied to the entire practice, as it were, of inscription, of employing paper and the physical organization it permitted, by way of pages, folding, binding, interleaving, and sequential placement. Scholarly and economic bookkeeping

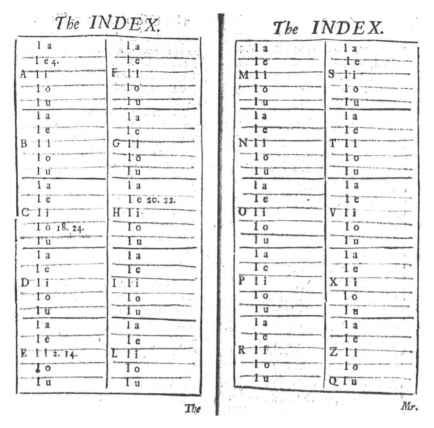

Figure 14.2. Index example from John Locke, *New Method of Making Common-Place-Books* (London, 1706).

extends well beyond the two-dimensionality of the paper page to involve spatial as well as written order.

That "learned bookkeeping" drew not only on paper and ink but also on containers, crates, and even cabinets had already been demonstrated by Vincent Placcius in his 1689 *De arte excerpendi*. Placcius offered precise instructions for recording and preserving notes and excerpts, and copperplate engravings illustrated the appropriate organization of one's literary gleanings (see figure 14.3). Notes, Placcius explained, should be made on small pieces of paper. These *schedae* are to be transferred to an excerpt cabinet, where they are pinned fast. Placcius demonstrates that the order of the excerpts was not restricted to the two-dimensionality of the surface; they were arranged in space as well. The example of the

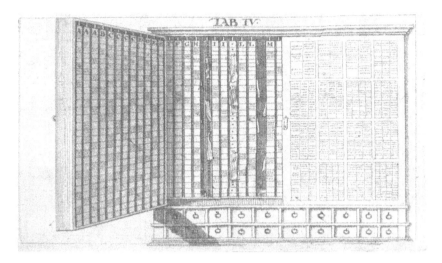

Figure 14.3. Excerpt cabinet from Vincent Placcius, *De arte excerpendi: Vom Gelahrten Buchhalten* (Stockholm and Hamburg, 1689).

excerpt cabinet blurs the division, apparently still so sharp in the case of Messerschmidt, between notes on paper and objects collected in boxes. On closer scrutiny, however, it becomes evident that one could also speak of spatialized indexes or lexicalizing objects.

As we have seen, Messerschmidt collected numerous seeds and seed-pods, among other specimens, and stored them during his journey in specially constructed capsule containers, out of which he formed his *seminarium*. His systematic activities were in a certain sense not unique; instructions for ordering seeds and plant parts had also been written for nontravelers. Many natural history collections of the day featured seed cabinets. They represented collections in their own right, although it was not always clear whether they were collections in progress or horticultural storehouses from which seeds could be drawn for use. Johann Georg Krünitz's *Ökonomisch-technische Enzyklopädie* (1824), for example, described such a seed cabinet as an object reminiscent of Placcius's excerpt cabinet (see figure 14.4). The engraving shows a cabinet similar in structure to cabinets that art collectors used to store medals or coins, although the seed cabinet's interior is markedly different:

As in the Figure, each drawer can be divided into 96 square sections, and these sections can be fashioned in wood by a cabinet-maker, or in thick cardboard by a bookbinder, if one wishes to undertake this latter task oneself, and neatly covered with white paper. In each section one must place a specific seed and inscribe a number, according to which one prepares a list of names. One can also

Figure 14.4. Seed cabinet from Johann Georg Krünitz, *Ökonomisch-technische Enzyklopädie oder allgemeines System der Staats-, Stadt-, Haus- und Landwirthschaft und der Kunstgeschichte in alphabetischer Ordnung*, vol. 135 (Berlin, 1824).

note the names on a small piece of paper in each section, which is doubtless the most appropriate procedure for a systematic order, and use it to make a catalogue.[41]

Krünitz's description of a seed cabinet reflects suggestions taken from the apothecary's trade, where collections of seeds, and above all *materia medica*, were stored in this manner.[42] The wealthy English physician and collector and president of the Royal Society (1727–41) Hans Sloane (1660–1753) took a more pragmatic approach to collecting horticultural specimens. His collection of "Vegetable and Vegetable Substances" offers a view of the inner life of a seed cabinet or *seminarium*. Objects were stored in ninety drawers housed in five cabinets of equal size. Numerous small boxes contained seeds and other medically relevant natural products, whose names were recorded on narrow labels pasted onto the boxes. Sloane labeled the little boxes and wrote the weighty accompanying index himself.[43] What Messerschmidt had stored in capsules Sloane organized as miniature peep shows, glassed in so that one could view the objects inside from all sides. These little boxes were then placed in the cabinets such that, size permitting, only the labels, not the objects within, were visible (see figure 14.5). Thus, initially, one cannot tell whether what one sees is a list or a compilation of little boxes. Sloane's seed collection took the form of a continuous list.

Both Messerschmidt and Sloane placed the objects they had collected on their journeys between the covers of books and within the walls of wooden boxes. Early eighteenth-century students of nature made use of an old practice of registration in order to record quotations, seeds, observations, and mineral samples. The parallel ordering practice involving writing and objects is related both to mercantile bookkeeping and to the learned notation system of excerpting. The accounting traditions of number and letter that they deployed were closely intertwined.

While Crusoe could look with satisfaction at the "Cave" he had turned into a "warehouse," Messerschmidt despaired at the flood of material he had accumulated. Despite the abundance of objects he had gathered, he felt that they represented but a minuscule portion of the natural history of the Siberian wilderness. Both men, the castaway without a concrete task and the explorer traveling terra incognita on an imperial mission, engaged the mental and habitual appropriation and ordering of nature with the help of particular techniques for managing paper and objects. Dual record keeping permitted the merchant and the natural historian alike to survey their emotional balance and bring it into equilibrium. The materiality of mercantile and scholarly practice allowed the protagonists to sustain themselves in the wilderness and thus to preserve their membership in civilization, by ordering the things around them.

What Defoe described as Crusoe's island, recognizable in fiction as a utopia, was in Messerschmidt's case his learned baggage—a real place that he inhabited, that he used, and that nevertheless separated him from everything he had known before. While justice could be estab-

Figure 14.5. Drawer from a cabinet in Sir Hans Sloane's collection of "Vegetable and Vegetable Substances," Collection of the Natural History Museum, Department of Botany, London.

lished on the island, the integrity of Messerschmidt's baggage was more fragile. The stock of objects he had collected over the years was confiscated on his return to Saint Petersburg in 1727, and the remnant finally allotted to him sank years later with a ship off the coast of Danzig. Unlike people, things cannot save themselves.

Chapter 15

Surgeons, Fakirs, Merchants, and Craftspeople
Making L'Empereur's Jardin *in Early Modern South Asia*

Kapil Raj

In recent years the strategies Europeans employed for gathering natural historical, ethnographic, and geographical knowledge beyond the confines of the metropolis have attracted substantial interest. Historians have, on the one hand, examined diverse modes of "field" inquiry (in contrast to the more common focus in science studies on knowledge making in the controlled setting of the laboratory[1]) and, on the other hand, analyzed the genre of "instructions to travelers"—often written by sedentary men of science in Europe—aimed at teaching what to observe in foreign lands, how to regulate and standardize techniques in collecting the requisite objects, and how to report them.[2]

Although both approaches have posed new and important questions for the history of science, each is riddled with serious difficulties. By opposing the heterogeneous space of the field sciences with the more uniform civilities of the laboratory sciences, the former approach fails to examine the relationship between the two spaces, between those "out there" and their sedentary colleagues who often played a crucial role in validating knowledge claims made in the field. And, by focusing exclusively on the collecting techniques that metropolitan savants recommended to work-a-day travelers (usually seamen, ships' surgeons, merchants, and sometimes missionaries), the latter set of studies suggests that scrupulously following the instructions was sufficient to acquire knowledge of the outside world. They thereby imply that the sought-after natural historical objects and knowledge were directly accessible to the travelers—that the whole project of collecting nature was akin to present-day space engineers programming planetary probes to retrieve relevant information from hostile environments. But the crucial difference between space probes and early modern travelers is that the latter mainly visited populated lands and had to negotiate with indigenous peoples to find out about and obtain objects that were often accessible only through their mediation.

Many European savants were well aware that the significance of natural historical objects, especially of flora and fauna, was embedded in the human cultures surrounding them.[3] Bringing to light these social—and economic—meanings was seen as an inevitable first step toward the commodification of these objects within the regional and global economies that the Europeans sought to enter. Because of its strategic importance, this type of information was itself highly prized merchandise.[4]

To be sure, the interactive nature of knowledge acquisition outside the metropolis has not escaped the notice of some historians, as attested to by recent research on the role of intermediaries in the construction of natural knowledge, although mainly in the context of the New World and the Pacific.[5] Little attention has, however, been focused on the other major contact zone—the Indian Ocean.

Despite many similarities, European encounters with the West and the East present significant differences, especially in the case of knowledge formation. Attracted to the East initially by the lucrative spice trade, Europeans discovered a world that was, ultimately, familiar to them, one already dominated by Muslims, their perennial, yet well-known, rivals. However, in this world Europeans formed but one small commercial group among many long-established trading communities of different racial, religious, and regional origins.[6] European survival in the region thus depended on the development of durable relationships with various regional agents—rulers, merchants, bankers, and interpreters, as well as skilled workmen and savants. In the Indian Ocean world, knowledge relating to botany and medicine was already formalized and circulated from the Arabian Peninsula to China within specialized communities, each with its own civility. Early modern European physicians, surgeons, and, later, naturalists in the region readily acknowledged this fact.

This has led at least one scholar to assert that early modern European botanizing in South Asia consisted essentially of compiling indigenous botanical texts, organized according to non-European precepts.[7] However enticing—and refreshing vis-à-vis the received notion of botany being a European preserve[8]—it may be, this interpretation begs important questions. How did Europeans and South Asians develop working relationships? What was the nature of the material, economic, and symbolic transactions that made the construction of specialized knowledge possible? How did these relate to the broader manufacturing and trading economies of the region? In what language(s) did they communicate? What was involved in the transformation of non-European knowledge into European forms? What was the relationship between this knowledge, its producers, and metropolitan savants and academies? Were there significant differences between the various European nations in their relationships toward non-European knowledge? Curiously, an unknown manuscript herbal held at the Muséum National d'Histoire Naturelle in Paris helps shed new and valuable light on these questions.

From a Forgotten Codex in a Paris Archive . . .

Under the title *Ellemans botanique des plante du Jardin de Lorixa leur vertu et quallite, tans conus que celle qui ne le sont pas avec leur fleur fruis et grainne traduit de louria an frances* (Botanical Elements of the Plants of the Garden of Orixa, Their Virtues and Qualities, Both Known and Unknown, with Their Flowers, Fruits and Seeds, Translated from the Oriya into French), the library of the Muséum National d'Histoire Naturelle holds a fourteen-volume folio herbal, twelve of which contain 725 double-folio paintings of 722 plants species. The remaining two volumes contain a description, in French, of each plant with an index of their vernacular names transcribed in the Roman script and a classification according to their medical or economic uses.[9] In addition, the first volume contains a "Preface," an "Avis au lecteur," and an intriguing frontispiece depicting five human figures, a potted tree in the foreground, and a Greco-Roman ruin in the background. The human figures are divided into two groups—three on the left, comprising an artist painting the tree, a man sitting next to him, and a woman carrying plants in a basket on her head; and two on the right: an ascetic holding a manuscript and a European standing behind him. The style of the frontispiece and the human figures it depicts, as well as that of the plant paintings leave no doubt as to the South Asian origins of the herbal (see figure 15.1).

But the library's catalog gives only two meager bits of information: its author is a certain L'Empereur—in all probability the European in the frontispiece—and it dates from the eighteenth century. If the catalog is laconic, the manuscript is more forthcoming. The title refers to a specific location in the Indian subcontinent: *Jardin de Lorixa* means "Flora of Orixa" (the common eighteenth-century spelling for present-day Orissa, a coastal region southwest of Bengal). It also claims that the work is a translation from the Oriya into French. The volumes yield further clues. Their similarity to accounting ledgers, the paintings, the French-watermarked paper, and Indian parchment binding lead one to surmise that the work was executed in a European trading settlement with the requisite infrastructure and indigenous craftspeople.

From the "Preface" and the "Note to the Reader" we learn that their author, although not a savant, is probably trained in medicine. L'Empe-

Figure 15.1. The frontispiece of the *Jardin de Lorixa*. © Bibliothèque Centrale, Muséum National d'Histoire Naturelle, Paris, 2003. Reproduced by permission.

reur modestly states that "it was not with the ambition of rendering it perfect" that he commissioned the work: "I only thought of making a start and leaving the glory of finishing it to whoever would like to take it up." He concludes, "I would be happy if, through my effort and expenditure, some poor invalid finds relief—that is the only aim I had in undertaking this botanical treatise"—statements obviously directed to appeal to a Catholic missionary order.[10]

Fortunately, both trade and religious leads prove fruitful. Among the many stories told by this and associated documents, the most remarkable concerns the conception, making, arrival in France, and ultimate fate of the *Jardin de Lorixa*. Briefly, we learn that it was started in Orissa in the late 1690s, completed in Bengal, and shipped to Paris in 1725. But, to fully appreciate the story, a few words about French presence in South Asia in the seventeenth and eighteenth centuries are necessary.

. . . To Eastern India in the Seventeenth Century

Formally arriving only in 1664, with the foundation of the Compagnie des Indes Orientales, the French were latecomers to Asia. Indeed, they were the last of the major European powers to enter the Indian Ocean, appearing over half a century after the Dutch and the English, and more than 150 years after the Portuguese. Unlike other European companies, the Compagnie des Indes was set up by royal edict, with capital raised from the royal family, courtiers, and financiers and only reluctantly from France's merchant communities. This factor played a crucial role in all domains, including that of knowledge making.[11]

The French company's purpose in finding a foothold in the Indian subcontinent was to obtain those goods already supplied to Europe by the Dutch and the English. Textiles, pepper, coffee, saltpeter, and a range of items covered by the term *drogues* formed the bulk of the cargoes bound for France, occasionally varied by wild beasts, such as rhinoceroses, for the royal menagerie, precious stones, books, and works of art. To obtain these commodities the French settled initially in Surat, in Gujarat, then in Pondicherry, and finally in Chandernagore in Bengal, close to the important Dutch and English townships of Chinsura and Calcutta. As these settlements were on the Hooghly, the main but hazardous distributary of the Ganges, the Europeans set up lodges in the 1630s at the mouth of the great river, in Balasore in Orissa, in order to house pilots to guide their ships upstream to their trading centers. It was there, soon after the French established themselves in 1686, that L'Empereur found employment as a surgeon to the Compagnie des Indes.[12]

The Origins of the *Jardin de Lorixa*

Nicolas L'Empereur was born in Normandy around 1660. His writings and correspondence suggest that he received a reasonable elementary education. There is no record showing that he trained as a doctor or surgeon in France. Instead, he must have enrolled as a surgeon's apprentice on an East Indiaman—a common way of entering the profession until the end of the eighteenth century.[13] At the end of a ten-year apprenticeship, around 1688, he finally earned the title of Surgeon Major. Unlike most of his fellow apprentices who returned to the French provinces, however, L'Empereur sought to make his living in the employ of the Compagnie des Indes and was posted at Balasore.

Here a decade later L'Empereur developed his plan for the herbal, first because the herbs and medicines Europeans carried with them deteriorated at sea and lost their efficacy by the time they arrived in India.

Second, Europeans met with a multitude of hitherto unknown diseases in these distant, tropical climes. Third, the number of medicinal plants traditionally known to Europeans was relatively small, leading them to look for new remedies overseas.[14] Maintaining health at sea was a major problem for Europeans until as late as the nineteenth century. Indeed, out of the 120,000 Frenchmen who sailed to the East between 1664 and 1789, 35,000 died during the voyage.[15] To aid in this crisis L'Empereur planned to buy "all the books on medicine that the people here have and find out how they use them. I plan to translate these into French so that we know all the cures, great and small, that are as yet unknown to Europeans. We will thereby be able to constitute a library of medical works for India as well as a pharmacy."[16] Later, he elaborated, "This work will be of considerable size and, once printed, nothing [of Indian medicine] will be left unknown to the European surgeon."[17]

L'Empereur was, of course, not the first European to conceive such a plan. Already during the sixteenth and seventeenth centuries Europeans had begun gathering material on Asian natural history for similar reasons. The best known of these were Garcia da Orta's 1563 *Coloquios dos simples e drogas . . . da India* and Hendrik Adriaan van Reede tot Drakenstein's *Hortus Indicus Malabaricus*, published in Amsterdam between 1678 and 1693. Indeed, van Reede's work and that of Paulus Hermann (1646–95)—another Dutchman—on Ceylon served as Linnaeus's primary sources for the flora of Asia.[18]

Dutch and English presence in Balasore played no small role in spurring L'Empereur to embark on his ambitious project. "While we [the French] . . . are the poorest," he complained to his friend Gabriel Delavigne, head of the powerful Catholic order La Société des Missions Étrangères de Paris, "the English flourish through their trade everywhere."[19] In 1706 L'Empereur moved as senior surgeon to Chandernagore, the most important French settlement in South Asia. He was now at the nerve center of European activity in eastern India and could report on it closely. "The English send a large quantity of calumba wood to England each year," he wrote to Antoine de Jussieu (1686–1758), professor at the Jardin du Roi, "as they have taken the trouble to test it and spare no means to obtain all that is curious." While it was relatively easy to report on rivals' exports, knowledge of the properties and uses of these botanical products was difficult to obtain from fellow Europeans, who did everything in their power to keep it secret or to mislead others. Ultimately, to garner natural knowledge, Europeans had to work their way into specialized local networks. Thus, da Orta, who besides practicing medicine was also a trader—chiefly in *materia medica* and jewels—and shipowner, depended chiefly on his Asian medical and trading partners for his knowledge and on a vast network of paid correspondents and agents, who sent him plants and seeds from all over Asia.[20]

Making the *Jardin de Lorixa*

In his "Note to the Reader," L'Empereur explains how he obtained his botanical knowledge: "There are fakirs who travel all their lives and many have a lot of wisdom. However, it is difficult to get them to share any of it, unless you know them intimately and offer them alms. Otherwise, . . . they inform you coldly that they are not interested in money. But I have been friendly with two of them for twelve or fifteen years and through them I meet other passing fakirs. Whenever I find a simple, they instruct me about its properties and uses."[21] In a letter he gives further details: "The fakirs who have the best remedies come every winter to bathe in the Ganges. By giving them something and speaking to them in [Hindustani], directly without interpreters, they let you into their secrets. It was a fakir who thus taught me the great remedy for epilepsy."[22]

In addition to his duties as surgeon major and member of the Council of Chandernagore, L'Empereur set himself up in private trade, selling uncut emeralds from South America bought for him in Europe; he became part owner of a small ship and bought and sold property for profit.[23] His daily experience with locally available simples convinced him of their efficacy, and he began purchasing indigenous books on medicine,[24] which circulated most commonly in the form of palm-leaf manuscripts in the various vernaculars of the subcontinent, from the Dravidian languages of the South to the Sanskrit-based ones of the North. Judging from the plant names in the *Jardin de Lorixa*, some, such as *china malli* (small jasmine), are clearly in Tamil. But L'Empereur seems not to have been aware of this linguistic diversity, considering all his material to be in Oriya, which, he declares, he translated into French.[25] However, the process was more complex. By his own admission, L'Empereur did not know Oriya: he got everything translated into Hindustani, the main language of intercourse between Europeans and South Asians in the region. This he then undertook to translate into French—"a tedious task," he wrote, "except for me since I speak Hindustani."[26]

Yet, not all the descriptions came from written texts. L'Empereur also employed a number of gardeners whom he sent at considerable cost to the mountains and forests, sometimes more than three hundred miles away, to bring back interesting plants.[27] He eventually established trade links with merchants as far away as Nepal, who sent him valuable plants. Some of these were unknown to the fakirs, leading him to start experimenting on local patients—both European and South Asian—with compounds he produced.[28]

L'Empereur organized his descriptions in a standardized form, starting with a physical description of each plant; its roots, flower, fruit, and seed; its habitat; and finally its properties and uses. But not all the plants were medicinal. Some were dyes, others aromatics, while a few had no apparent use at all. Some were even exotic—such as papaya, chili, custard apple, and potato—introduced from South America in the sixteenth century by the Portuguese. L'Empereur, like other contemporary Europeans, however, did not distinguish between local and exotic varieties: he gives the distinct impression that they formed part of the region's traditional flora.

L'Empereur also employed local artists to draw and paint each plant, with its flowers and fruits, and a cross-sectional representation of the seeds at the bottom. Chandernagore being a major trading port, tens of thousands of Asian merchants, interpreters, bankers, and craftspeople worked for the European export market.[29] Many were painters who earned their livelihood executing floral designs on calicoes that formed some of the main Indian exports to Europe. L'Empereur thus found it "easy to get natives to draw the plants."[30]

As the above account suggests, the *Jardin de Lorixa* is not a translation of indigenous texts in a purely linguistic sense. Furthermore, it differs from Indian palm-leaf *materia medica* in that the latter do not describe the plants but enumerate their properties and uses and, above all, contain no illustrations. There was, of course, an established tradition, since at least the sixteenth century, of illustrating natural history memoirs for South Asian nobility. The floral borders and stylized plant representations from these soon found their way into pictorial arts, from cloth printing and wall paintings to illustrations of popular tales and religious epics, but not into medical practitioners' vade mecums.[31]

The typically Indian style of the paintings and L'Empereur's own claims to having simply translated Oriya works notwithstanding, the *Jardin* is also a recognizably European botanical work in its general organization and presentation: it is a hybrid work containing disparate elements reconfigured into a new homogeneity. Of course, L'Empereur was no medical neophyte and knew the conventions of European medico-botanical treatises, but, as noted earlier, many such works had been produced in Asia, and it is interesting to compare the *Jardin* to them.

The *Jardin de Lorixa* and the *Hortus Malabaricus*

The most obvious candidate is van Reede's renowned *Hortus Malabaricus*, the last volume of which appeared just a few years before L'Empereur embarked on his own scheme. In addition to their remarkably similar formats and number of plant descriptions, they bear an uncanny resemblance in other ways.

The first similarity concerns the heterogeneity of the agents involved in their construction. Like L'Empereur, van Reede employed many different specialists—a council of at least four physicians from the Malabar coast (to supervise the collection of plants, help identify them, and provide information on their medicinal uses), local arboriculturists and gardeners, a Luso-Indian translator, and a team of Dutch draftsmen to execute its numerous illustrations.[32] Of particular interest is the frontispiece, portraying a vast tropical garden. In the center stands an ornamental summerhouse with two caryatids bearing an entablature whose tympanum is inscribed with the title of the work. In the foreground, beneath an arched pergola, sits the (apocryphal) goddess of Indian botany holding a rake, with a pruning knife lying at her feet. Four kneeling Malayali cherubs (on the left) offer her a potted tree (see figure 15.2).

At first sight, this engraving looks very different from the painted frontispiece of the *Jardin de Lorixa.* Any similarity seems to stop at the way the plants and human groups are placed in both: the central potted tree, the Malayali cherubs/group of artists, the goddess of botany/fakir, all under a pergola/an arch made by two flowering trees, a classical summerhouse/Greco-Roman ruin. However, we discover that the woman carrying plants in a basket on her head is a replica of the left caryatid of the summerhouse in the *Hortus Malabaricus.* L'Empereur's artists clearly had access to the *Hortus,* but they tell a very different story from the allegory imagined by the Dutch engravers sitting in faraway Holland. In the same way as the caryatid carrying a sheaf of corn is brought to life as a real woman carrying plants to be painted and described, so too all the other figures of the *Hortus* come alive as the different actors involved in the making of the *Jardin.* The kneeling Malayali cherubs metamorphose into artists; the goddess transforms into a fakir wielding his palm-leaf manuscripts instead of a rake and a pruning knife. Other elements change as well. The pergola becomes an arch formed by flowering trees, inspired from traditional Indian paintings and embroidery; and the ornamental summerhouse transforms into a Greco-Roman ruin. The central tree now planted in a Chinese pot—a witness to the lively intra-Asian trade—serves to demarcate the different groups, the manual workers to the left and the "cerebrals" to the right. L'Empereur, in his role as patron, finds himself in front of the ruin just above the ascetic.

A close examination of the *Hortus* further confirms that it provided the template for European botanical conventions for the *Jardin.* For, while painting floral motifs was the main livelihood of Indian artists,

Figure 15.2. The frontispiece of van Reede's *Hortus Malabaricus.* © Bibliothèque Centrale, Muséum National d'Histoire Naturelle, Paris, 2003. Reproduced by permission.

their painted calicoes did not, and were not meant to, respect any botanical conventions. These latter required, for instance, that seeds be shown apart, whole and laterally dissected, that flowers be drawn separately, and that roots be depicted with the plant. Not only do the paintings in the *Jardin* respect these conventions, but some of them are more or less directly inspired by the engravings of the *Hortus Malabaricus,* as, for example, the banana, the papaya, and the jackfruit (see figures 15.3a and 15.3b).[33] L'Empereur's artists did not, however, mechanically copy the illustrations from the printed book: the very fact that they colored their illustrations, getting the colors of all parts right each time—the *Hortus Malabaricus* was in black and white—removes all doubt on the matter. Besides, many illustrations such as *Nux vomica,* are very differently represented in the two works (see figures 15.4a and 15.4b).[34] Moreover, the vast majority of the plants described respectively in the two works refer

Figure 15.3a. The banana tree in the *Jardin de Lorixa.* © Bibliothèque Centrale, Muséum National d'Histoire Naturelle, Paris, 2003. Reproduced by permission.

Figure 15.3b. The banana tree in the *Hortus Malabaricus.* © Bibliothèque Centrale, Muséum National d'Histoire Naturelle, Paris, 2003. Reproduced by permission.

Figure 15.4a. *Nux vomica* as represented in L'Empereur. © Bibliothèque Centrale, Muséum National d'Histoire Naturelle, Paris, 2003. Reproduced by permission.

to two regions of different climes and over a thousand miles apart. Once the local artists had understood what was wanted of them, they followed the drift without having to copy directly from a "pattern-book." This conforms to what is already commonly known of the capacity of Indian weavers and painters to execute floral patterns shown them by their foreign clientele into the chintzes and palampores that formed the staple export of Bengal and the Coromandel Coast in the seventeenth and eighteenth centuries (see figure 15.5).[35] In this respect, it is interesting to note the similarity in the versatility of these artists and that of the engravers working for European publishing houses.

The verbal descriptions in the two herbals only loosely resemble each other in that they systematically describe the habitat and different parts of the plants before giving their properties and uses—by then an already well-established convention in European botany. L'Empereur, however,

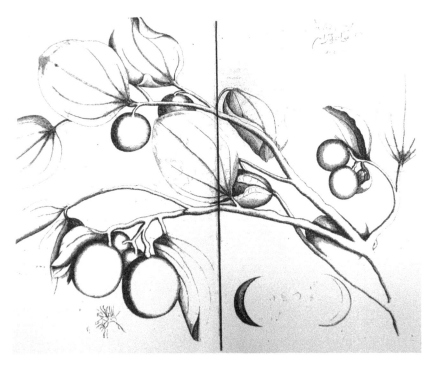

Figure 15.4b. *Nux vomica* as represented in van Reede. © Bibliothèque Centrale, Muséum National d'Histoire Naturelle, Paris, 2003. Reproduced by permission.

systematically gives the dimensions of each plant in feet and inches, and his account of the properties and uses is markedly different from that of van Reede. But the latter transcribes the local names of the plants in the Roman script and also gives both the Malayalam names, in the Aryaezuthu and Arabic scripts, and the Konkani names in Nagari; L'Empereur, as noted earlier, records only the local names he gathers (mostly, but not always, in Oriya) in the Latin script. Neither herbal gives the exact composition of remedies, stopping at a description of the uses to which the respective indigenous populations put the different parts of each plant. L'Empereur explicitly states that he "do[es] not describe the dosage because they [Indian physicians] do not weigh the drugs. Only experience can teach one how to administer the medicines."[36]

L'Empereur's *Jardin* Comes to Paris . . .

L'Empereur was confident of being suitably rewarded for his entrepreneurship. Already in 1698 he sent samples of his work to Paris with De-

Figure 15.5. Early eighteenth-century painted tent panel from eastern India with a floral pattern similar to the painted bedspreads that formed the staple export of the region to Europe.

lavigne, hoping he would find a patron. The latter did not succeed in making much headway, and by 1701 L'Empereur's impatience was palpable. "No one shall have my work unless I make a suitable profit from it," he wrote to Delavigne, informing him that he had started looking for alternative patronage: he was sending some samples of his work to "Monsieur Petit" in London and to his brother's friend in Dol, a retired canon who had once been tutor to the sons of Guy-Crescent Fagon (1638–1718).[37] L'Empereur's indefatigable efforts paid off: in 1719 L'Empereur received two gold medals as tokens of royal patronage and "a promise to be suitably rewarded upon completing the work."[38]

The flora was finally completed in 1725, and L'Empereur shipped it to the Académie Royale des Sciences in Paris along with a wonder remedy for epilepsy. He was now sixty-five years old and eagerly awaiting a sizable reward. L'Empereur had lost his job with the Compagnie des Indes and had spent his last penny on this gigantic work—so much so that he was reduced to bankruptcy and begging.[39] The volumes and remedy reached Paris safely and were handed over to Antoine de Jussieu. L'Empereur, however, received no acknowledgment, let alone any remuneration. After a couple of unanswered letters to de Jussieu,[40] he began complaining about the latter's behavior to the directors of the company and to his numerous correspondents in Parisian high society.[41] In 1733 the Compagnie des Indes reinstated L'Empereur as visitor of its godowns in Chandernagore, but in spite of intercessions on his behalf, de Jussieu stubbornly refused to pay him his due, although de Jussieu did admit that the various remedies were indeed effective.[42]

Nicolas L'Empereur died in anonymity in Chandernagore on 13 February 1742, age eighty, cared for to the last by his Bengali doctor, to whom he left part of his meager savings.[43]

. . . And Gets Anonymized in the *Jardin du Roi*

Why the *Jardin de Lorixa* should have suffered a fate so different from that of the *Hortus Malabaricus* cannot be explained in terms of its contents or structure. Exotic herbals were highly coveted as much for their usefulness to naturalists and physicians as for their commercial potential. By the eighteenth century illustrated herbals were also highly prized on the European art market. Personal animosity on the part of de Jussieu (L'Empereur's explanation of de Jussieu's reluctance to reward his work) cannot entirely explain the *Jardin*'s failure.[44]

The *Jardin de Lorixa*'s ill success did not result from de Jussieu's disinterest in Asian flora for, as the botanical expert to the Compagnie des Indes, he was well aware of the Dutch monopoly over the European drug market, a monopoly which, by his own admission, "they acquired by gaining a thorough knowledge of the natural history and uses of drugs in the lands they visit." Consequently, if the French hoped to compete in the drug market, they had to encourage the company's servants overseas to collect useful plants, "send them for study to Paris," and eventually "transplant the most useful of them in our newly founded colonies."[45] De Jussieu maintained a regular correspondence with Frenchmen in the East and even secretly sent a certain Jean-Claude Barbé to botanize in Chandernagore in 1725. (On the latter's sudden death in 1729, L'Empereur rummaged through his affairs and was furious to discover de Jussieu's duplicity.[46])

De Jussieu seems to have thought that knowledge outside of Europe could be gained without great expense. L'Empereur did not lose the opportunity to point out to him the folly of this assumption: "With 1200 *livres*, a botanist cannot do much considering that food, clothing, wine at 30 *sous* a bottle, the local doctor, housing, servants, interpreters, other domestics, and presents cost a lot more. Moreover, your botanists come during the monsoons, the worst time to botanize. In addition, they have to learn the language and buy books from the natives."[47] The Asian world was as commercially organized and segmented as the European one was. De Jussieu, being ignorant of this, was not to make much headway in collecting nature at a distance in the East.

One of de Jussieu's writings, however, does throw light on his appreciation of L'Empereur's herbal. Short though it may be, "Des avantages que nous pouvons tirer d'un commerce littéraire avec les botanistes etrangers" (The Advantages We Can Gain from a Literary Trade with Foreign Botanists), most probably written in 1732, provides insight into de Jussieu's notion of the botanical enterprise. "It is neither simple curiosity," he states, "nor the desire to adorn one's garden with exotic and hitherto unknown plants that are the main reasons for corresponding with botanists abroad—no, if botany is to have any place in the progress of medicine and other arts, then it must establish comparisons between European and exotic flora. It is only thus that one can identify plants of the same type, know their uses in medicine and the arts, and finally improve the quality of the European flora." According to de Jussieu, it is this correspondence that helped establish that the ipecacuanha was none other than the common violet and that the plants from which Japanese paper is made are merely a species of white mulberry and althaea.

De Jussieu then goes on to give five practical examples to show the utility of such correspondence—the second being none other than that of Nicolas L'Empereur:

The second letter [*sic*], dated 20 January 1729, is from Mr. L'Empereur, formerly surgeon at Chandernagore in the kingdom of Bengal. It contains a number of observations on the plants of that country drawn and painted by him in twelve folio volumes that he sent to the Academy and that are now with me. The observations are mainly on the uses in Bengal of most of the plants described in this collection, which is almost a corpus of medicine in this distant kingdom. However, an examination of the plants has led me to remark that most of those that grow there naturally and are thus wild, are to be found here among our vegetables, which are cultivated and so have a different flavor.[48]

Surprisingly, de Jussieu and L'Empereur concur on the performativity of botany, but what they mean by it is very different. For the former, knowledge of foreign flora was of interest only inasmuch as it helped establish concordances between foreign and local plants in order for France to find import substitutes and protect both its markets and powerful professional groups from Dutch competition. This was certainly not L'Empereur's purpose. His was a scheme, inspired from the Dutch model in the Indian Ocean, of gaining knowledge of regional pharmacopoeias in order to commodify them. At any rate, it was de Jussieu who, as a senior savant-expert in the network of French royal institutions, had the last word. He thus sounded the death knell of the *Jardin de Lorixa*. Exiled from the world of certified knowledge, it lay in his personal library at the Jardin du Roi before ending up as an anonymous, exotic curiosity in the *Muséum*'s library in the course of the nineteenth century.

Conclusion

The L'Empereur corpus, as also the works of other Europeans in Asia, throws considerable light on the triangular relationship between Europeans at large, their indigenous interlocutors, and their armchair, metropolitan colleagues. It highlights the relationships between knowledge practices and broader economic and political contexts that constitute some of the principal themes of this volume.

Early modern South Asia and the Indian Ocean were spaces in which knowledge was intellectually and socially constituted prior to European contact. The knowledge that circulated there was not some form of popular knowledge but the prerogative of discrete, well-defined groups. This was clearly acknowledged by Europeans both outside and inside Europe. L'Empereur's enterprise, as much as those of van Reede and da Orta, thus consisted not in gathering information held by undifferentiated, autochthonous groups, but in *reconfiguring* and *constructing* knowledge, skills, and specialized practices—for the regional as much as for the European knowledge market.

Using the market metaphor here is not out of place: it is used by the actors themselves—L'Empereur refers to trade and profit as does Jussieu ("literary trade," for example). It brings to the fore the material dimensions of knowledge formation and circulation. Science has so far been commonly presented as a special "symbolic" economy distinct from other dimensions of human intercourse. Instead of freely circulating in an idyllic and seamless republic of letters, science in Europe when observed from the vantage point of the Indian Ocean, moved in spaces bounded by national political and economic interests, and shaped by different regimes of performativity within which alone the meaningfulness of knowledge can be determined.[49] L'Empereur's example argues for an understanding of early modern science as part of the market economy that partakes of the larger political economies of burgeoning nation-states, of early modern mercantilism, and of nascent European colonialism.[50] It is only by considering it thus that we can begin to clarify the complex nexus between knowledge and power.

In addition, even more than L'Empereur's experience, those of da Orta, van Reede, Hermann, and others plead for studying knowledge construction in the Asian context not as an extension of its construction within Europe, but as a phenomenon in its own right. Their experiences bring to light the fact that these men gained their credibility not in providing information to European armchair savants, but by making knowledge through negotiations with local Asian groups. They either published their works in Asia, as in the cases of da Orta, or else, as in the cases of van Reede and Hermann, made their names by circulating their works in manuscript form within the Indian Ocean world, mainly through Batavia, without appealing to European metropolitan authority.

The change in historiographical perspective suggested here makes it possible to begin to investigate the site of knowledge production and the dynamics of knowledge making set in motion there. As we have seen, the translation from South Asian vernaculars to European ones was only one of the many translations that L'Empereur's enterprise involved. Indeed, L'Empereur translated a motley assortment of medical, religious, economic, social, and cultural skills and practices through a series of complex negotiations into a single work that obeyed no single preset idiom.[51] And if L'Empereur failed to make it at the far end of the chain—in France—he did succeed at the near end, in South Asia, in pulling together a complex network of savants, merchants, missionaries, and craftspeople. His enterprise, like that of other Europeans, had a long-term effect on the local communities with which it interacted. Indigenous painters found natural history drawing to be an increasingly lucrative business for the European market throughout the eighteenth century and began specializing in this art form. At first they did this on an individual basis, but with the British colonization of Bengal and the Coromandel Coast a few decades later, a whole institutional space opened up with the founding of botanical gardens and the various natural historical and geographical surveys. Indian painters and draftsmen were now employed on a massive scale in these colonial institutions to execute maps, landscapes, and some of the great herbals of the late eighteenth and early nineteenth centuries.[52] Likewise, European naturalists were to start finding employment in South Asian princely courts to establish botanical and medicinal gardens.[53] This intercultural interaction also had long-term effects on medical and botanical practices in the region itself. Although difficult to apprehend because of the difficulty of finding adequate sources, bringing these effects to light would prove a valuable contribution to the history of Indian medicine.

Chapter 16

Measurable Difference
Botany, Climate, and the Gardener's Thermometer in Eighteenth-Century France

Marie-Noëlle Bourguet

In 1788 the chief gardener of the Jardin du Roi in Paris, André Thouin (1747–1824), equipped his apprentice Joseph Martin with instructions concerning the cargo of plants he was to escort from Le Havre to the Isle de France (Mauritius): "Two thermometers will be installed in the greenhouse, one tied to the glass, the other driven into the ground to the zero graduation." The instruments would, Thouin explained, allow Martin to measure the difference between the ground temperature and the surrounding air: "We believe that, for European plants to be safe, the temperature of the soil should not exceed 36 to 40 degrees." Similarly, on the return trip, designed to bring back to France a collection of plants from the Torrid Zone, the young gardener was instructed to maintain the plants' native atmosphere, measured on the graduated scale of a Réaumur thermometer, as the ship passed through different latitudes and climates: "As soon as the temperature goes below 10 degrees at night," small greenhouses should be installed on the deck; if a frost occurs, Martin should not hesitate "to take the cases of plants away from the deck and put them in some room where a fire can be lit and the heat be kept at 8 degrees, at a minimum."[1]

In the early modern period European nations engaged in a chesslike game of acquiring botanical resources; in a key maneuver the transfer of plants substituted for the exchange of currency. While breaking down trade monopolies, the shipment and acclimatization of plants were intended to allow Europeans to grow exotic spices and other useful foreign plants in their colonial territories, if not at home. Beyond the economic and political stakes, the "mobilization" of plants posed both a conceptual and a technical challenge for the botanists who, as gardeners of the earth, endeavored to reshape nature's distribution of floras by moving seeds and plants across the seas.[2] Indeed, no living plant or seed

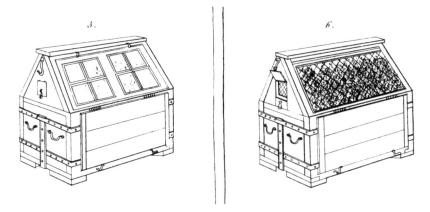

Figure 16.1. "Chests. Tools for gardening and naturalization. Wooden case or portable greenhouse. Very well suited for transporting and preserving living trees from one hemisphere to the other. These various models have traveled around the world or brought back from our most distant colonies many precious plants." From André Thouin, *Cours de culture et de naturalisation des végétaux*, 3 vols., and atlas (Paris: Mme Huzard/Déterville, 1827), vol. 1, 431; atlas, pl. 15, figs. 5, 6 (details). © Bibliothèque Nationale de France, Paris. Reproduced by permission.

can be extracted from its native habitat and transferred elsewhere without some elements of its environment (type of soil, degree of heat or moisture) and some precise information (about mode of cultivation, techniques of preparation, or practices of consumption) being imported along with the sample to its new terrain—whether a greenhouse, a garden, or a plantation. Should one of these elements be missing, the whole transfer and naturalization process might fail.[3] (See figure 16.1.)

This tension between mobility and embedding lay at the core of eighteenth-century botanical science and practice. As Thouin's memorandum vividly conveys, the gathering of information on climatic conditions and the use of meteorological instruments together played a key role in the process of plant transfer. Thermometers and barometers were used as mediating devices, the function of which was to translate the singularity of a local climate into calibrated, portable, and comparable information. The practice of taking meteorological instruments on board ships or installing them inside greenhouses or gardens progressively became part and parcel of the story of voyaging plants. By the late eighteenth century botany came thus to be allied with instrumental practices, and with the development of an ethics of quantification and precision. (See Figure 16.2.)

This chapter considers eighteenth-century botanists' unusual interest

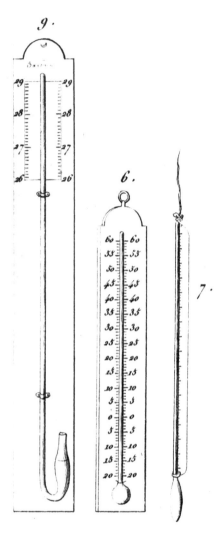

Figure 16.2. Thermometers and barometer "proper for the knowledge of atmospheric conditions that can affect vegetation." Number 6 is the Réaumur alcohol thermometer, described as being "useful in the cultivation of delicate plants, for registering the temperature outside or inside the greenhouse and for maintaining temperatures inside the greenhouses suitable for plants from various climates." Number 7 is a dipping thermometer (*Thermomètre à bain*) "used in hothouses, to know the temperature of the hotbeds and liquids." Number 9, a mercury barometer, indicates "atmospheric pressure variations, and helps to guard against those variations that could jeopardize the cultivation and especially the harvest of the plants." From Thouin, *Cours de culture*, vol. 1, 478–83; atlas, pl. 28, figs. 6, 7, 9 (details). © Bibliothèque Nationale de France, Paris. Reproduced by permission.

in the precise environmental conditions required for growing particular plants and the link these botanists established between botany and meteorology when confronted in the field with the local singularities of plants and climates. Focused on French examples for the most part, and threaded through with the theme of thermometric measurements, it will address several questions: What understanding of the relation between plants and climate was formulated by means of meteorological records? How were precision and measurement deployed in the debates about colonial botany and the cultivation of foreign plants? To what extent were the meteorological data gathered by colonial observers or naturalist travelers related to the concern for the geography of plants that emerged at the turn of the nineteenth century in connection with the mapping of climates and the history of the earth?

From Latitude to Temperature: Climates as Numbers

For an understanding of how eighteenth-century naturalists approached the question of climate, the account by the Minim friar Louis Feuillée (1660–1732) of his expedition to Spanish America as a mathematician and botanist for Louis XIV can be helpful. In his *Journal*, Feuillée diligently recorded the information he collected daily about longitude and latitude, winds, barometric height, compass variation, and the like.[4] All were standard scientific topics at the time: starting at the end of the seventeenth century, as portable devices became available, travelers were instructed to take measurements en route. Meteorology was a prime object of attention. As the English naturalist and collector John Woodward (1665–1728) explained, only through a world survey of temperatures could a true science of climates come into being: "By this means the heat or cold of all places in the same climate or under the same latitude may be compared and known, for any or all seasons of the year."[5] Together with quantified data, travelers were expected to collect curiosities and natural history samples. Thus, a whole section of Feuillée's report is devoted to the botanical resources of the places he visited in Peru and Chile, with special attention to medicinal and economic plants: "Because nature had furbished this land with a variety of animals and plants, I thought it useful for the public to describe them and inquire about their use."[6] Each specimen is represented by a drawing and a short description of its aspect, place, and uses in indigenous pharmacopoeia. A plant he picked up "on the edge of a swamp," for instance, he described as *Gramen bromoides catharticum, vulgo Guilno*, "one of the best purgative plants, most frequently used by the people of Chile." Yet, except for a brief mention of its habitat—"this plant grows in humid places"—the plant's native environment is delineated only through geographical coordinates: "I discovered this sample at a latitude of 36° 46′ in the southern hemisphere."[7] Latitude was enough for the traveler to define the plant's "climate." Such a disjunction between botanical observations and meteorological measurements recalls the geographical tradition, inherited from antiquity, that characterized climates in spatial terms, as "stripes" or "zones" on the earth's surface, bounded by two parallels.[8]

This system of representation prevailed throughout the eighteenth century, framing the assumptions of many travelers and naturalists. In 1747 the Swedish botanist Carl Linnaeus (1707–78) sent his student Pehr Kalm (1716–79) to North America to collect plants in the Hudson Bay area because the latitude of Sweden and Canada—between 55° and 60°—led him to expect the climate to be similar in both countries (Kalm was to register the daily temperature with a thermometer); Linnaeus assumed that plants from the northern parts of America would easily adapt to his native land. According to this principle of "latitudinal homogeneity," any given plant was sufficiently characterized by its native habitat or zonal location: "*Canna indica* . . . habitat inter tropicos, Asiae, Africae, Americae. . . . *Poinciana pulcherrima* . . . habitat in Indiis."[9] Conversely, Joseph Boissieu de La Martinière (c. 1750–88?), a botanist who embarked with Jean-François de Galaup, comte de La Pérouse's expedition around the world, expressed his astonishment when he discovered the differences in Pacific coast vegetation between plants that grew in the Northern Hemisphere and those in the Southern Hemisphere: "Because Monterrey in California lies approximately at the same latitude and longitude as Concepción in Chile, I was expecting the vegetations to be alike. I was wrong."[10]

It was as late as the 1730s that the correlation between climate and temperature, between botany and meteorology, assumed the status of a research program in France. It did so through the efforts of the naturalist René-Antoine Ferchault de Réaumur (1683–1757), a member of the Académie Royale des Sciences in Paris, who managed to organize a small network of observers who were scattered to the limits of the French colonial domain. Among the scientists on the move who participated in this information-gathering network were the academicians Pierre Bouguer (1698–1758) and Charles-Marie de La Condamine (1701–74) in Peru; the naturalist traveler Claude Granger (?–1737) in Syria and Persia; an agent of the India Company, Pierre Barthélemy David (1711?–95), in Senegal; and members of the colonial elite and administration such as the physician Jacques-François Artur (1708–79) in Cayenne and the military engineer Jean-François Charpentier de Cossigny (1692–1778) in the Mascarene Islands. Having equipped his correspondents with thermometers, calibrated according to his own instructions (see figure

16.3), Réaumur asked them to register the daily maximum and minimum heat, and he published a decade's worth of results in the *Mémoires de l'Académie royale des sciences*.[11] By collecting degrees of extreme heat and cold, Réaumur did more than satisfy his audience's curiosity for rare and singular phenomena. He also intended his survey to serve the utility of the state: the tables were to give the king of France a map of the potential expansion of his power through long-distance commerce, colonial settlements, and the introduction of useful plants.[12] Such assessments made the relations between botany, climate (often merely conceived as temperature), and empire a central concern for naturalist travelers and gave to their scientific undertaking a centrifugal dynamic.

The botanical and meteorological survey organized in the 1740s by the physicist and agronomist Henri-Louis Duhamel du Monceau (1700–1782), Réaumur's colleague at the Académie, concretely illustrates these connections. The project was designed as a comparative study of climates and vegetation, in a colonial context: "It would be useful . . . for both agriculture and physics to know more positively the relations between temperature and plant productions."[13] While he was himself registering daily thermometric variations, together with other seasonal events (blossoming trees, harvest time, migrating birds, and so forth), in his family domain of Denainvilliers near Pithiviers, Duhamel asked the royal physician at Quebec, at that time Jean-François Gauthier (1708–56), to make the same observations in Canada. But the comparison was not an easy task: in fact, Gauthier had to give up using a Réaumur thermometer, the graduations of which did not allow for registering extremely cold temperatures (on many occasions the liquid fell into the tank of the instrument). Instead, he used a thermometer that the geographer Joseph-Nicolas Delisle (1688–1768) had designed in 1732 for his own meteorological observations in Russia and Siberia.[14] In these early attempts at quantification, standardizing thermometers and making them speak the same language at a distance was a real accomplishment.[15] Since the thermometer allowed botanists and agronomists to synthesize local peculiarities or differences between distant places in tables of calibrated and comparable data, the instrument became a crucial device for bringing space under control and mastering nature.[16]

A Science for the Colonies? Meteorology and Economic Botany in the West Indies

The loss of Canada by the French brought to an end this early attempt at scientific colonial botany. In order to study further the development of meteorology in relation to colonial agriculture, one has to look elsewhere: to the French West Indies. In the eighteenth century the region,

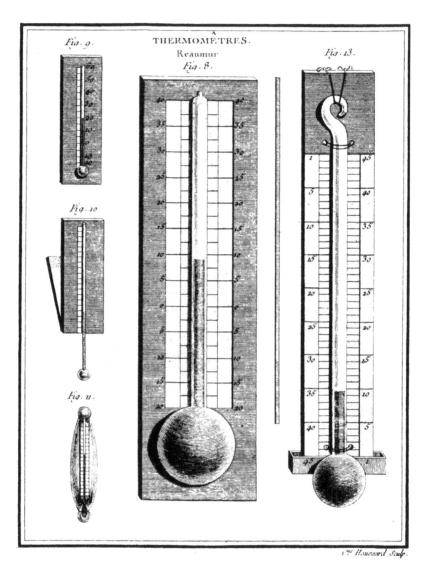

Figure 16.3. Réaumur thermometers. "Mr. de Réaumur's thermometer unites all the advantages that one might wish from such an instrument. By following step by step the physicist's wise instructions . . . , one can build such thermometers, that function comparably, anywhere and any time, as they are calibrated according to fixed degrees of cold and heat, and directly readable to the observer." Number *8* represents "the great thermometer, as Mr. de Réaumur first built it"; Number *9*, a small thermometer; Numbers 10 and 11, dipping thermometers. From Father Louis Cotte, *Traité de météorologie* (Paris: Imprimerie royale, 1774), 117–18 and pl. V, figs. 8, 9, 10, 11. © Bibliothèque Nationale de France, Paris. Reproduced by permission.

with the exception of Guiana, was no longer terra incognita. Along with the development of a settled colonial society, scientific activities came to depend more heavily on local practices and were supported by the arrival of colonial officers, civil engineers, physicians, and naturalists.[17] In the case of Saint-Domingue, the distribution and use of instruments among the settlers can be minutely documented thanks to the encyclopedic survey of Médéric Louis Élie Moreau de Saint-Méry (1750–1819): from the 1760s onward meteorological devices—thermometers, barometers, and a few hygrometers—appeared in various places (in 1764 in the plain of "La Limonade"; in 1774 at Bombardopolis, in the island's western part; in 1775 in the North, at "La Petite Anse," and so forth); in the late 1770s thermometers were present throughout the island.[18]

The same was true in the other islands as well: a former student of Réaumur, the young colonial administrator Jean-Baptiste Thibault de Chanvalon (1725–88), took with him to Martinique one barometer and two thermometers; from July to December 1751, with a friend's help, he assiduously made three observations each day, both in the sun and in the shade. In Guadeloupe a planter named Barthélemy de Badier (1740?–89) left in his cabinet, when he died, "3 packets of barometer tubes, estimated 66 pounds, . . . a fixed barometer, worth 12 pounds, . . . 6 thermometers with their cases, estimated 6 pounds." Such a collection suggests that its owner was not a mere amateur but a diligent observer of the local weather: Badier, a "Naturaliste du Roi" and corresponding member of the Société Royale d'Agriculture in Paris, was actively involved in making his own garden an experimental site for the introduction and acclimatization of new varieties of potatoes, cotton plants, banana trees, and the like, which he had imported from other Caribbean islands and from the South American continent. Gathering meteorological data and experimenting with new crops were important to his botanical and agronomic endeavor.[19]

Still, in most cases recording weather data remained a private and isolated practice in the colonies, and its connection with larger projects was somewhat problematic. Only a few records on the West Indies actually reached metropolitan institutions. Thibault de Chanvalon, who emphasized the need for a comprehensive picture of the colonies' climate and resources, presented his work to the Académie Royale des Sciences in 1761 and two years later published some seventy pages of tables in his *Voyage à la Martinique*.[20] Two colonists of Saint-Domingue, Baussan and Lefebvre-Deshayes, were in contact with the astronomer Joseph-Jérôme Lalande (1732–1807), who transmitted their records to Father Louis Cotte (1740–1815) and to the Société Royale de Médecine. Some physicians filled the printed forms circulated by the Society: in 1782 Charles Arthaud (1748–93?) sent observations collected at Léogane (Saint-Domingue); his colleague Le Gaux collated monthly tables at Basse-Terre (Guadeloupe) (see figure 16.4). In Sainte-Lucie, Jean-Baptiste Cassan strove to give a complete account of the Torrid Zone climate, based on his meteorological and medical observations.[21]

If these attempts, taken together, attest to developing scientific activity in the colonies, they also demonstrate the difficulties, both technical and cognitive, that observers encountered when attempting to transfer either instruments or concepts from the metropolis into their local environment. Thibault de Chanvalon had to give up reading his thermometer "at the same hours as in Europe" when he noticed that the maximum temperature occurred in Martinique not at 3 P.M. as in Europe, but around 1 P.M. local time. Furthermore, having observed that mercury thermometers were more reliable in a tropical climate, he deplored European physicists' continued use of alcohol instruments, which were "suited for a few countries only," instead of mercury ones, which were "suitable for all."[22] Cassan, for his part, noted that the atmosphere's constant humidity made electric machines or instruments such as the hygrometer, invented by the Genevan physicist Horace-Bénedict de Saussure (1740–99), nearly useless in the Caribbean. Even the barometer, he asserted, was "hardly of any use in the colonies" since the daily variations were "so insignificant as to be scarcely worthy of attention."[23]

All such difficulties aside, a local network of observers did emerge in the French West Indies in the 1780s that paralleled those organized in France under the patronage of the Société Royale de Médecine and the Société Royale d'Agriculture. In January 1780, in a public announcement printed in the *Affiches américaines*, the Saint-Domingue newspaper, Nicolas-Joseph Thiery de Menonville (1739–80), the introducer of the Mexican nopal cactus to the colony and the head of the newly founded royal botanical garden in Port-au-Prince, declared his intention to promote regular weather recording among the settlers. In exchange for "a meteorological memoir with an accurate and truthful account of the quantity of rain, from January 20, 1780 to January 20, 1781," he offered each observer specimens of living nopal cactus, wild and Mexican cochineal insects, Guatemalan indigo, and other economically useful plants.[24] Meanwhile, Charles Mozard (1755–1810), the editor of the *Affiches américaines* and a passionate amateur of meteorology, sold thermometers and blank forms in his bookstore in Port-au-Prince: in the mid-1780s he collated meteorological logbooks that his correspondents were requested to send through the intendant's postal privilege.[25] As a whole, however, meteorology in the West Indies lagged behind developments in Europe. As James E. McClellan III has shown in his study of the Cercle des Philadelphes, the learned and scientific society founded in Port-au-Prince in 1784, the utility of such an empirical science as meteorology

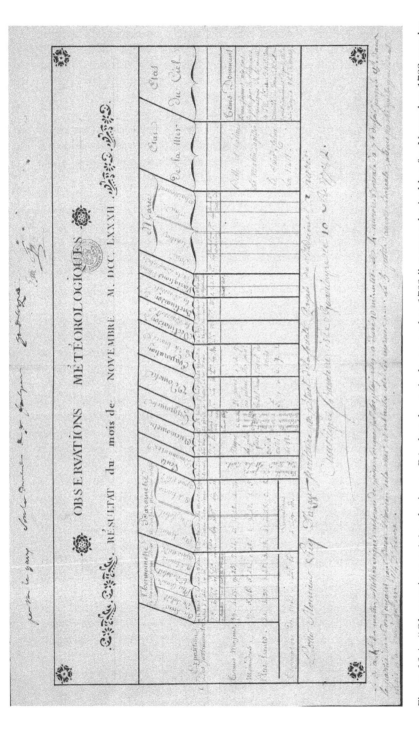

Figure 16.4. "Observations météorologiques. Résultat du mois de novembre 1782," meteorological log for November 1782 sent by Le Gaux, a physician at Basse-Terre (Guadeloupe), to Vicq d'Azyr, secretary of the Société Royale de Médecine, 10 December 1782. This manuscript table is organized according to the printed forms sent by the Société Royale de Médecine, although some entries (on magnetic variations and tides) have not been filled in. From the Bibliothèque de l'Académie Nationale de Médecine, Archives de la Société Royale de Médecine, carton 191, dossier 18, pièce 2. © Académie Nationale de Médecine, Paris. Reproduced by permission.

was deemed by most settlers "not yet sufficiently demonstrated" because it could not provide "the ability to predict conditions and different revolutions in the atmosphere"; a reliable weather forecast was the only motive for a planter's interest in meteorology: "Without that capacity, . . . no power to control, no ability to prevent or to remedy anything, results."[26]

An interesting event occurred in 1787, at a time when César-Henri de La Luzerne (1737–99), a former governor of Saint-Domingue, newly appointed as "Secrétaire d'État" for the navy and the colonies, was supporting the introduction of plants from the Indian Ocean into the Caribbean islands. Mozard seized this opportunity to publish a report in the *Affiches* on the climate of the Mascarenes and to cross-reference it with the records available for Saint-Domingue. He hoped that the comparison would encourage local planters to welcome a cargo of exotic trees that was on its way: the climatic similarities suggested that precious plants from the East Indies, such as cinnamon, clove, or nutmeg, might be easily "naturalized" in the West Indies. In Mozard's view, the thermometric tables he had gathered were the legitimate basis on which to found a politics of transfer and acclimatization: "It is now well established that the temperature is more or less the same in the Isle de France and in Saint-Domingue." Both scientific and economic reasons thus justified his plea for the establishment of a direct correspondence between the colonies of the eastern and western parts of the Torrid Zone, with no regard for their hierarchical subordination to the metropolis. Despite the distance, such a connection emphasized the colonies' shared features and common concerns. In the long run, the cultural identity and political consciousness of the colonies were as much at stake as their respective economic interests were.[27]

In 1788 the arrival of the botanist Hypolite Nectoux (1759–1836) in Saint-Domingue boosted the trend toward scientific agriculture. A well-traveled naturalist (he had explored Guiana with his uncle, the botanist Jean-Baptiste Leblond) and a skilled gardener (in Cayenne he had managed to rescue some spice plants arrived from the Isle de France), Nectoux was charged with reorganizing the royal botanical garden in Port-au-Prince after the death of its former director, the physician R.-N. Joubert de la Motte (?–1787). Here Nectoux began to grow a local species of cinchona, developed the cochineal production, and succeeded in acclimatizing the Polynesian breadfruit. His connections with botanists such as Louis-Claude Richard (1754–1821) and André Thouin provided him with up-to-date scientific news from Europe, instruments (including a portable thermometer), and books, among them Antoine-Laurent de Jussieu's *Genera plantarum* (1789). Despite his remove and the fact that he was a better gardener than a taxonomist, Nectoux was anxious to participate in the greater scientific community through the many letters and memoirs he addressed to the Société Royale d'Agriculture and, later on, to the Institut National. One such memoir, on the cultivation of coffee, reveals how he attempted to position himself in the domain of tropical botany and colonial agriculture.[28]

Taking the Caribbean islands as a field to be surveyed and the botanical garden as a kind of laboratory, Nectoux called for a scientific approach to the introduction of foreign plants. Accurate and precise measurements served to assess the diversity of the Caribbean landscape and climate: "Our colonies offer for the spice plants all kinds of exposures and all possible temperatures, from the very hot plains to the freezing summit of the mountains"; and each exotic species had to be assigned its specific habitat: "The coffee plant is not indigenous to the American colonies, nor robust enough to grow in all the various climates of this mountainous region."[29] During his excursions Nectoux visited coffee plantations and observed specimens that had grown wild in various places. A thermometer in hand, he paid special attention to the soil, to the location, and "most important" in his view, to the temperature. "The temperatures that seem best suited to the cultivation of the coffee plant are between 10° to 22°." In order to define the plant's topographical boundaries on the island, he transcribed these thermometric figures into height: coffee, he assured, would grow and develop best between "400–500 and 600–800 meters."[30] If planted arbitrarily "above or below these height limits," a coffee plant would not develop "at its best" and would bring no profit. Herein lay the core of the gardener's plea for a scientific agriculture. Anyone interested in growing coffee should first equip himself with "a good thermometer" and make three observations a day, at sunrise, midday, and sunset, so as to choose the best possible site for his plantation.[31] By fashioning himself as a scientific gardener, while introducing a language of quantification and precision in colonial practices, Nectoux envisioned tropical agriculture as a domain at the junction of botany and meteorology—in other words, as a science about the climates of plants.

Proof, Persuasion, and Expertise: The Rhetoric of Numbers

In the meantime the French Revolution, soon followed by the war at sea and the rebellion in the West Indies, threatened the supply of exotic foodstuffs from colonial territories and rendered the dream of acclimatization and self-sufficiency a crucial political issue in metropolitan France. Should the new nation conquer some colonial domain across the Mediterranean Sea—as, indeed, the Directory was to do later in Egypt? Or could a scientifically managed agriculture transform a portion of the national territory into a domestic "Indies"? The many projects on "foreign plants" that deluged governmental committees reveal the extent to which the fascination with exotic plants functioned in revolutionary France as a collective utopia. As Emma Spary has shown, at this time scientific questions concerning the acclimatization of exotic plants and animals were tightly interwoven with moral and political issues surrounding the regeneration of society.[32] Not surprisingly, arguments about locality and climate played a momentous role in these debates, which involved a variety of actors: exiled planters, landowners, scientific experts, politicians, and the like. In order to understand how the discussion developed and to untangle the issues, we need to examine the criteria defining expertise and the extent to which instrumental skills and quantification carried epistemological weight in scientific and social controversies.

In years II and III, a heated controversy emerged between a promoter of acclimatization, a physician from Nîmes, the *citoyen* Basset, and the experts of the Commission d'Agriculture et des Arts in Paris. After an adventurous life as a planter in Saint-Domingue and South Carolina, Basset served as a military physician in the revolutionary army near Menton, in southern France. Not very literate, as his handwriting testifies, he bombarded Parisian institutions with proposals to create an experimental farm for the cultivation of exotic plants, such as rice, sugar cane, cotton, indigo, spices, and so forth. In his view, all such plants would "succeed in the vicinity of Menton, in the Alpes-Maritimes *département*, as well as on the islands facing the city of Hyères." To lend credibility to his proposal, Basset provided all kinds of proof that he intended to be scientific. He sent lists of useful plants with which he claimed familiarity despite his obvious ignorance of their scientific names (the nopal cactus is called "the racket that bears the cochineal insect"). To add authority to his pronouncements, he sprinkled his pamphlet with references to botanists he had met during his travels, including "the physician Joubert" (probably Joubert de la Motte, head of the royal garden in Port-au-Prince from 1780 to 1787), with whom he allegedly produced a pair of silk gloves as a gift to Louis XVI, "the last Capet." Sketchy comparisons and analogies permitted him to assume, for instance, that since orange trees could grow in southern France, so too could sugarcane. He considered his best argument ("Let us come now to irrefutable proofs") to be the decision, taken by the authorities of the Alpes-Maritimes, to welcome his project on the grounds that "the temperature in their *département* was hardly different from that of America."[33]

The Commission d'Agriculture sent a lengthy rebuttal, both derisory and polemical, of Basset's memoir signed by three experts—the nurseryman Philippe Vilmorin (1746–1840), the botanist Jacques Cels (1740–

1806), and the veterinarian François-Hilaire Gilbert (1757–1800). With irony they pointed out his ignorance of systematic botany and dismissed his assertions about rice cultivation in the New World by quoting their own informant, the botanist André Michaux (1746–1802), who had visited the Carolinas in the same years.[34] In the end, their core argument was—like Basset's—built on climate. But accuracy and precision were now the issue: "In order to give credit to his project, the citizen Basset should have been more precise about the degree of heat which was needed, and about the duration of the vegetation cycle. He should . . . have told how much sugar the plant can produce at a given latitude and height above the sea-level." Grounded in numbers and measures, the experts' conclusion amounted to a definitive dismissal: "The sugarcane cannot grow except with a mean temperature of at least 12 degrees, maintained over a period of nine months."[35] Unable to decide whether he was a naive ignoramus or an impostor, the commission advised Basset to make better use of his imagination the next time he wanted to serve the public good.

This derogatory attitude is all the more significant since the commission, far from being hostile to the kind of project Basset had in mind, was contemporaneously supporting a similar project. In the vicinity of Menton a farmer named Bermond was growing indigo, cotton, sugarcane, coffee, and other exotic plants with the support of the Muséum d'Histoire Naturelle and of the Commission d'Agriculture, who provided him with living specimens. Truly, the Parisian experts did not expect much from this trial, with the exception of coffee. Still, since Bermond was considered an "enlightened farmer," his experiment was widely publicized in the *Décade philosophique*.[36] Contrasting the negative reception of Basset's proposal with the support granted to Bermond reveals how, in those debates, the definition of a legitimate discourse of science and the ability to speak or write its language were as much issues as was the reality of plant acclimatization. Precision and quantification were now necessary for scientific credibility.

From the Geography of Plants to the History of the Earth

By the late eighteenth century instrumental skills and quantification had thus become the guarantees of expertise for botanists and gardeners involved in the transfer of plants. Individual plants existed in a matrix of observations and calculations that were intended to identify the climatic environment in which they thrived and to allow for their replication in another terrain, whether a greenhouse or garden in Europe, or a plantation in the overseas tropical colonies. Yet, whereas precise information was deemed indispensable for the management of plant life, the examples cited above illustrate that such knowledge was nonetheless considered occasional or merely practical, and eighteenth-century botanists rarely integrated it into their division and description of taxa. One further question remains to be addressed, albeit briefly: Did, at some point, the quantifying practices that developed in meteorology and the physical description of nature converge with systematic botany such that the relations between the geographic distribution of plants and their taxonomic arrangement constituted a new field of inquiry?

On 10 Fructidor, year IX, one could read in the *Décade philosophique* that some specimens of "Tahitian mulberry" had been successfully acclimatized in the Jardin des Plantes. The author affirmed that another exotic plant, the breadfruit tree, could also be naturalized in Paris. First, he argued, the two plants were native to the same area in the South Sea Islands. Second, they both belonged to the same family: "Their foliage looks alike, and their wood has nearly the same density." Both their common geographic origin and their shared morphological and physiological features were solid clues that each of the two plants could be acclimatized with equal success.[37]

A sudden cold spell in early fall 1805 offered André Thouin the opportunity to investigate further the link between plant physiology and atmospheric conditions. "On September 20th, the thermometer indicated 2.75 degrees below zero at night, in . . . the lowest and most humid corner of the Jardin." The meteorological accident had taken the gardeners by surprise, and some exotic plants had not yet been brought into their winter greenhouses. Training his taxonomic eye on the plants, Thouin studied in great detail how each specimen responded to the cold: he took notice that the frost had had similar effects—the leaves fell off, the wood showed dark marks on the outside and got dry inside—on some plants (fig trees, mulberries, and one Polynesian plant named *Broussonetia*) that all belonged to a single natural family, identified by Thouin as *Orties* (now known as *Urticaceae*). In Thouin's view, that pattern strongly supported a taxonomy based on natural families: "The effect of the frost on these plants is a new demonstration of the relations that link them all together. The fact that they all underwent a similar decaying process seems to prove that the external aspect of a plant is the product of its inner organization." De Jussieu's system was, therefore, a precious heuristic tool for botanists and gardeners, as well as for chemists and physicists: the study of the plants' physical and physiological organization could help practitioners and scientists alike anticipate their reactions to atmospheric and meteorological conditions, and could throw new light on their geographic distribution.[38]

A new field of interest and research was thus emerging at the junction of systematic botany and geo-climatic information. Already in 1788

Thouin had urged the young Joseph Martin to register in his logbook the daily temperatures and their effects on the plants on board while sailing to and from the Torrid Zone. In 1806, exiting the experimental site of a ship or botanical garden, he expanded his request to encompass the entire natural world and invited all voyaging botanists to investigate whether "the internal arrangement of plants makes them more or less sensitive to cold, and more or less able to survive heat." The correlation between the spatial distribution of plants and their taxonomic groupings was to be scrutinized: "One is tempted to believe that there are entire families, or even genera, of plants, which grow only in specific places in the world."[39]

This very question would animate early nineteenth-century botany, and constitute a central concern for many traveling naturalists in their attempts to articulate field observations in relation to a comprehensive view of botany and to identify hitherto uncharted regularities in the operations of nature.[40] A most exemplary case is offered by Alexander von Humboldt (1769–1859), who left for Spanish America in 1799 with a whole array of portable instruments with which he measured and charted the distribution of life over the surface of the earth, while building up the catalog of its botanical diversity. Whereas contemporary naturalists such as the abbé Jean-Louis Giraud-Soulavie (1752–1813) in the province of Vivarais, Louis-François Ramond de Carbonnières (1755–1827) in the Pyrenees, and Francisco José de Caldas (1771–1816) in the Andes had limited their scope to local floras, Humboldt introduced a global dimension to the study of the geography of plants. His systematic measurements enabled him to demonstrate that individual plant species were arranged in large communities and distributed according to climatic and environmental variables (for example, height, moisture, temperature, soil). The descriptions of plants, the precision of thermometric and barometric records, and the observation of environments were now merged in a single and all-encompassing endeavor.[41]

Just as they helped to unravel patterns in the distribution of natural kinds, numbers also enabled the discovery of unexpected phenomena. A major event, in this regard, was Robert Brown's survey of Australia in the early 1800s. In order to compare the composition and distribution of flora in the "southern continent" with those of other regions of the world, Brown (1773–1858) investigated the correlation between geography, climate, and taxonomy through a precise tabulation: "It is well known that Dicotyledonous plants greatly exceed Monocotyledonous in number; I am not however aware that the relative proportion of these two primary divisions have anywhere been given, or that it has been enquired how far they depend on climate." To answer these questions, Brown calculated the changing proportions of plant species according to latitude: "From the Equator to the 30° of latitude, in the northern hemisphere at least, the species of Dicotyledonous plants are to the Monocotyledones as about 5 to 1; . . . in the higher latitudes, a gradual diminution of Dicotyledones takes place, until in about 60° N latitude and 55° S latitude, they scarcely equal half their intra-tropical proportion." Yet Brown's statistical method—or "botanical arithmetic"—revealed strange irregularities when applied to the southern continent: a plant such as the eucalyptus, albeit "generally spread over the whole of *Terra Australis* (nearly one hundred species have been already observed there)," appeared to be strictly confined to its southern habitat and "hardly found beyond this country" even under similar climates. This was an intriguing observation indeed: "I confess I can perceive nothing . . . in the nature of the soil or climate of *Terra Australis* . . . to account for these remarkable exceptions to the general proportions of the two classes in the corresponding latitudes of other countries."[42]

To make sense of such discontinuities in the geography of plants, variables other than temperature or barometric height would have to be called on: the history of the earth, its geological or climatic changes, and their impact on ancient plant distribution would have to be taken into account. In other words, time as well as space was to be explored: as much as a natural science, the study of plant distribution was to become a historical discipline. Naturalists did not, however, have to renounce their interest in plant habitats or give up their measuring practices in order to delve into the past. As Giraud-Soulavie had anticipated in a somewhat visionary program, "weights and measures" were as needed in the exploration of nature's past as they were in the description of its present organization: "The observation of fossil plants . . . will inform us about the meteorology of ancient climates. . . . Fossils will be our thermometers."[43] Through the graduations of his thermometer, whether a fossil plant or a mercury tube, the naturalist could dream of encompassing the entire history and deep unity of nature.

Notes

Introduction

1. Charles-Marie de La Condamine, "Sur l'arbre du quinquina" (28 Mai 1737) *Histoire mémoires de l'Académie Royale des Sciences* (Amsterdam, 1706–55): 319–46, esp. 326.

2. Charles-Marie de La Condamine, *Relation abrégée d'un voyage fait dans l'interieur de l'Amérique Méridionale* (Paris, 1745), 26–27.

3. Lisbet Koerner, *Linnaeus: Nature and Nation* (Cambridge: Harvard University Press, 1999), 136–39. See also Müller-Wille's chapter in this volume.

4. See Francisco Guerra, "Drugs from the Indies and the Political Economy of the Sixteenth Century," *Analecta Medico-Historica* 1 (1966): 29–54.

5. Alfred Crosby, *The Columbian Exchange: Biological and Cultural Consequences of 1492* (Westport, Conn.: Greenwood, 1972); Lucile Brockway, "Plant Science and Colonial Expansion: The Botanical Chess Game," in *Seeds and Sovereignty: The Use and Control of Plant Resources,* ed. Jack Kloppenburg Jr. (Durham: Duke University Press, 1988), 49–66.

6. Alice Stroup, *A Company of Scientists: Botany, Patronage, and Community at the Seventeenth-Century Parisian Royal Academy of Sciences* (Berkeley: University of California Press, 1990); Harold J. Cook, "The Cutting Edge of a Revolution? Medicine and Natural History Near the Shores of the North Sea," in *Renaissance and Revolution: Humanists, Scholars, Craftsmen and Natural Philosophers in Early Modern Europe,* ed. J. V. Field and Frank A. J. L. James (Cambridge: Cambridge University Press, 1993), 45–61; Steven Harris, "Long-Distance Corporations, Big Sciences, and the Geography of Knowledge," *Configurations* 6 (1998): 269–304.

7. William Stearn, "Botanical Exploration to the Time of Linnaeus," *Proceedings of the Linnean Society of London* 169 (1958): 173–96, esp. 175; Julius von Sachs, *Geschichte der Botanik vom XVI: Jahrhundert bis 1860* (Munich, 1876); Edward Lee Greene, *Landmarks of Botanical History,* 2 vols. (Washington, D.C.: Smithsonian Institution, 1909); Roger Williams, *Botanophilia in Eighteenth-Century France* (Dordrecht: Kluwer Academic Publishers, 2001).

8. In addition to works cited separately, see Marie-Cécile Bénassy-Berling, ed., *Nouveau monde et renouveau de l'histoire naturelle,* 3 vols. (Paris: Presses de la Sorbonne nouvelle, 1986–94); John MacKenzie, ed., *Imperialism and the Natural World* (Manchester: University of Manchester, 1990); Mary Louise Pratt, *Imperial Eyes: Travel Writing and Transculturation* (London: Routledge, 1992); N. Jardine, J. A. Secord, and E. C. Spary, eds., *Cultures of Natural History: From Curiosity to Crisis* (Cambridge: Cambridge University Press, 1995); Yves Laissus, ed., *Les naturalistes français en Amerique de sud* (Paris: Edition du CTHS, 1995); David Miller and Peter Reill, eds., *Visions of Empire: Voyages, Botany, and Representations of Nature* (Cambridge: Cambridge University Press, 1996); Tony Rice, *Voyages: Three Centuries of Natural History Exploration* (London: Museum of Natural History, 1999);

Marie-Noëlle Bourguet and Christophe Bonneuil, *De L'Inventaire du monde à la mise en valeur du globe: Botanique et colonization (fin XVIIe siècle-début XXe siècle)*, special issue of *Revue française d'histoire d'outre-mer* 86, no. 322–23 (1999); Emma Spary, *Utopia's Garden: French National History from the Old Regime to Revolution* (Chicago: University of Chicago Press, 2000).

9. Philip Curtin, *The Rise and Fall of the Plantation Complex* (Cambridge: Cambridge University Press, 1990); Juan Pimentel, "The Iberian Vision: Science and Empire in the Framework of a Universal Monarchy, 1500–1800," in *Nature and Empire: Science and the Colonial Enterprise*, ed. Roy MacLeod, special issue of *Osiris* 15 (2000); 17–30, esp. 26.

10. Wim Klooster, *Illicit Riches: Dutch Trade in the Caribbean, 1648–1795* (Leiden: KITLV Press, 1998), 20; C. R. Boxer, *The Dutch Seaborne Empire: 1600–1800* (New York: Knopf, 1965), 49, 94.

11. Curtin, *Rise and Fall of the Plantation Complex*.

12. See Antonio Barrera, "Local Herbs, Global Medicines: Commerce, Knowledge, and Commodities in Spanish America," in *Merchants and Marvels: Commerce, Science, and Art in Early Modern Europe*, ed. Pamela Smith and Paula Findlen (New York: Routledge, 2002), 163–81, esp. 164.

13. See Klaas van Berkel, "Een onwillige Mecenas?," in *VOC en Cultuur: Wetenschappelijke en Culturele Relaties tussen Europa en Azië ten tijde van de Verenigde Oostindische Compagnie*, ed. J. Bethlehem and A. C. Meijer (Amsterdam: Schiphouwer & Brinkman, 1993), 39–58.

14. See, most recently, Rebecca Parker Brienen, "Art and Natural History at a Colonial Court: Albert Eckhout and Georg Marcgraf in Seventeenth-Century Brazil" (Ph.D. diss, Northwestern University, 2002).

15. Stewart Mims, *Colbert's West India Policy* (New Haven: Yale University Press, 1912).

16. James E. McClellan III and François Regourd, "The Colonial Machine: French Science and Colonization in the Ancien Régime," in *Nature and Empire*, ed. MacLeod, 31–50, esp. 32.

17. John Gascoigne, *Science in the Service of Empire* (Cambridge: Cambridge University Press, 1998).

18. Richard Drayton, *Nature's Government: Science, Imperial Britain, and the Improvement of the World* (New Haven: Yale University Press, 2000); Richard Drayton, "European Imperialism and the Enlightenment: Botany and the Science of Government, c. 1750–1815," paper delivered at the "Botany in Colonial Connection," Einstein Forum, Potsdam, May 2001. We regret not having a chapter on Great Britain in this volume. Interested readers are directed to the wealth of literature on this topic, including David Mackay, *In the Wake of Cook: Exploration, Science, and Empire, 1780–1801* (London: Croom Helm, 1985; and Gascoigne, *Science in the Service of Empire*.

19. Crosby, *Columbian Exchange*.

20. Richard Ligon, *History of the Island of Barbados* (London, 1657), 99.

21. Kloppenburg, *Seeds and Sovereignty*.

22. Judith Carney, *Black Rice: The African Origins of Rice Cultivation in the Americas* (Cambridge: Harvard University Press, 2001).

23. Judith Carney, "African Traditional Plant Knowledge in the Circum-Caribbean Region," *Journal of Ethnobiology* 23, no. 2 (2003): 167–85.

24. Richard Grove, *Green Imperialism: Colonial Expansion, Tropical Island Edens and the Origins of Environmentalism, 1600–1860* (Cambridge: Cambridge University Press, 1995).

25. *Grand Robert*, 1985, s.v. "botaniste"; *Oxford English Dictionary*, 2004, s.v. "botany."

26. Harold Cook has made the argument that "new" histories of the Scientific Revolution have effectively dropped medicine and natural history from their charts; see Cook, "Cutting Edge of a Revolution?"

27. Londa Schiebinger, *Plants and Empire: Colonial Bioprospecting in the Atlantic World* (Cambridge: Harvard University Press, 2004).

28. Agnes Arber, *Herbals; Their Origin and Evolution, a Chapter in the History of Botany, 1470–1670* (Cambridge: Cambridge University Press, 1912).

29. Agnes Arber, "From Medieval Herbalism to the Birth of Modern Botany," in *Science, Medicine, and History: Essays on the Evolution of Scientific Thought and Medical Practice Written in Honour of Charles Singer*, ed. E. Ashworth Underwood, 2 vols. (London: Oxford University Press, 1953), vol. 1, 317–36, esp. 317–18. See also Eileen Reeves, "Old Wives' Tales and the New World System: Gilbert, Galileo, and Kepler," *Configurations* 7 (1999): 301–54.

30. See Harold J. Cook's essay in this volume.

31. Drayton, *Nature's Government*, 92.

32. James McClellan III, *Colonialism and Science: Saint Domingue in the Old Regime* (Baltimore: Johns Hopkins University Press, 1992), 148.

33. Thomas Dancer, *Some Observations Respecting the Botanical Garden* (Jamaica, 1804), 3.

34. William M. Ivins Jr., *Prints and Visual Communication* (London: Routledge & K. Paul 1953), 36.

35. See Brienen, "Art and Natural History," iii; Maurits's letter is cited in note 7.

36. L'Empereur's images are rare but not unique; see Stuart Cary Welch, "A Confluence of East and West, of Art and Science," in *A Selection of Late 18th and Early 19th Century Indian Botanical Paintings* (Pittsburgh: Hunt Institute for Botanical Documentation, 1980), xxx. See also Beth Fowkes Tobin, *Picturing Imperial Power: Colonial Subjects in Eighteenth-Century British Painting* (Durham: Duke University Press, 1999).

37. MacLeod, ed., *Nature and Empire*, 6.

Chapter 1. Dominion, Demonstration, and Domination

1. See Richard Drayton, *Nature's Government: Science, Imperial Britain, and the 'Improvement' of the World* (New Haven: Yale University Press, 2000); Chandra Mukerji, *Territorial Ambitions and the Garden of Versailles* (Cambridge: Cambridge University Press, 1997).

2. Marguerite Duval, *The King's Garden*, trans. Annette Tomarken and Claudine Cowen (Charlottesville: University Press of Virginia, 1982), 9–18; Davy de Virville, *Histoire de la Botanique en France* (Paris: Société d'Édition d'Enseignment Supérieure, 1954), 31.

3. Duval, *King's Garden*, 1–7, 19–30, 55–56, 61; André Bailly, *Défricheurs d'inconnu: Peiresc, Tournefort, Adanson, Saporta* (Aix-en-Provence: Edisud, 1992), 68–69.

4. Mukerji, *Territorial Ambitions*, 71–77, 151–61, 166–71; Jean de la Quintinie, *Instructions pour les jardins fruitiers et potagers* (Amsterdam, 1692).

5. Mukerji, *Territorial Ambitions*, 9–18.

6. Henri Carré, *Sully: Sa vie et son oeuvre, 1559–1641* (1932; Paris: Payot, 1980),

185–203; Chandra Mukerji, "Bourgeois Culture and French Gardening in the sixteenth and seventeenth Centuries," in *Bourgeois Influences in Garden Design*, ed. Michel Conan (Washington, D.C.: Dumbarton Oaks Press, 2002), 173–88.

7. Quoted in Henry Morley, *Palissy the Potter: The Life of Bernard Palissy, of Saintes, His Labours and Discoveries in Art and Science*, 2 vols. (London: Chapman and Hall, 1852), vol. 2, 241–42.

8. Olivier de Serres, *Théâtre d'agriculture et mesnage des champs* (Geneva, 1611), 28–29.

9. Olivier de Serres, *The Perfect Use of Silk-Wormes* (London, 1607).

10. John Prest, *The Garden of Eden: The Botanic Garden and the Re-Creation of Paradise* (New Haven: Yale University Press, 1981).

11. Jacques Boyceau, *Traité du jardinage selon les raisons de la nature et de l'art* (Paris, 1638).

12. Mukerji, *Territorial Ambitions*, 27–29, 41–42.

13. André Félibien, *Description de la grotte de Versailles* (Paris, 1672).

14. James McClellan III, *Colonialism and Science: Saint Domingue in the Old Regime* (Baltimore: Johns Hopkins University Press, 1992), 21–74.

15. Alice Stroup, *A Company of Scientists: Botany, Patronage, and Community at the Seventeenth-Century Parisian Royal Academy of Sciences* (Berkeley: University of California Press, 1990), 185–97.

16. Raymond Paskvan, "The Jardin du Roi: The Growth of Its Plant Collection 1715–1750" (Ph.D. diss., University of Minnesota, 1971), 108–11.

17. Ibid., 75–79, 93–96.

18. The university was first founded by Jews and used Arab medicine as a basis for study. See Ellison Hawks, *Pioneers of Plant Study* (1928; Freeport, N.Y.: Books for Libraries Press, 1969), 98; Arthur Steele, *Flowers for the King* (Durham: Duke University Press, 1964), 11.

19. Duval, *King's Garden*, 5–7; Lucia Tomasi, "Projects for Botanical and Other Gardens: A Sixteenth-Century Manual," *Journal of Garden History* 3 (1983): 1–34; Geofroy Linocier, *L'histoire des plantes* (Paris, 1584).

20. Paskvan, "Jardin du Roi," 11–15; Duval, *King's Garden*, 5–7.

21. Duval, *King's Garden*, 1–8.

22. Jacques Mathieu, *Le premier livre de plantes du Canada* (Saint-Foy: Les Presses de l'Université Laval, 1998), 60–75; Virville, *Histoire*, 33.

23. Paskvan, "Jardin du Roi," 20–54.

24. Mukerji, *Territorial Ambitions*, 73–82, 116–23, 171–81.

25. Ibid., 150, 159–60; Serres, *Théâtre d'agriculture*, 302–6.

26. Ibid., 131–43, 153–55, 177–78, 259–63.

27. Louis Marin, *Portrait du Roi* (Paris: Éditions de Minuit, 1981), 251–60.

28. J.-A. Le Roi, ed., *Journal de la santé du roi Louis XIV de l'année 1647 à l'année 1711* (Paris: Durard, 1862).

29. Paskvan, "Jardin du Roi," 66–73, 75–79, 93–96.

30. Duval, *King's Garden*, 55.

31. Ibid.

32. Léon Lapeyssonnie, *La médecine coloniale* (Paris: Seghers, 1988), 71–73.

33. Arthur Vallée, *Un biologiste Canadien: Michel Sarrazin 1659–1735* (Quebec: LS-A. Proulx, 1927), 1–25.

34. Ibid., 35–45, esp. 40–41.

35. Ibid., 49–58.

36. Ibid., 59–62.

37. Paskvan, "Jardin du Roi," 111–13.

38. Ibid., 20–54.

39. Mukerji, *Territorial Ambitions*, 199–203.

40. Roger L. Williams, *Botanophilia in Eighteenth-Century France: The Spirit of the Enlightenment* (Dordrecht: Kluwer Academic Publishers, 2001).

41. Steven Shapin and Simon Schaffer, *Leviathan and the Airpump* (Princeton: Princeton University Press, 1985).

42. Charles Plumier, *Description des Plantes de l'Amerique* (Paris, 1693).

43. Stroup, *Company of Scientists*, 73–74; Jacques Roger, *Les sciences de la vie dans la pensée française du XVIIIe siècle* (Paris: Armand Colin, 1971), 176–77.

44. Stroup, *Company of Scientists*, chap. 1.

45. Duval, *King's Garden*, 43–44.

46. Bailly, *Défricheurs d'inconnu*, 77–99.

47. Ibid., 71–77.

48. Williams, *Botanophilia*, 31–44.

49. Buffon quoted in Duval, *King's Garden*, 66–67.

50. Jean-Baptiste Thibaut de Chanvalon, *Voyage à la Martinique* (1761; Paris, 1763).

51. Ibid., dedication.

52. Ibid., 83–85.

53. Pierre Sonnerat, *Voyage à la Nouvelle Guinée* (Paris, 1776), viii–x.

Chapter 2. *Walnuts at Hudson Bay, Coral Reefs in Gotland*

1. On the role of Linnaean nomenclature in colonialism, see Mary Louise Pratt, *Imperial Eyes: Travel Writing and Transculturation* (London: Routledge, 1992), 24–37; James E. McClellan III, *Colonialism and Science: Saint Domingue in the Old Regime* (Baltimore: Johns Hopkins University Press, 1992), 118–19; David Arnold, *Science, Technology and Medicine in Colonial India* (Cambridge: Cambridge University Press, 2000), 38; Richard Drayton, *Nature's Government: Science, Imperial Britain, and the "Improvement" of the World* (New Haven: Yale University Press, 2000), 18–19; Zaheer Barber, *The Science of Empire: Scientific Knowledge, Civilization and Colonial Rule in India* (New York: State University of New York Press, 1996), 175.

2. See Margareta Revera, "The Making of a Civilized Nation: Nation Building, Aristocratic Culture and Social Change," in *New Sweden in America*, ed. Carol E. Hoffecker, Richard Waldron, and Lorraine E. Williams (Newark: University of Delaware Press, 1995), 26–55.

3. Lisbet Koerner, *Linnaeus: Nature and Nation* (Cambridge: Harvard University Press, 1999), 188.

4. Ibid., chap. 2.

5. Carl Linnaeus, "An Oration Concerning the Necessity of Travelling in One's Own Countrey," in *Miscellaneous Tracts Relating to Natural History*, ed. Benjamin Stillingfleet (1741; London, 1762), 11–12.

6. Ibid., 19–20, 26–27.

7. Åke Berg and Arvid Hj. Uggla, eds., *Herbationes upsalienses: Protokoll över Linnés exkursioner i Uppsalatrakten I. Herbationerna 1747* (Uppsala: Almqvist & Wiksell, 1952), 13.

8. Linnaeus, "Oration Concerning the Necessity of Travelling," 9.

9. Bruno Latour, *Science in Action: How to Follow Scientists and Engineers through Society* (Cambridge: Harvard University Press, 1987); John Law, "On the Meth-

ods of Long-Distance Control: Vessels, Navigation and the Portuguese Route to India," *Sociological Review Monographs* 32 (1986): 234–63.

10. On the Swedish East India Company, see Paul Hallberg and Christian Koninckx, eds., *A Passage to China: Colin Campbell's Diary of the First Swedish East India Company Expedition to Canton, 1732–1733* (Göteborg: Royal Society of Arts and Sciences, 1996). For Linnaeus's relation to it, see Sverker Sörlin, "Scientific Travel—The Linnaean Tradition," in *Science in Sweden: The Royal Swedish Academy of Sciences 1739–1989*, ed. Tore Frängsmyr (Canton, Mass.: Science History Publications, 1989), 96–123. On Linnaeus's correspondence network, see William T. Stearn, introduction to *Carl Linnaeus: Species plantarum: A Facsimile of the First Edition 1753*, 2 vols. (London: Ray Society, 1957), vol. 1, 1–176.

11. For information on Kalm's life and work I rely throughout on the excellent analysis of Martti Kerkkonen, *Peter Kalm's North American Journey: Its Ideological Background and Results* (Helsinki: Finnish Historical Society, 1959).

12. Carl Linnaeus, Memorandum to the Royal Academy of Sciences, 10 January 1746, in *Bref och skrifvelser af och till Carl von Linné*, ed. Th. M. Fries (Stockholm: Ljus, 1908), part I, vol. 2, 58–59.

13. Linnaeus to Elvius (secretary of the Royal Academy of Sciences), 30 January 1749, in *Bref och skrifvelser*, ed. Fries, part I, vol. 2, 126–27.

14. Kalm to Linnaeus, 28 May 1749, in *Bref och skrifvelser*, ed. J. M. Hulth (Uppsala: Akademiska Bokhandeln, 1922), part I, vol. 8, 44.

15. Martti Kerkkonen, John E. Roos, and Harry Krogerus, eds., *Pehr Kalm: Resejournal över resan till norra Amerika*, 4 vols. (Helsinki: Svenska Litteratursällskapet i Finland, 1966–88), vol. 2, 331.

16. Kalm to Linnaeus, 30 August 1749, in *Bref och skrifvelser*, ed. Hulth, part I, vol. 8, 47.

17. Quoted according to Kerkkonen, *Peter Kalm's North American Journey*, 107.

18. Linnaeus to P. Wargentin, 16 February 1750, in *Bref och skrifvelser*, ed. Fries, part I, vol. 2, 138–39.

19. Kalm to Linnaeus, 10 October 1748, in *Bref och skrifvelser*, ed. Hulth, part I, vol. 8, 41.

20. Linnaeus to Wargentin, 11 December 1750, in *Bref och skrifvelser*, ed. Fries, part I, vol. 2, 161.

21. Linnaeus to Abraham Bäck, 28 June 1751, in *Bref och skrifvelser*, ed. Th. M. Fries (Stockholm: Ljus, 1910), part I, vol. 4, 154.

22. Kalm to Linnaeus, 5 December 1750, in *Bref och skrifvelser*, ed. Hulth, part I, vol. 8, 59–61.

23. A translation of this report by Esther Louise Larsen was published in *Agricultural History* 13 (1939): 33–64.

24. Linnaeus to Bäck, 28 June 1751, in *Bref och skrifvelser*, ed. Fries, part I, vol. 4, 154.

25. Carl Linnaeus, *Nova plantarum genera* (Uppsala, 1751). Such dissertations were based on private lectures given by Linnaeus, and published under his name in *Ammoenitates academicae*, 7 vols. (Holmiae, 1749–69). Linnaeus can therefore be considered their author.

26. Cf. Staffan Müller-Wille, "Joining Lapland and the Topinambes in Flourishing Holland: Center and Periphery in Linnaean Botany," *Science in Context* 76 (2003), 467–88.

27. On the economy of botanical exchange see Staffan Müller-Wille, "Nature As a Market-Place: The Political Economy of Linnaean Botany," in *Œconomies in the Age of Newton*, ed. Neil de Marchi and Margaret Schabas, supplement to *His-*

tory of Political Economy 35 (2003): 154–72; Paula Findlen, *Possessing Nature: Museums, Collecting, and Scientific Culture in Early Modern Italy* (Berkeley: University of California Press, 1994), esp. 334.

28. Pratt, *Imperial Eyes*, 32–33; Kavita Phillip, "Global Botanical Networks, Environmentalist Discourses, and the Political Economy of Cinchona Transplantation to British India," in *De l'inventaire du monde à la mise en valeur du globe: Botanique et colonisation*, ed. Marie-Noëlle Bourguet and Christophe Bonneuil, special issue of *Revue française d'histoire d'Outre-mer* 86 (1999): 119–42; Kapil Raj, "Colonial Encounters and the Forging of New Knowledge and National Identities: Great Britain and India, 1760–1850," in *Nature and Empire: Science and the Colonial Enterprise*, ed. Roy Mcleod, special issue of *Osiris* 15 (2000): 119–34, esp. 133.

Chapter 3. Mission Gardens

I would like to express my gratitude to the Royal Society for a History of Science research grant (2002) to study the history of Moravian natural history. Thanks also go to Cristina Grasseni and my editors for their detailed comments on successive drafts. I would like to acknowledge the generosity of John Mason in sharing his extensive knowledge of Moravian missions and offering extensive and valuable comments, as well as the importance of ongoing collaborations with Michael Harbsmeier and Dorinda Outram on Moravian natural history. Paul Peucker very kindly tracked down the Saint Croix garden illustration. Jan Timbrook of the Santa Barbara Natural History Museum generously drew my attention to the fascinating and complex ethnobotanical practices of the Californian Indians as an avenue for further research. His extensive writings about Chumash ethnobotany include "Chumash Ethnobotany: A Preliminary Report," *Journal of Ethnobiology* 4 (1984): 141–69; and "Ethnobotany of Chumash Indians," *Economic Botany* 44 (1990): 236–53.

1. Jean François de Galaup, comte de Lapérouse, *A Voyage Round the World . . . 1785–1788*, ed. and trans. M. L. A. Milet-Mureau, 3 vols. (London, 1798). Citations are taken from a more fluid translation that covers only the period of Lapérouse in California; see Jean François de Galaup, comte de Lapérouse, *Monterey in 1786, Life in a California Mission: The Journals of Jean François de Lapérouse*, introduction and annotation by Malcolm Margolin, rev. trans. by Glenn Farris (Berkeley, Calif.: Heydey Books, 1989).

2. Lapérouse, *Voyage Round the World*, vol. 2, 233; Lapérouse, *Monterey in 1786*, 108–9.

3. Jan Timbrook, personal communication with author, September 2003.

4. Lapérouse, *Voyage Round the World*, vol. 2, 216; Lapérouse, *Monterey in 1786*, 87.

5. Steven J. Harris, "Long-Distance Corporations, Big Sciences, and the Geography of Knowledge," *Configurations* 6 (1998): 288–93; see also S. J. Harris, "Confession-Building, Long-Distance Networks, and the Organization of Jesuit Science," *Early Science and Medicine* 1 (1996): 299–304.

6. One exception to which I shall return in passing is J. C. S. Mason, "Moravians in Labrador: The Labrador Affair, 1764–1784," in J. C. S. Mason, *The Moravian Church and the Missionary Awakening in England 1760–1800* (London: Boydell Press, 2001), 28–58. For a discussion of tensions between missionaries and other colonial botanists in South Africa in the nineteenth century, see Richard H. Grove, "Scottish Missionaries, Evangelical Discourses, and Conservation

Thinking in Southern Africa 1820–1900," *Journal of South African Studies* 15 (1999): 163–87. For a discussion of nineteenth-century missionary botany and agrarian ideology in the context of South Africa and the London Missionary Society and the South Pacific, see Sujit Sivasundaram, "Natural History Spiritualized: Civilizing Islanders, Cultivating Breadfruit and Collecting Souls," *History of Science* 32 (2001): 417–43.

7. For a concise discussion of Joseph Banks's relationship with the Moravians and his debt to them as collectors, see Mason, *Moravian Church*, 52–53.

8. The natural history network of Saxony that I have seen has been mainly concerned with ethnography. See, for example, Stephan Augustin and Lydia Icke-Schwalbe, "Kuntstsachen" *von Cooks Reisen: Die Sammlung und ihre Geschichte im Völkerkundemuseum Herrnhut* (Münster: LIT, 1993).

9. Tim Ingold, *The Perception of the Environment: Essays on Livelihood, Dwelling and Skill* (London: Routledge, 2000). Ingold describes the properties of skill in "Of String Bags and Birds' Nests: Skill and the Construction of Artefacts," in ibid., 349–61. His argument here is that the historical change from artisanship in the early modern era to a technological industrial society should be understood in terms of an externalization of social relations and of disembedding the technical from the social, not progressive evolution, as is often argued.

10. Max Weber, *The Protestant Ethic and the Spirit of Capitalism*, trans. Talcott Parsons (London: G. Allen & Unwin, Ltd., 1930).

11. My account here of Linnaeus's cameralist project and his construction of the Sàmi people as a version of noble savages is indebted to Lisbet Koerner, *Linnaeus: Nature and Nation* (Cambridge: Harvard University Press, 1999). Chapter 3, " 'The Lapp Is Our Teacher': Medicine and Ethnography," is particularly incisive and ironic, esp. 78. See also Staffan Müller-Wille's contribution to this volume.

12. Frederick IV had previously granted Hans Egede, a Norwegian Lutheran from Bergen, permission to establish a mission in Greenland (1721).

13. Mason provides a well-researched overview of the history of the Moravian missions in "The Moravian Church and Its Missions: Zinzendorf to Spangenberg," *Moravian Church*, 5–27.

14. Mason, personal communication with author, September 2003.

15. Mason, *Moravian Church*, 98–103.

16. Christian Oldendorp, *Geschichte der Mission der Caribischen Inseln Sanct Thomas, Sanct Crux und Sanct Jan*, ed. Johann Bossart (Barby, 1777). Quotations and page references are taken from the English translation, Christian Oldendorp, *A Caribbean Mission: C. G. A. Oldendorp's History of the Mission of the Evangelical Brethren on the Caribbean Islands . . .* , ed. and trans. Arnold Highfield and Vladimir Barac (Ann Arbor: Karoma Publishers, 1987). Also relevant is Christian Oldendorp, *Historie der Caribischen Inseln Sanct Thomas, Sanct Crux und Sanct Jan*, ed. Gudrun Meier, Stephan Palmié, Peter Stein, and Horst Ulbricht (Berlin: Verlag für Wissenschaft und Bildung, 2000).

17. Oldendorp, *Caribbean Mission*, 225.

18. Mason explains how the Moravians' "choir system, 'speakings,' and 'helpers'" were adopted. See Mason, *Moravian Church*, 96–98.

19. Oldendorp, *Caribbean Mission*, 393–95.

20. 1769 Synod record, fol. 242, at the Church Mission House, London; cf. Mason, *Moravian Church*, 103.

21. Oldendorp, *Caribbean Mission*, 225.

22. Ibid., 155.

23. Ibid., 17.

24. David Cranz, *Historie von Grönland* (Barby and Leipzig, 1765). Translations include David Cranz, *The History of Greenland: Containing a Description of the Country and Its Inhabitants: And Particularly a Relation of the Mission*, 2 vols. (London, 1767).

25. The expectations and events surrounding the publication of the English translation of Cranz's *Historie von Grönland* are nicely dissected in Mason, *Moravian Church*, 28–59.

26. They initially set up their missions near the Lutheran camps of Hans Egede, but following some disagreements they moved away toward Kangeq, a nearby island, and in subsequent missions moved further south to Lichtenfels and Lichtenau.

27. Cranz, *History of Greenland*, 60.

28. Ibid., 64.

29. Ibid., 65–67.

30. Ibid., 150.

31. Oldendorp, *Caribbean Mission*, 115.

Chapter 4. Gathering for the Republic

1. H. B. Trout to Benjamin Smith Barton, 20 January 1814, Barton Papers, Correspondence, American Philosophical Society (hereafter APS), Philadelphia, Pa. Trout referred to Benjamin Smith Barton, *Elements of Botany: Or, Outlines of the Natural History of Vegetables* (Philadelphia, 1803). The modern classification for the opium poppy is *Papaver somniferum.*

2. Francis Hopkinson, "An Address to the American Philosophical Society, Held at Philadelphia, for Promoting Useful Knowledge: Delivered January 16, 1784," in Francis Hopkinson, *The Miscellaneous Essays and Occasional Writings of Francis Hopkinson, Esq.* (Philadelphia, 1792), 364–66.

3. Nicholas Collin, "An Essay on Those Inquiries in Natural Philosophy, Which at Present Are Most Beneficial to the United States of North America," *Transactions of the American Philosophical Society* 3 (1793): v.

4. Humphry Marshall, *Arbustrum Americanum: The American Grove, or, an Alphabetical Catalogue of Forest Trees and Shrubs, Native of the American United States, Arranged According to the Linnaean System* (Philadelphia, 1785), v–ix.

5. Charles Willson Peale, *Introduction to a Course of Lectures on Natural History: Delivered in the University of Pennsylvania, Nov. 16, 1799* (Philadelphia, 1800), 10–12, 14.

6. Thomas Jefferson, *Notes on the State of Virginia*, ed. William Peden (1787; Chapel Hill: University of North Carolina Press, 1954).

7. Jeremy Belknap, *The History of New-Hampshire* (Boston, 1792), 96.

8. Among the most important of these "histories of place" are Belknap, *History of New-Hampshire*; another important local natural history was Samuel Williams, *The Natural and Civil History of Vermont* (Walpole, N.H., 1794).

9. William Paul Crillon Barton, *Vegetable Materia Medica of the United States; of Medical Botany: Containing a Botanical, General, and Medical History, of Medicinal Plants Indigenous to the United States*, 2 vols. (Philadelphia, 1817).

10. Benjamin Smith Barton, *Fragments of the Natural History of Pennsylvania* (Philadelphia, 1799), viii.

11. This notice is reprinted in James Mease, *The Picture of Philadelphia, Giving*

an Account of Its Origin, Increase, and Improvements in Arts, Sciences, Manufactures, Commerce and Revenue (Philadelphia, 1811), 305.

12. Gabriel Crane to Secretary of the American Philosophical Society, 27 May 1817, Archives, APS.

13. Peter Curtis to Benjamin Smith Barton, 15 June 1804, Verbal Communications, APS; John D. Gillespie to John Vaughan, 23 November 1807, MSS Communications, APS; David Thomas to David Rittenhouse and the APS, 24 January 1792, Archives, APS; William Currie, "A Sketch of the Errors Which Have Been Discovered in Some of the Philosophical Opinions of the Illustrious Sir Isaac Newton, and Other Philosophers of Acknowledged Genious and Talents," Archives, APS.

14. Lorraine Daston and Katherine Park, *Wonders and the Order of Nature, 1150–1750* (New York: Zone Books, 1998), esp. chap. 9.

15. Edward Johnson to Benjamin Smith Barton, 4 November 1807, Barton Papers, Correspondence, APS.

16. Unfortunately, the Archives of the American Philosophical Society contain only a few outbound letters and a few reports on these matters.

17. Charles Creswell to American Philosophical Society, 1809, MSS Correspondence, Archives, APS; C. Brown to Benjamin Smith Barton, 30 November 1792, Barton Papers, Correspondence, APS; D. R. Patterson to Benjamin Smith Barton, 15 March 1808, Barton Papers, Correspondence, APS.

18. William Thorton to John Vaughan, 13 December 1805, MSS Correspondence, Archives, APS; Joseph Richardson to John Vaughan, 6 June 1805, MSS Correspondence, Archives, APS; North Carolina Gold-Mine Company, Broadside, n.d., Archives, APS. For information on the North Carolina gold economy, see Richard F. Knapp, "Golden Promise in the Piedmont: The Story of John Reed's Mine," *North Carolina Historical Review* 52 (1975): 1–19; Richard D. Knapp and Brent D. Glass, *Gold Mining in North Carolina: A Bicentennial History* (Raleigh: North Carolina Division of Archaeology and History, 1999); and Jeff Forret, "Slave Labor in North Carolina's Antebellum Gold Mines," *North Carolina Historical Review* 76 (1999): 135–62.

19. Benjamin Smith Barton, *Collections for an Essay towards a Materia Medica of the United-States* (Philadelphia, 1798), 43–45.

20. Dr. S. P. Hildreth, "Information Concerning the Frasera Carolinensis, Otherwise Called the American Colombo Plant: In a Letter from Dr. S. P. Hildreth, of Marietta, in Ohio, Dated Marietta, July 30, 1810 (with a Figure)," *The Medical Repository, and Review of American Publications on Medicine, Surgery, and the Auxiliary Branches of Philosophy* 15, no. 2 (1811); James Woodhouse, "An Account of a New, Pleasant, and Strong Bitter, and Yellow Dye, Prepared from the Stem and Root of the Xanthorhiza Tinctoria, or Shrub Yellow Root; with a Chemical Analysis of This Vegetable: Communicated by James Woodhouse, M.D. Professor of Chemistry in the University of Pennsylvania, &C.," *The Medical Repository* 5, no. 2 (1802): 159–64.

21. Barton, *Collections for an Essay*, 13.

22. John Beatty to Benjamin Smith Barton, 19 April 1809, Barton Papers, Correspondence, APS.

23. Thomas T. Hewson to Benjamin Smith Barton, 11 April [1809], Barton Papers, Correspondence, APS.

24. Richard Brown to Benjamin Smith Barton, 30 October 1807, Barton Papers, Correspondence, APS.

Chapter 5. Books, Bodies, and Fields

I am grateful to Anthony Grafton, Ken Mills, Alex Haw, Claudia Swan, Londa Schiebinger, Katie Holt, Chris Garces, and the students and faculty in the Program in the History of Science at Princeton University for their comments on earlier versions of this essay.

1. Particularly noteworthy are John H. Elliott, *The Old World and the New 1492–1650* (Cambridge, U.K.: Cambridge University Press, 1970); Serge Gruzinski, *The Conquest of Mexico: The Incorporation of Indian Societies into the Western World, 16th–18th Centuries*, trans. Eileen Corrigan (Cambridge: Polity Press, 1993), and *La pensée métisse* (Paris: Fayard, 1999); Margaret T. Hodgen, *Early Anthropology in the Sixteenth and Seventeenth Centuries* (Philadelphia: University of Pennsylvania Press, 1971); Hugh Honour, *The European Vision of America* (Cleveland: Cleveland Museum of Art and Kent State University Press, 1975); Karen Ordahl Kupperman, ed., *America in European Consciousness, 1493–1750* (Chapel Hill: University of North Carolina Press, 1995); James Lockhart, *Nahuas and Spaniards* (Stanford, Calif.: Stanford University Press, 1991); Anthony Pagden, *The Fall of Natural Man: The American Indian and the Origins of Comparative Ethnology* (Cambridge: Cambridge University Press, 1982), and *European Encounters with the New World: From Renaissance to Romanticism* (New Haven: Yale University Press, 1993); Tzvetan Todorov, *The Conquest of America: The Question of the Other* (New York: Harper & Row, 1984).

2. See John Brewer and Roy Porter, eds., *Consumption and the World of Goods* (London: Routledge, 1994); Paula Findlen and Pamela Smith, eds., *Merchants and Marvels: Commerce, Science, and Art in Early Modern Europe* (New York: Routledge, 2002), esp. Antonio Barrera, "Local Herbs, Global Medicines: Commerce, Knowledge, and Commodities in Spanish America," 163–81; Lisa Jardine, *Worldly Goods* (London: Macmillan, 1996).

3. Nicolás Monardes, *Primera y segunda y tercera partes de la historia medicinal de las Cosas que se traen de nuestras India Occidentales que sirven en medicina* (Seville, 1565–74) (hereafter *HM*; citations refer to the facsimile edition [Seville: Padilla Libros, 1988]). All translations are mine; for a contemporary English edition, see John Frampton, *Joyfull Newes Out of the Newe Founde Worlde*, 2 vols. (1577; London: Constable and Co. Ltd., 1925).

4. See Miguel Ángel Ladero Quesada, *El primer oro de América: Los comienzos de la Casa de la Contratación de las Yndias* (Madrid: Real Academia de la Historia, 2002); José Cervera Pery, *La Casa de Contratación y el Consejo de Indias* (Madrid: Ministerio de Defensa, 1997); Alison D. Sandman, "Cosmographers vs. Pilots: Navigation, Cosmography, and the State in Early Modern Spain" (Ph.D. diss., University of Wisconsin at Madison, 2001).

5. *Real Cédula dando licencia al Doctor Monardes, médico, para pasar a Indias 200 esclavos negros, un tercio hembras*, Archivo General de Indias (Seville), 28 July 1561, Sección Indiferente, 425, legajo 24, 36v–37v and 15 September 1561, Sección Indiferente, 425, legajo 24, 58r–59r.

6. See Francisco Guerra, *Nicolás Bautista Monardes: Su vida y su obra* (Mexico: Compañía Fundidora de Fierro y Acero de Monterrey, S.A., 1961), 80–81; and David C. Goodman, *Power and Penury: Government, Technology, and Science in Philip II's Spain* (Cambridge: Cambridge University Press, 1988), 238.

7. Guerra, *Nicolás Bautista Monardes*, 91–109.

8. The classic study on Clusius is Friedrich W. T. Hunger, *Charles de l'Escluse (Carolus Clusius), Nederlandsch kruidkundige* (The Hague: M. Nijhoff, 1927–43).

See also Josep Lluís Barona and Xavier Gómez Font, *La correspondencia de Carolus Clusius con los científicos españoles* (Valencia: Seminari d'estudis sobre la ciència, 1998).

9. For an overview of European literature on the New World, see Anthony Grafton, *New Worlds, Ancient Texts: The Power of Tradition and the Shock of Discovery* (Cambridge: Harvard University Press, 1992); and Dennis Channing Landis, *The Literature of the Encounter: A Selection of Books from European Americana* (Providence, R.I.: John Carter Brown Library, 1991).

10. See Emily Walcott Emmart, trans., *The Badianus Manuscript (Codex Barberini, Latin 241), Vatican Library: An Aztec Herbal of 1552* (Baltimore: Johns Hopkins University Press, 1940); José María López Piñero and María Luz López Terrada, *La influencia española en la introducción en Europa de las plantas americanas (1493–1623)* (Valencia: CSIC, 1997); Simon Varey, ed., *The Mexican Treasury: The Writings of Dr. Francisco Hernández* (Stanford, Calif.: Stanford University Press, 2000); Simon Varey, Rafael Chabrán, and Dora V. Weiner, eds., *Searching for the Secrets of Nature: The Life and Works of Dr. Francisco Hernández* (Stanford, Calif.: Stanford University Press, 2000).

11. *HM*, 30r–30v.

12. Ibid., 30v–31r.

13. Ibid., 79r, 91v–92r.

14. There are many similar stories, for instance the entry on pepper; see ibid., 86v–87r.

15. *HM*, 78r.

16. Ibid., 39r.

17. In colonial contexts, the word "profit" commonly carried religious connotations, alluding to potential gains in spiritual wealth. At times this represented a critique of the worldly excesses that accompanied economic profit. Antonio de la Calancha's *Crónica moralizada* (Barcelona, 1638), for example, argued that the true wealth and profit of the silver mining town of Potosí, Peru, were spiritual and not material. I thank Ken Mills for this observation.

18. *HM*, 31r.

19. *HM*, 31r. There were some exceptions to this lack of interest, such as the Franciscan friars who transported Michoacan and successfully grew the plant outside the infirmary of their Seville monastery, where Monardes saw it (ibid., 31v).

20. *HM*, 31r–31v.

21. Ibid., 29r–30r.

22. Ibid., 9v, 11v–12r.

23. Ibid., 71r. Osma refers to the first part of the *Historia medicinal* (1565).

24. *HM*, 71v–74r.

25. Ibid., 75r.

26. Ibid., 75v.

27. Ibid., 76r. See Kenneth J. Andrien and Rolena Adorno, eds., *Transatlantic Encounters: Europeans and Andeans in the Sixteenth Century* (Berkeley: University of California Press, 1991), and Susan E. Ramirez, *The World Upside Down: Cross-Cultural Contact and Conflict in Sixteenth-Century Peru* (Stanford, Calif.: Stanford University Press, 1996).

28. *HM*, 74v.

29. On sixteenth-century medicine in Spain and the New World, see Goodman, *Power and Penury*; José María López Piñero, "The Medical Profession in 16th Century Spain," in *The Town and State Physician in Europe from the Middle Ages*

to the Enlightenment, ed. A. W. Russell (Wolfenbüttel: Herzog August Bibliothek, 1981); José María López Piñero and José Luis Fresquet Febrer, eds., *El mestizaje cultural y la medicina novohispana del siglo XVI* (Valencia: CSIC, 1995).

30. Although there are no grounds for doubting Monardes's veracity, it is possible that Osma's letter was not an actual missive but a fabricated publicity stunt, a common practice at the time. Regardless of its authorship, the letter was exploited rhetorically by Monardes. His contemporaries did not suspect its authenticity, and it was included in numerous translations and editions of his work.

31. *HM*, 77r.

32. Ibid., 47r–51v.

33. Ibid., 90r.

34. Ibid., 36r.

35. Ibid., 36v.

36. Ibid., 73v–74r.

37. See Nicholas Griffiths and Fernando Cervantes, eds., *Spiritual Encounters: Interactions between Christianity and Native Religions in Colonial America* (Lincoln: University of Nebraska Press, 1999).

38. See Fernando Cervantes, *The Devil in the New World* (New Haven: Yale University Press, 1994), 60.

Chapter 6. Global Economies and Local Knowledge in the East Indies

1. Simon Schaffer, "Herschel in Bedlam: Natural History and Stellar Astronomy," *British Journal of the History of Science* 13 (1980): 211–39; John V. Pickstone, *Ways of Knowing: A New History of Science, Technology and Medicine* (Manchester: Manchester University Press, 2000).

2. I used the phrase in Harold J. Cook, "The Cutting Edge of a Revolution? Medicine and Natural History Near the Shores of the North Sea," in *Renaissance and Revolution: Humanists, Scholars, Craftsmen and Natural Philosophers in Early Modern Europe*, ed. J. V. Field and Frank A. J. L. James (Cambridge: Cambridge University Press, 1993), 45–61.

3. On "matters of fact," see esp. Steven Shapin and Simon Schaffer, *Leviathan and the Air Pump: Hobbes, Boyle, and the Experimental Life* (Princeton: Princeton University Press, 1986), following the lead given by Barbara J. Shapiro, *Probability and Certainty in Seventeenth-Century England: A Study of the Relationships between Natural Science, Religion, History, Law, and Literature* (Princeton: Princeton University Press, 1983).

4. See William Eamon, *Science and the Secrets of Nature: Books of Secrets in Medieval and Early Modern Culture* (Princeton: Princeton University Press, 1994); Pamela O. Long, *Openness, Secrecy, Authorship: Technical Arts and the Culture of Knowledge from Antiquity to the Renaissance* (Baltimore: Johns Hopkins University Press, 2001); and Pamela Smith, *The Body of the Artisan* (Chicago: Chicago University Press, 2004).

5. Niels Steensgaard, *The Asian Trade Revolution of the Seventeenth Century: The East India Companies and the Decline of the Caravan Trade* (Chicago: University of Chicago Press, 1974), 43.

6. For the critical role of information in the early modern Dutch economy, see Jan De Vries and Ad van der Woude, *The First Modern Economy: Success, Failure,*

and Perseverance of the Dutch Economy, 1500–1815 (Cambridge: Cambridge University Press, 1997), 147–50, passim.

7. Simon Schaffer, "Scientific Discoveries and the End of Natural Philosophy," *Social Studies of Science* 16 (1986): 387–420.

8. K. van Berkel, "Een Onwillige Mecenas? De Rol Van de VOC Bij Het Natuurwetenschappeijk Onderzoek in de Zeventiende Eeuw," in *VOC en Cultuur: Wetenschappelijke en Culturele Relaties Tussen Europa an Azië Ten Tijde Vande Verenigde Oostindische Compagnie,* ed. J. Bethlehem and A. C. Meijer (Amsterdam: Schipbhouwer en Brinkman, 1993), 39–58.

9. L. S. A. M. von Römer, *Dr. Jacobus Bontius,* Bijblad op het Geneeskundig Tijdschrift voor Nederlandsch-Indië (Batavia: G. Kolff & Co., 1932); Frits de Haan, "Uit Oude Notarispapieren, II: Andries Cleyer," *Tijdschrift Voor Indische Taal-, Land- en Volkenkunde* 46 (1903): 423–68; J. M. H. van Dorssen, "Willem Ten Rhijne," *Geneeskunde Tijdschrift Voor Nederlands Indië* 51 (1911): 134–228.

10. The comment is contained in two of the three letters of Bontius dated 10 March 1631, printed (unpaginated) in L. S. A. M. von Römer, ed., *Epistolæ Jacobi Bontii* (Batavia: Gualtherium Kolff, 1921).

11. G. F. Pop, *De Geneeskunde Bij Het Nederlandsche Zeewezen (Geschiedkundige Nasporingen)* (Weltevreden-Batavia: G. Kolff, 1922), 333–34.

12. Römer, *Epistolæ Jacobi Bontii,* letter 2.

13. Römer, *Dr. Jacobus Bontius,* 4–5.

14. First published as the fourth part (but with its own title page) in Jacobus Bontius, *De Medicina Indorum* (Lugduni Batav: Apud Franciscum Hackium, 1642), pp. 115–17.

15. Jacobus Bontius, *Tropische Geneeskunde/On Tropical Medicine,* ed. M. A. van Andel, *Opuscula Selecta Neerlandicorum De Arte Medica,* no. 10 (Amstelodami: Sumptibus Societatis, 1931), 128–29. I have throughout cited page numbers to this accessible edition. I have, however, revised the English to make it more closely reflect the original meaning since the 1931 edition of the medical works reproduces the free and sometimes anachronistic English translation of 1769.

16. Ibid., 102–5, 128–29.

17. Published in Bontius, *De Medicina,* 59–106.

18. Garcia da Orta, *Colloquies on the Simples and Drugs of India* (Lisbon, 1895), trans. Clements Markham (London: Henry Sotheran and Co., 1913), vii–ix.

19. Guy Attewell, "India and the Arabic Learning of the Renaissance: The Case of Garcia D'Orta" (M.A. thesis, Warburg Institute, University of London, 1997), 7–13, 17–18, 22–23.

20. Carolus Clusius, *Aromatum, et Simplicium Aliquot Medicamentorum Apud Indos Nascentium Historia* (1567), intro. M. de Jong and D. A. Wittop Koning (Nieuwkoop: B. de Graaf, 1963).

21. L. J. Rather, "The 'Six Things Non-Natural': A Note on the Origin and Fate of a Doctrine and a Phrase," *Clio Medica* 3 (1968): 337–47; P. H. Niebyl, "The Non-Naturals," *Bulletin of the History of Medicine* 45 (1971): 486–92; Heikki Mikkeli, *Hygiene in the Early Modern Medical Tradition,* Humaniora Series no. 305, Finnish Academy of Science and Letters (Helsinki, 1999).

22. Bontius, *Tropische Geneeskunde,* 80–83.

23. While the work (published in Bontius, *De Medicina,* 1–55) is undated, in a letter to his brother of 18 February 1631 Bontius wrote about the "animadversions" that were being sent with the letter, along with an explanation of why he felt compelled to correct da Orta, clearly indicating that it was this treatise he had just finished; see Bontius, *Tropische Geneeskunde,* xlviii–xlix.

24. Ibid., 4–5, 2–3, 10–11.

25. In commenting in his "Dialogues" on the fruit locally called *focqui*, he noted that "both the depiction and description of it are to be seen in my *Exotics of the Indies*, which you will have in about a year"; see ibid., 94–95.

26. While the letter is undated as printed in ibid., xlviii–li, it is dated by the editor, Van Andel, in his introduction, xvi–xvii.

27. Willem Piso, *De Indiæ Utriusque Re Naturali et Medica Libri Quatuordecim* (Amsterdam: Apud Lodovicum et Denielem Elzevirios, 1658).

28. MS Sherard 186, listed as "Drawings of animals and plants with explanatory text apparently prepared for the press," now in the Plants Sciences Library of the University of Oxford, Oxford, U.K.

29. Bontius, *Tropische Geneeskunde*, 224–25, 226–27, 232–33, 234–35, 246–47.

30. This last story is contained in ibid., 276–77.

31. Ibid., 224–25, 232–33, 234–35, 246–47, 226–27, 380–83; the chapter on flowers is not in MS Sherard 186 and is otherwise unlike Bontius's chapters.

32. Römer, *Epistolæ Jacobi Bontii*, letters 1 and 2.

33. Bontius, *Tropische Geneeskunde*, 230–31.

34. Ibid., 292–95; MS Sherard 186, 79v–80.

35. Bontius, *Tropische Geneeskunde*, 294–301.

36. Ibid., 102–3.

37. In the second letter printed in Römer, *Epistolæ Jacobi Bontii*, he writes "Ick heb een magnifycke bibliotheek wel by de twee duysent boecken meest uytgelesen autheuren, so dat ickde beste occasye heb om iet fræijs te doen als ick oyt in Europa soude connen becomen hebben," but while implying that he had his books with him, the context does not make it certain. Leonard Blussé informed me that at least one of Bontius's contemporaries in Batavia bought books from him.

38. Dirk Schoute, *De Geneeskunde in den Dienst der Oost-Indische Compagnie in Nederlandsch-Indie* (Amsterdam: J. H. de Bussy, 1929), 118.

39. Bontius, *Tropische Geneeskunde*, 2–3, 38–41, 10–11, 14–15.

40. Ibid., 248–49; the bird is illustrated in the Piso edition but not drawn in the MS Sherard.

41. Bontius, *Tropische Geneeskunde*, 40–43, 32–35.

42. See Jean Gelman Taylor, *The Social World of Batavia: European and Eurasian in Dutch Asia* (Madison: University of Wisconsin Press, 1983); and Leonard Blussé, *Strange Company: Chinese Settlers, Mestizo Women and the Dutch in VOC Batavia* (Dordrecht and Riverton: Foris, 1986).

43. His will is printed in Bontius, *Tropische Geneeskunde*, xliv–xlvii.

44. Ibid., 220–21, xliv–xlvii.

45. Ibid., 28–29.

46. The reference is to Pieter Pauw (1564–1617); for more on Pauw, see the chapter by Swan in this volume.

47. Bontius, *Tropische Geneeskunde*, 92–95, 24–25, 366–67.

48. Marcel Mauss, *The Gift: Forms and Functions of Exchange in Archaic Studies*, trans. Ian Cunnison, intro. E. E. Evans-Pritchard (New York: W. W. Norton, 1967).

49. For some sympathetic but critical treatments, see Jonathan Parry, "On the Moral Perils of Exchange," in *Money and the Morality of Exchange*, ed. J. Parry and M. Block (Cambridge: Cambridge University Press, 1989), 64–93; James G. Carrier, *Gifts and Commodities: Exchange and Western Capitalism since 1700* (London and New York: Routledge, 1995), 11, passim; Janet Hoskins, *Biographical Objects:*

How Things Tell the Stories of People's Lives (New York and London: Routledge, 1998), esp. 7–12, 192–98.

50. Maurice Godelier, *The Enigma of the Gift*, trans. Nora Scott (Oxford: Polity Press, 1999), 102.

51. I am referring here to another ideal type of the function of commodities in a global economy, not about how they are often personalized, much less valued differently in different cultures and circumstances; on the latter point, see esp. Arjun Appadurai, ed., *The Social Life of Things: Commodities in Cultural Perspective* (Cambridge: Cambridge University Press, 1986); Carrier, *Gifts and Commodities*; and Hoskins, *Biographical Objects*.

52. Bontius, *Tropische Geneeskunde*, 396–97.

53. Joel Kaye, *Economy and Money in the Fourteenth Century: Money, Market Exchange, and the Emergence of Scientific Thought* (Cambridge: Cambridge University Press, 1998), 11, 116–62.

54. Richard Hadden, *On the Shoulders of Merchants: Exchange and the Mathematical Conception of Nature in Early Modern Europe* (Albany: State University of New York Press, 1994).

55. Mark Harrison, "Medicine and Orientalism: Perspectives on Europe's Encounter with Indian Medical Systems," in *Health, Medicine and Empire: Perspectives on Colonial India*, ed. Biswamoy Pati and Mark Harrison (Hyderabad: Orient Longman, 2001), 37–87; Michael Adas, *Machines as the Measure of Men: Science, Technology, and Ideologies of Western Dominance* (Ithaca: Cornell University Press, 1989).

Chapter 7. Prospecting for Drugs

1. Denis Diderot and Jean Le Rond d'Alembert, eds., *Encyclopédie: ou, Dictionnaire raisonné des sciences, des arts et des métiers* (Paris, 1751–76), s.v. "Amerique."

2. [Nicolas Bourgeois], *Voyages intéressans dans différentes colonies Françaises, Espagnoles, Anglaises, etc.* (London, 1788), 459.

3. Hans Sloane, *A Voyage to the Islands Madera, Barbadoes, Nieves, St Christophers, and Jamaica; with the Natural History . . .* , 2 vols. (London, 1707–25), vol. 1, preface, n.p.

4. Figures vary from sixty thousand to four million. See Noble David Cook, *Born to Die: Disease and New World Conquest, 1492–1650* (Cambridge: Cambridge University Press, 1998), 23–24.

5. David Watts, *The West Indies: Patterns of Development, Culture and Environmental Change since 1492* (Cambridge: Cambridge University Press, 1987).

6. Jean-Baptiste-René Pouppé-Desportes, *Histoire des Maladies de Saint Domingue*, 3 vols. (Paris, 1770), vol. 3, 59.

7. Londa Schiebinger, *Plants and Empire: Colonial Bioprospecting in the Atlantic World* (Cambridge: Harvard University Press, 2004), chap. 5.

8. Pouppé-Desportes, *Histoire*, vol. 3, 59.

9. Ibid.

10. [Bourgeois], *Voyages*, 67.

11. [Anonymous], *Histoire des désastres de Saint-Domingue* (Paris, 1795), 47.

12. John Stedman, *Stedman's Surinam: Life in Eighteenth-Century Slave Society*, ed. Richard Price and Sally Price (1796; Baltimore: Johns Hopkins University Press, 1992), 63. For the African diaspora of seeds, plant knowledge, and cultivation techniques, see Judith Carney's chapter in this volume.

13. [Bourgeois], *Voyages*, 458, 470.

14. Ibid., 470.

15. Ibid., 468, 470.

16. Galvin de Beer, *Sir Hans Sloane and the British Museum* (London: Oxford University Press, 1953), 41–42.

17. Pierre Barrère, *Essai sur l'histoire naturelle de la France Equinoxiale* (Paris, 1741), 204.

18. David de Nassy, *Essai historique sur la colonie de Surinam* (Paramaribo, 1788), 64.

19. Sloane, *Voyage*, vol. 1, xiii–xiv.

20. Charles-Marie de La Condamine, "Sur l'arbre du quinquina," (28 May 1737), in *Histoire mémoires de l'Académie Royale des Sciences* (Amsterdam, 1706–55): 319–46, esp. 330. For French territorial ambitions, see the chapters by Chandra Mukerji and Emma Spary in this volume.

21. Pouppé-Desportes, *Histoire*, vol. 3, 81.

22. Edward Long, *The History of Jamaica*, 3 vols. (London, 1774), vol. 2, 381.

23. See also Antonio Barrera, "Local Herbs, Global Medicines: Commerce, Knowledge, and Commodities in Spanish America," in *Merchants and Marvels: Commerce, Science, and Art in Early Modern Europe*, ed. Pamela Smith and Paula Findlen (New York: Routledge, 2002), 163–81, esp. 174.

24. Emma Spary, *Utopia's Garden: French Natural History from the Old Regime to Revolution* (Chicago: University of Chicago Press, 2000), 84.

25. Mary Louise Pratt, *Imperial Eyes: Travel Writing and Transculturation* (London: Routledge, 1992), 6–7.

26. Long called the kidnappers "mantraders" for stealing young men and forcing them into service (*History*, vol. 2, 287); Carl Thunberg called the Dutch East India Company "man-stealers" in *Travels in Europe, Africa and Asia, Performed between the Years 1770 and 1779*, 3 vols. (London, 1795), vol. 1, 73–75.

27. Maria Sibylla Merian, *Metamorphosis insectorum Surinamensium*, ed. Helmut Deckert (1705; Leipzig: Insel Verlag, 1975), commentary to pls. 7, 25, 13.

28. Edward Bancroft, *An Essay on the Natural History of Guiana, in South America* (London, 1769), 3.

29. La Condamine, "Sur l'arbre du quinquina," 340.

30. Charles-Marie de La Condamine, *Relation abrégée d'un voyage* (Paris, 1745), 53–55.

31. Alexander von Humboldt and Aimé Bonpland, *Personal Narrative of Travels*, trans. Helen Williams, 7 vols. (London, 1821), vol. 3, 301–3.

32. Cited in Robin Blackburn, *The Making of New World Slavery: From the Baroque to the Modern: 1492–1800* (London: Verso, 1997), 281.

33. Charles de Rochefort, *Histoire naturelle et morale des Iles Antilles de l'Amerique* (Rotterdam, 1665), 449.

34. Guenter Risse, "Transcending Cultural Barriers: The European Reception of Medicinal Plants from the Americas," in *Botanical Drugs of the Americas in the Old and New Worlds*, ed. Wolfgang-Hagen Hein (Stuttgart: Wissenschaftliche Verlagsgesellschaft, 1984), 31–42, esp. 32.

35. Spary, *Utopia's Garden*, 87. See also Anthony Pagden, *European Encounters with the New World* (New Haven: Yale University Press, 1993), 21.

36. Humboldt (and Bonpland), *Personal Narrative*, vol. 5, 256.

37. Ibid., vol. 5, 132.

38. La Condamine, *Relation abrégée*, 74–75.

39. Sloane, *Voyage*, vol. 1, xviii.

40. [Bourgeois], *Voyages*, 487.

41. Philippe Fermin, *Description generale, historique, geographique et physique de la colonie de Surinam*, 2 vols. (Amsterdam, 1769), vol. 1, preface. In their chapters in this volume both Harold Cook and Judith Carney emphasize the contributions of women in other cultures to European pharmacopoeia.

42. [Bourgeois], *Voyages*, 487.

43. Fermin, *Description generale*, vol. 1, 209.

44. Sloane, *Voyage*, vol. 1, liv–v. For Spanish naturalists in New Spain, see Daniela Bleichmar's chapter in this volume.

45. Jaime Jaramillo-Arango, *The Conquest of Malaria* (London: Heinemann, 1950), 79.

46. Hans Sloane, *An Account of a Most Efficacious Medicine for Soreness, Weakness, and Several Other Distempers of the Eyes* (London, 1745), 1.

47. Francisco Guerra, "Drugs from the Indies and the Political Economy of the Sixteenth Century," in *Analecta Medico-Historia* (Oxford: Pergamon Press, 1966), 29–54, esp. 29.

48. C. R. Boxer, *The Dutch Seaborne Empire: 1600–1800* (New York: Knopf, 1965).

49. Thunberg, *Travels*, vol. 2, 286.

50. Sverker Sörlin, "Ordering the World for Europe: Science as Intelligence and Information as Seen from the Northern Periphery," in *Nature and Empire: Science and the Colonial Enterprise*, ed. Roy MacLeod, special issue of *Osiris* 15 (2000): 51–69, esp. 55.

51. Jean-Paul Nicolas, "Adanson et le mouvement colonial," in *Michel Adanson's "Familles des plantes"*, ed. George Lawrence, 2 vols. (Pittsburgh: Carnegie Institute of Technology, 1963), vol. 2, 440.

52. Lisbet Koerner, "Women and Utility in Enlightenment Science," *Configurations* 2 (1995): 233–55, esp. 251.

53. David Mackay, *In the Wake of Cook: Exploration, Science, and Empire, 1780–1801* (London: Croom Helm, 1985), 15.

54. R. G. Lathan, ed., *The Works of Thomas Sydenham*, 2 vols. (London: Sydenham Society, 1848), vol. 1, 82.

55. S. W. Zwicker, *Breviarium apodemicum methodice concinnatum* (Danzig, 1638), cited in Justin Stagl, *A History of Curiosity: The Theory of Travel 1500–1800* (Chur: Harwood Academic Publishers, 1995), 78.

56. Her cure was published in "Mrs. Stephen's Cure for the Stone," *London Gazette*, 16 June 1739, n.p.

57. Zachary Cope, *William Cheselden, 1688–1752* (Edinburgh: Livingstone Ltd., 1953), 24–25.

58. Londa Schiebinger, *The Mind Has No Sex? Women in the Origins of Modern Science* (Cambridge: Harvard University Press, 1989), 237–38.

59. Koerner, "Women and Utility," 250–51.

60. John Douglas, *A Short Account of the State of Midwifery in London, Westminster* (London, 1736), 19.

61. Pierre-Henri-Hippolyte Bodard, *Cours de botanique médicale comparée*, 2 vols. (Paris, 1810), vol. 1, xviii.

62. Ibid., vol. 1, xxx.

63. [Bourgeois], *Voyages*, 460.

64. Pouppé-Desportes, *Histoire*, vol. 3, 59.

Chapter 8. Linnaean Botany and Spanish Imperial Biopolitics

The authors thank the Centro de Estudios Hispanos e Iberoamericanos for its support of the project "Redes, objetos y prácticas en la Ilustración española y americana" CEHI 02/03.

1. John Law, "On the Methods of Long-Distance Control: Vessels, Navigation, and the Portuguese Route to India," in *Power, Action and Belief: A New Sociology of Knowledge?*, ed. John Law (London: Routledge, 1986), 234–62.

2. Paul Carter, *The Road to Botany Bay: An Essay in Spatial History* (London: Faber, 1987).

3. In the reign of Carlos III (1760–88) thirty-three expeditions to the colonies were organized; twenty-four more were organized in the time of Carlos IV (1789–1808). Most of them would contribute to a better knowledge of American flora. See Francisco Javier Puerto Sarmiento, *La ilusión quebrada: Botánica, sanidad y política científica en la España ilustrada* (Barcelona: Serbal, 1988).

4. Antonio Lafuente, "Institucionalización metropolitana de la ciencia española en el siglo XVIII," in *Ciencia colonial en América*, ed. Antonio Lafuente and José Sala Catalá (Madrid: Alianza, 1992), 91–118.

5. Miguel A. Puig-Samper, "Difusión e institucionalización del sistema linneano en España y América," in *Mundialización de la ciencia y cultura nacional*, ed. Antonio Lafuente, Alberto Elena, and María Luisa Ortega (Aranjuez: Doce Calles, 1993), 349–59.

6. Casimiro Gómez Ortega, "Curso Elemental de Botánica," in *Ciencia de Cámara*, ed. Francisco Javier Puerto Sarmiento (1785; Madrid: CSIC, 1992), 262.

7. Francisco Bruno Fernández, *Instrucciones para el bien público y común de la conservación, y aumento de las poblaciones* (Madrid, 1769), 3.

8. Gaston Bachelard, *La formation de l'esprit scientifique: Contribution à une psychanalyse de la connaissance objective* (Paris: Librairie Philosophique J. Vrin, 1972).

9. Bruno Latour, *L'espoir de Pandore: Pour une version réaliste de l'activité scientifique* (Paris: La Découverte, 2001).

10. Donna Haraway's conclusion deals with both the scarcity of data and the ideological dependence on the notion of economic rentability that pervades the history of natural history and biology. See her *Simians, Cyborgs, and Women: The Reinvention of Nature* (New York: Routledge, 1991), 108.

11. Johannes Fabian, *Time and the Other: How Anthropology Makes Its Object* (New York: Columbia University Press, 1983).

12. New Spain refers to modern-day Mexico, Central America, the Caribbean islands, and the states of California, Nevada, Arizona, New Mexico, and Utah.

13. José Antonio Alzate, "Carta satisfactoria dirigida a un literato por . . . sobre lo contenido en la Gaceta de México de 16 de Mayo de 1788," in *Linneo en México: Las controversias sobre el sistema binario sexual, 1788–1798*, ed. Roberto Moreno (Mexico D.F.: UNAM, 1989), 23.

14. The accusation of obscenity against the Linnaean system was fairly common at the time but by no means general. See Londa Schiebinger, "Gender and Natural History," in *Cultures of Natural History*, ed. N. Jardine, J. A. Secord, and E. C. Spary (Cambridge: Cambridge University Press, 1996), 163–77.

15. "I propose devoting myself to their most elegant language in order to show that the [native] names infer their uses, in the same way that Greek does. If, as I hope, I can master it [the Nahuatl language], for this advantage alone I shall be able to be a member of the proposed botanical expedition" (Letter from Martín Sessé to Gómez Ortega, Mexico, 1 May 1786, cited in Xavier Lozoya

Legorreta, ed., *Plantas y luces en México: La Real Expedición Científica a Nueva España, 1787–1803* [Barcelona: Serbal, 1984], 30).

16. José Antonio Alzate, "Botánica," in *Linneo en México*, ed. Moreno, 4.

17. José Hipólito Unanue, "Una idea general del Perú," in José Hipólito Unanue, *Obras científicas y literarias*, 2 vols. (Lima: CEM, 1975), vol. 2, 296.

18. On this controversy, see Jorge Cañizares-Esguerra, *How to Write the History of the New World* (Stanford, Calif.: Stanford University Press, 2001).

19. These theses were defended by Vicente Cervantes's students in public exercises. See "Ejercicios públicos de botánica que tendrán en esta Real y Pontificia Universidad el Bachiller don José Vicente de la Peña, Don Francisco Giles y Arellano y Don José Timoteo Arsinas, dirigiéndolos don Vicente Cervantes, . . . el jueves 11 de Diciembre a las tres de la tarde," in *Linneo en México*, ed. Moreno, 49–52.

20. Cited in José L. Maldonado Polo, *Las huellas de la razón: La expedición científica de Centroamérica (1795–1803)* (Madrid: CSIC, 2001), 289.

21. See Robert J. Shafer, *The Economic Societies in the Spanish World: 1763–1821* (Syracuse, N.Y.: Syracuse University Press, 1958).

22. José Antonio Goycoechea, "Introduction to the Notes," in José Mariano Moliño, *Tratado del Xiquilite y Añil de Guatemala*, annotated by J. A. Goycoechea (1799; San Salvador: Ministerio de Educación, Dirección de Publicaciones, 1976), 36. His complete reply implied that even though there was agreement on the means—the practices and the methods—only those characteristics specified could guarantee the maintenance of uniqueness.

23. Ibid., 40.

24. José Mociño, *Flora de Guatemala de José Mociño*, ed. José L. Maldonado Polo (Madrid: CSIC, 1996), 141.

25. Hipólito Ruiz, *Relación histórica del Viaje que hizo a los Reynos del Peru y Chile . . .* , 2 vols. (Madrid: RACEFN, 1952), vol. 1, 349.

26. Ibid., 113.

27. Juan J. Izquierdo, *Montaña y los orígenes del movimiento social y científico de México D.F.* (Mexico: Ciencia, 1955).

28. Gonzalo Hernández de Alba, *Quinas amargas: El sabio Mutis y la discusión naturalista del siglo XVIII* (Bogotá: Academia de Historia, 1996). See also Jorge Cañizares-Esguerra's chapter in this volume.

29. Antonio Lafuente, "Enlightenment in an Imperial Context: Local Science in the Late-Eighteenth-Century Hispanic World," in *Nature and Empire: Science and the Colonial Enterprise*, ed. Roy MacLeod, special issue of *Osiris* 15 (2000): 155–73.

30. Luis Urteaga, *La tierra esquilmada: Las ideas sobre la conservación de la naturaleza en la cultura española del siglo XVIII* (Madrid: Serbal-CSIC, 1987).

31. Mª Luisa Laviana, "Los intentos de controlar la explotación forestal en Guayaquil: Pugna entre el cabildo y el gobierno colonial," in *Ciencia, vida y espacio en Iberoamérica*, ed. José L. Peset, 3 vols. (Madrid: CSIC, 1989), vol. 2, 406–7.

32. Consuelo Naranjo, "Los reconocimientos madereros en Cuba (1780–1810)," in *El bosque ilustrado: Estudios sobre la política forestal española en América*, ed. Manuel Lucena Giraldo (Madrid: ICONA, 1991), 110.

33. Reproduced in Manuel Rubio Sánchez, *Historia del añil o xiquilite en Centro América*, 2 vols. (San Salvador: Ministerio de Educación, 1976), vol. 1, 127.

34. On the work of Caldas and his relation with Humboldt, see Jorge Cañizares-Esguerra's chapter in this volume.

35. Unanue was emphatic on this point: "In this part of the Torrid Zone which runs along the coast of Peru from the Equator to the Tropic of Capricorn,

we see rising to the east the enormous hills of the Andes range, from whose foothills rise in ranks, one after the other, all the climates of the Universe" (José Hipólito Unanue, "Observaciones sobre el clima de Lima," in *Obras científicas*, vol. 1, 12). Francisco José de Caldas wrote, "It is enough to descend 5000 *varas* [14,000 feet] to go from the moss of the pole to equatorial jungle. Two inches more on the barometer change the face of the empire of flora" (Francisco José de Caldas, "Del influjo del clima sobre los seres organizados," in *Francisco José de Caldas: Un peregrino de las ciencias*, ed. Jeanne Chenu [Madrid: Historia 16, 1992], 311).

36. Caldas, "Del influjo del clima," 311.

37. Francisco José de Caldas, "Estado de la geografía del Virreinato de Santa Fé," in *Francisco José de Caldas*, ed. Chenu, 295.

38. Unanue, "Observaciones," 12.

39. Ibid., 54.

40. Ibid., 66.

41. José Hipólito Unanue, "Disertación sobre la Coca," in *Obras científicas*, vol. 2.

42. Caldas, "Del influjo del clima," 301.

Chapter 9. How Derivative Was Humboldt?

1. Douglass Botting, *Humboldt and the Cosmos* (New York: Harper & Row, 1973); L. Kellner, *Alexander von Humboldt* (New York: Oxford University Press, 1963).

2. Janet Browne, *The Secular Ark: Studies in the History of Biogeography* (New Haven: Yale University Press, 1983).

3. David Brading, *First America, the Spanish Monarchy, Creole Patriots, and the Liberal State, 1492–1867* (Cambridge: Cambridge University Press, 1991), 526–32.

4. See Jorge Cañizares-Esguerra, *How to Write the History of the New World. Histories, Epistemologies, and Identities in the Eighteenth-Century Atlantic World* (Stanford, Calif.: Stanford University Press, 2001).

5. Pablo Vila, "Caldas y los orígenes eurocriollos de la geobotánica," *Revista de la Academia Colombiana de Ciencias* 11 (1960): 16–20.

6. On Caldas, see John Wilton Appel, *Francisco José de Caldas: A Scientist at Work in New Granada* (Philadelphia: American Philosophical Society, 1994). On Mutis, see Marcelo Frías Núñez, *Tras el Dorado vegetal: José Celestino Mutis y la Real Expedición Botánica del Nuevo Reino de Granada (1783–1808)* (Seville: Diputación Provincial de Sevilla, 1994).

7. All these unpublished maps and memoirs by Caldas are kept at the Archive of the Royal Botanical Garden in Madrid, División III, Serie Botánica, Fondo Mutis. The full title of Humboldt's essay is *Essai sur la géographie des plantes: Accompagné d'un tableau physique des régions équinoxiales, fondé sur des mesures exécutées, depuis le dixième degré de latitude boréale jusqu'au dixième degré de latitude australe, pendant les années 1799, 1800, 1801, 1802 et 1803* (Paris: Levrault, Schoell et Compagnie, 1805).

8. Letters by Caldas to Santiago Arroyo, 20 July, 21 September, and 6 October, 1801, in *Francisco José de Caldas: Un peregrino de las ciencias*, ed. Jeanne Chenu (Madrid: Historia 16, 1992) [hereafter *FJC*], 107, 131, 133.

9. For a recent example of this type of patriotic literature, see Jorge Arias de Greif, "Encuentro de Humboldt con la ciencia en la España Americana: Transferencias en dos sentidos," in *El Regreso de Humboldt: Exposición en el Museo de la*

Ciudad de Quito Junio-Agosto del 2001, ed. Frank Holl (Quito: Municipio de Quito, 2001), 33–41.

10. John Murra, *Formaciones económicas y políticas del mundo andino* (Lima: Instituto de Estudios Peruanos, 1975).

11. John M. Prest, *The Garden of Eden. The Botanic Garden and the Re-creation of Paradise* (New Haven: Yale University Press, 1981).

12. Carl Linnaeus, "On the Increase of the Habitable Earth," in *Select Dissertations from the Amoenitates Academicae*, trans. F. J. Brand, 2 vols. (1781; New York: Arno Press, 1977), vol. 1, 71–127.

13. Christopher Columbus, "Tercer Viage," in *The Four Voyages of Columbus: A History in Eight Documents, Including Five by Christopher Columbus in the Original Spanish with English Translations*, ed. and trans. Cecil Jane, 2 vols. (New York: Dover, 1988), vol. 2, 29–47. On paradise in the supralunar sphere, see Charles S. Singleton, "Stars over Eden," *Annual Report of the Dante Society* 75 (1975): 1–18.

14. José de Acosta, *Historia natural y moral de las Indias*, ed. José Alcina Franch (Madrid: Historia 16, 1987), book II, passim, esp. chaps. 12, 14; book III, chaps. 19–20.

15. On marvels and curiosities in the early modern period, see Lorraine Daston and Katherine Park, *Wonders and the Order of Nature, 1150–1750* (New York: Zone Books, 1998).

16. For his views of the Andes as a privileged space, see Antonio de León Pinelo, *El paraiso en el Nuevo Mundo*, ed. Raul Porras Barrenechea, 2 vols. (Lima: Comité del IV Centenario del Descubrimiento del Amazonas, 1943), vol. 1, 307–13, 383–96.

17. Lisbet Koerner, *Linnaeus: Nature and Nation* (Cambridge: Harvard University Press, 1999).

18. Antonio Lafuente and A. Mazuecos, *Los Caballeros del punto fijo* (Barcelona and Madrid: Ediciones del Serbal and CSIC, 1987); Arthur R. Steele, *Flowers for the King* (Durham, N.C.: Duke University Press, 1964); Antonio González Bueno, ed., *Expedición botánica al virreinato del Peru (1777–1788)* (Barcelona: Lunwerg Editores, 1988); Francisco Javier Puerto Sarmiento, *La ilusión quebrada: Botánica, sanidad y política* (Barcelona: Ediciones del Serbal, 1988); Xavier Lozoya, *Plantas y Luces en México* (Barcelona: Ediciones del Serbal, 1984); Iris H. W. Engstrand, *Spanish Scientists in the New World: The Eighteenth-Century Expeditions* (Seattle: University of Washington Press, 1981); Juan Pimentel, *La física de la monarquía* (Madrid: Doce Calles, 1998); Frías Núñez, *Tras el dorado vegetal*.

19. Quoted in Francisco Javier Puerto Sarmiento, *Ciencia de cámara: Casimiro Gómez Ortega (1741–1818): El científico cortesano* (Madrid: Consejo Superior de Investigaciones Científicas, 1992), 155–56.

20. On this expedition, see Frías Nuñez, *Tras el dorado vegetal*.

21. See Mutis's article on quinine in *Diario de Madrid*, 11 November 1880, no. 315; the article is reproduced in *Flora de la Real Expedición Botánica del Nuevo Reino de Granada*, 47 vols. (Madrid: Ediciones de Cultura Hispánica, 1954–   ), vol. 44, 42–43.

22. José Celestino Mutis, "Te de Bogotá," in *Escritos científicos de Don José Celestino Mutis*, ed. Guillermo Hernández de Alba, 2 vols. (Bogotá: Instituto Colombiano de Cultura Hispánica, 1983), vol. 1, 177. On this project, see Marcelo Frías Núñez, "El té de Bogotá: Un intento de alternativa al té de China," in *Nouveau Monde et renouveau de l'histoire naturelle*, ed. Marie-Cécile Bénassy-Berling, 3 vols. (Paris: Presses de la Sorbonne Nouvelle, 1986–94), vol. 3, 201–19.

23. Pedro Fermín de Vargas, "Memoria sobre la población del reino de

Nueva Granada" (c. 1790), in *Pensamientos políticos y memoria sobre la población del Nuevo Reino de Granada* (Bogotá: Biblioteca Popular de Cultura Colombiana, 1944), 6 (quotation).

24. Francisco Antonio Zea, "Avisos de Hebephilo a los jóvenes de los dos colegios sobre la inutilidad de sus estudios presentes, necesidad de reformarlos, elección y buen gusto en los que deben abrazar," *Papel periódico de la ciudad de Santafé de Bogotá,* no. 9, 8 April 1791, 68.

25. Sabio patriota, "Al señor autor del periódico," *Papel periódico,* no. 11, 22 April 1791, 81. Many other articles emphasized the same theme of New Granada as microcosm of the world; for example, see Observador amigo del país, "Discurso," ibid., 86; Luis de Atigarraga, "Disertación sobre la agricultura dirigida a los habitantes del Nuevo Reyno de Granada," ibid., no. 56, 9 March 1792, 36–37; Diego Martín Tanco, "Discurso por el cual se manifiestan los medios de aumentar la población de este reyno," ibid., no. 76, 27 July 1792, 197; "Idea del nuevo Reyno de Granada," ibid., no. 256, 12 August 1796, 1537–38.

26. Francisco José de Caldas, "Estado de la geografía en el virreinato de Santa Fe de Bogotá con relación a la economía y al comercio (1808)," *FJC,* 276.

27. Francisco José de Caldas, "Influjo del clima sobre los seres organizados (1808)," in *Obras completas* (Bogotá: Universidad de Colombia, 1966), 112 and Caldas, "Ensayo sobre el estado de la geografía," *FJC,* 275.

28. Caldas, "Estado de la geografía en el virreinato," *FJC,* 276–77.

29. Tadeo Lozano quoted in Gonzalo Hernández de Alba, *Quinas Amargas; El sabio Mutis y la discusión naturalista del siglo XVIII* (Bogotá: Academia de Historia de Bogotá and Tercer Mundo Editores, 1991), 148.

30. Hipólito Unanue, "Geografía física del Peru," *Mercurio Peruano,* no. 4 (1792): 11; see also ibid., 16. Peru as a temple of God has its facade to the north: its dome is the celestial vault at the equator; its columns are the mountains; and its perpetual lamps are the volcanoes.

31. Unanue, "Geografía física," 21. Like Caldas, Unanue thought that the Andes were a privileged laboratory to study the influence of climate on humans. See *Observaciones sobre el cuma de Lima y su influencia en los seres organizados en especial el hombre,* in *Los ideólogos: Hipolito Unanue,* ed. Jorge Arias-Schreiber Pezet, 8 vols. (Lima, 1974), vol. 8, 47–171.

32. Unanue, "Geografia física," 22–26. For a detailed analysis of Unanue's views, see Jorge Cañizares-Esguerra, "La utopía de Hipólito Unanue: Comercio, naturaleza, y religión en el Perú," in *Saberes andinos: Ciencia y tecnología en Bolivia, Ecuador y Peru,* ed. Marcos Cueto (Lima: Instituto de Estudios Peruanos, 1995), 91–108.

33. Hipólito Unanue, "Disertación sobre el aspecto, cultivo, comercio y virtudes de la famosa planta del Perú nombrada Coca," *Mercurio Peruano,* no. 11 (1794): 241–45.

34. Pedro Antonio Cerviño, "El tridente de neptuno es el cetro del mundo: Discurso inaugural de la Academia Naútica, del 25 noviembre de 1799," reproduced in José Carlos Chiaramonte, *La Ilustración en el Río de la Plata: Cultura eclesiástica y cultura laica durante el Virreinato* (Buenos Aires: Punto Sur Editores, 1989), 295.

35. Francisco Antonio Caballe, "Continua la idea general del comercio de las provincias del Rio de la Plata," *Telégrafo Mercantil, Rural, Politico-economico, e Historiografo del Rio de la Plata,* vol. 4 (11 April 1801), cited in Chiaramonte, *Ilustración,* 227, 229 (on the privileged central position of Buenos Aires).

36. Francisco Antonio Caballe, "Comercio," *Seminario de Agricultura,* vol. 4 (13 October 1802), Chiaramonte, *Ilustración,* 266–67; Francisco Antonio Ca-

balle, "Agricultura," *Seminario de Agricultura*, vol. 1 (1 September 1802), Chiara-monte, *Ilustración*, 254.

37. Juan Manuel de San Vicente, *Exacta descripción de la magnífica corte mexi-cana, cabeza del nuevo americano mundo* (Cadiz, 1768); reproduced in *Anales del Museo Nacional de Antropología de Mexico*, 3ra época, vol. 5 (1913), 32, 34.

38. Juan Manuel Venegas, *Compendio de la medicina o medicina práctica* (Mexico City, 1788), advertencia (n.p.).

39. José Mariano Mozino, quoted in E. Trabulse, *Historia de la ciencia en México (version abreviada)* (Mexico City: Fondo de Culture Económica 1994), 116–17.

Chapter 10. The Conquest of Spice and the Dutch Colonial Imaginary

1. Kristof Glamman, *Dutch-Asiatic Trade, 1620–1740* (Oxford: Clarendon Press, 1989). See also Kees Zandvliet, *The Dutch Encounter with Asia 1600–1950* (Zwolle: Waanders, 2002).

2. For a directory to primary documents, see John Landwehr, *VOC: A Bibliog-raphy of Publications Relating to the Dutch East India Company, 1602–1800* (Utrecht: HES, 1991). For the catalog of the exhibition commemorating the company's four-hundredth anniversary, see Leo Akveld and Els M. Jacobs, eds., *The Colour-ful World of the VOC* (The Hague: Thoth Publishers Bussum, 2002).

3. "Punten en Artikelen in Form van Generale Instructie," Amsterdam, 26 April 1650, in *Verzameling van Instructien, Ordonnancien en Reglementen voor de Re-gering van Nederlandsch Indië, 1609–1836*, ed. P. Myer (Batavia, 1848), 71–116, cited in translation by Charles Ralph Boxer in his *The Dutch Seaborne Empire: 1600–1800* (London: Hutchinson, 1965), 95.

4. On still lifes in Dutch household inventories, see Julie Berger Hochstras-ser, "Imag(in)ing Prosperity: Still Life and Material Culture in the Seventeenth-century Dutch Household," in *Wooncultuur in de Nederlanden/The Art of Home in the Netherlands 1500–1800, Netherlands Yearbook for History of Art*, vol. 51, ed. Jan de Jong, Bart Ramakers, Herman Roodenburg, Frits Scholten, and Mariët Wester-mann (Zwolle: Waanders Uitgevers, 2000).

5. Johan van Beverwyck, *Schat der Gesontheyt* (Dordrecht: Hendrick van Esch, 1636), 135–36; republished numerous times over the course of the century.

6. For a more comprehensive account of this and other examples, see Julie Berger Hochstrasser, "Life and Still Life: A Cultural Inquiry into Seventeenth-Century Dutch Still-Life Painting" (Ph.D. thesis, University of California, Berke-ley, 1995), chap. 1.

7. For example, Floris van Schooten, Frans Halsmuseum, Haarlem, and Floris van Dijck, signed and dated 1622, private collection, Amsterdam; reproduced in N. R. A. Vroom, *A Modest Message as Intimated by the Painters of the Monochrome Banketje*, trans. Peter Gidman, 2 vols. (Schiedam, the Netherlands: Interbook In-ternational, 1980). Claesz. alone painted more than twenty-one banquets with pepper signed and dated between 1624 and 1657—roughly spanning the period of high and even peak prices for pepper on the market and (interestingly) ceas-ing when the prices bottomed out. A particularly grand early example is signed and dated 1628 (Rijksmuseum, Amsterdam). Another Heda banquet with pep-per, dated the same year as our illustrated example in 1635 (National Gallery, Washington, D.C.), closely mirrors the pyramidal composition of the Rijksmu-seum Heda, again pairing oysters and pepper. See Vroom, *Banketje*, for reproductions.

8. One of two versions of Kalf's painting is in the Musée Picardie, Amiens France.

9. See, for instance, Norman Bryson, *Looking at the Overlooked: Four Essays on Still Life Painting* (Cambridge: Harvard University Press, 1990), chap. 4, "Abundance."

10. Josiah Child, *A New Discourse of Trade*, 4th ed. (London, n.d.), first published in 1694 but originally drafted in 1669, as Child states in the preface.

11. Denis Diderot, *Voyages en Hollande* (Paris: F. Maspero, 1982), 68.

12. Quoted in Boxer, *Dutch Seaborne Empire*, 28.

13. Jacob Cats, in van Beverwyck, *Schat der Gesontheyt* (1672 ed.), 133–34. All translations are mine.

14. Ibid.

15. Ibid.

16. Ibid.

17. For a useful overview of the extensive literature, see the annotated bibliography by Jeanie M. Welch, *The Spice Trade: A Bibliographic Guide to Sources of Historical and Economic Information* (Westport, Conn.: Greenwood Press, 1994), esp. 53–116, covering the age of exploration and colonialism.

18. Quoted in Boxer, *Dutch Seaborne Empire*, 22–23.

19. Ibid. Compare the painting by Hendrick Cornelis Vroom (1566–1640), *Return of the Homecoming Fleet, July 1599* (Amsterdam Historical Museum).

20. Jan de Vries and Ad van der Woude, *The First Modern Economy: Success, Failure, and Perseverance of the Dutch Economy, 1500–1815* (Cambridge: Cambridge University Press, 1997), 453, where their estimate was one in three. In the Dutch edition their estimate was a more severe one in two; see Jan de Vries and Ad van der Woude, *Nederland 1500–1815: De Eerste Ronde van Moderne Economische Groei* (Amsterdam: Uitgeverij Balans, 1995), 525.

21. Pieter Geyl, *The Netherlands in the Seventeenth Century, Part One, 1609–1648* (London: Ernest Benn, 1964), 168. He reports that Arnold de Vlamingh van Outshoorn had carried out the ruthless campaign that first concentrated the cultivation of cloves on Ambon and surrounding islands, exterminating trees and population alike on other islands.

22. Coen served as governor-general from 1618 to 1623 and again from 1627 to 1629.

23. Boxer, *Dutch Seaborne Empire*, 98.

24. Ibid.

25. Compare Anthony Reid, *Southeast Asia in the Age of Commerce, 1450–1680*, vol. 2, *Expansion and Crisis* (New Haven: Yale University Press, 1993), 274; and Akveld and Jacobs, who add that one potential motive for Coen's ruthlessness was revenge: as a young man he witnessed the murder of his commander Pieter Verhoeff in an ambush on Banda during negotiations. *Colourful World of the VOC*, 113.

26. Quoted in Geyl, *Netherlands*, 178. Compare the devastating assessment of Boxer, *Dutch Seaborne Empire*, 102, and the equally grim, if less sensationalized account of J. L. van Zanden in *The Rise and Decline of Holland's Economy: Merchant Capitalism and the Labour Market* (Manchester: Manchester University Press, 1993), sect. IV, "Forced Labour in the Periphery: The V.O.C. in the Moluccas," 67–87. On the islands of Lontor, Banda Neira, and Ai, the Dutch constructed nutmeg plantations with slaves from Java, Celebes, and Timor taught by the Bandanese slaves. See Akveld and Jacobs, *Colourful World of the VOC*, 114.

27. Long dominated by China, trade at Bantam was also conducted by English and Danish merchants; Sultan Ageng of Bantam had repeatedly attacked

Batavia in an effort to drive off the Dutch. In a 1680 power struggle, the VOC instead supported crown prince Haji, promising military support provided he would close the port to their European rivals. By 1684 they had concluded a contract closing Bantam to European competitors and requiring the new Sultan to provide all pepper to the VOC at low prices. See Akveld and Jacobs, *Colourful World of the VOC,* 117.

28. A year after the English had been driven out of Bantam, they established a new outpost at Benkulu on the southwest coast of Sumatra. The VOC also pursued futile military initiatives among the widely scattered pepper-producing regions along the Malabar Coast of India. Elsewhere in India, as in China, Japan, Siam (Thailand), and Persia (Iran), the power of local rulers obliged the Dutch instead to negotiate. See Akveld and Jacobs, *Colourful World of the VOC,* 18.

29. It proved "extremely costly in men and money" (Boxer, *Dutch Seaborne Empire,* 98). In the end the Malabar Coast fell to the English as well, although the tiny English outpost on the nutmeg-producing island of Run in the Banda archipelago was finally and famously traded back to the Dutch in exchange for New Netherland (present-day Manhattan). See also Giles Milton, *Nathaniel's Nutmeg, or, the True and Incredible Adventures of the Spice Trader Who Changed the Course of History* (New York: Farrar, Straus and Giroux, 1999).

30. Geyl, *Netherlands,* 182. Scathing indictments from Indonesian sources confirm Reael's allegation; see those cited in Reid, *Southeast Asia,* vol. 2, 270–71.

31. Van Beverwyck, *Schat der Gesontheyt* (1636), 116.

32. The most recent of numerous editions of Linschoten's *Itinerario* is *Jan Huygen van Linschoten and the Moral Map of Asia: The Plates and Text of the Itinerario and Icones* (London: for members of the Roxburghe Club, 1999), including a useful introduction by Ernst van den Boogaart now reprinted as *Civil and Corrupt Asia: Image and Text in the Itinerario and the Icones of Jan Huygen van Linschoten* (Chicago: University of Chicago Press, 2003).

33. Linschoten, *Itinerario,* 57.

34. Linschoten, *Itinerario,* 81. Perhaps this explains the wanton exploitation Westerners exercised upon their arrival: the VOC resorted to wholesale destruction of plants in their attempt to maintain monopoly control of spices. On the other hand, Richard Grove maintains that the economic motivation on the part of colonial powers (including the Dutch) to preserve and manage the microecologies of colonial settlements, most especially on islands, stimulated the awakening of environmental consciousness; see Richard Grove, *Green Imperialism: Colonial Expansion, Tropical Island Edens and the Origins of Environmentalism, 1600–1860* (Cambridge: Cambridge University Press, 1995).

35. Oddly, these so-called figs are actually bananas, complete with peels; likewise, the tree labeled fig is a banana tree, obviously lacking the deeply lobed leaves of a fig. Compare van den Boogaart, *Jan Huygen van Linschoten and the Moral Map of Asia,* 225.

36. Perhaps this Edenic vision swayed even the botanical identification of the "Indian fig."

37. Johan Nieuhof, *Beschryving van't Gesandschap der Nederlandsche Oost-Indische Compagnie aen Den Grooten Tartarischen Cham nu Keyser van China* (Amsterdam: Jacob van Meurs, 1665). English edition: *An Embassy from the East India Company of the United Provinces to the Grand Tartar Cham Emperor of China,* trans. John Ogilby (London: John Macock, 1669; reprint Menston: Scolar Press Limited, 1975).

38. Nieuhof, *China,* 259.

39. Ibid.

40. Johan Nieuhof, *Gedenkwaerdige Zee en Lantreize door de Voornaemste Landschappen van West en Oostindien* (Amsterdam: widow of Jacob van Meurs, 1682); in English, see Nieuhof, *Voyages and Travels to the East Indies, 1653–1670,* with an introduction by Anthony Reid (London: Awnsham and John Churchill, 1704, reprinted 1732; reprint, Oxford: Oxford University Press, 1988), for example 161–66.

41. The clavier was painted by Pieter Isaacsz, possibly from a design by Karel van Mander, 74.4 × 165 cm (Rijksmuseum, Amsterdam), reproduced in Akveld and Jacobs, *Colourful World of the VOC,* 39–41.

42. The cog-ship, the vessel used in Dutch trade in the Middle Ages, appears also in the windvane on the Town Hall's cupola. Van Campen's design for the back pediment dates from c. 1648 (Rijksprentenkabinet, Amsterdam), while the sculptures were completed between 1650 and 1665. See Katharine Fremantle and Willy Halsema-Kubes, eds., *Beelden Kijken: De Kunst van Quellien in het Paleis op de Dam/ Focus on Sculpture: Quellien's Art in the Palace on the Dam* (Hilversum: De Jong & Co., 1977), esp. 4; and Katherine Fremantle, *The Baroque Town Hall of Amsterdam* (Utrecht: Haentjens Dekker & Gumbert, 1959).

43. Olfert Dapper, *Historische Beschryvinghe van Amsterdam* (Amsterdam: Jacob van Meurs, 1663). Compare also Filips von Zesen's German *Beschreibung der Stadt Amsterdam* (Amsterdam: M. W. Doornik, 1664). For a parallel discussion of allegorical images of America, reproducing the title page of Abraham Ortelius, *Theatrum orbis terrarum* (Antwerp: Plantin, 1570), in which the "maid" representing Europe rules like a seated monarch over figures representing the other continents, see Benjamin Schmidt, *Innocence Abroad: The Dutch Imagination and the New World, 1570–1670* (Cambridge and New York: Cambridge University Press, 2001), 127, fig. 18.

44. Richard Grove makes a parallel argument with regard to mapping: "Cartographically, what was left out of the survey became just as important as what was included in the map. On Tobago the Caribs were left out of the map-making process altogether and within twenty years had disappeared as a separate population" (Grove, *Green Imperialism,* 283, with thanks to Londa Schiebinger).

45. Elsewhere I too forward other parallel arguments, for example regarding Frans Post's renderings of sugar mills in Dutch Brazil, which avoid the grisly realities of mangling and mortality of African slaves; see Julie Berger Hochstrasser, *Still Life and Trade in the Dutch Golden Age* (London: Yale University Press, 2004).

46. Akveld and Jacobs make the former claim (*Colourful World of the VOC,* 117); on failed profits, see de Vries and van der Woude, *First Modern Economy,* 536–37.

47. Boxer, *Dutch Seaborne Empire,* 102.

48. See van Zanden, *Rise and Decline of Holland's Economy,* 171, passim.

49. Karl Marx, *Capital: A Critique of Political Economy,* ed. Frederick Engels, trans. Samuel Moore and Edward Aveling (New York: Charles H. Kerr, 1967 [1867]), 535.

50. On this, see Charles Corn, *The Scents of Eden: A Narrative of the Spice Trade* (Tokyo: Kodansha International, 1998). Less even-handed in its open partiality to the English is Milton, *Nathaniel's Nutmeg,* also narrated in the style of historical fiction. Foremost in the scholarly history is the extensive work of Reid, particularly that cited above, note 25. Literary criticism too has unmasked the political and economic agendas permeating the spice trade, see Timothy Morton, *The Poetics of Spice: Romantic Consumerism and the Exotic* (Cambridge: Cambridge University Press, 2000).

51. See for example Tina Rosenberg, "The Free-Trade Fix," *The New York*

Times Magazine, 18 August 2002, section 6 cover, and 28–33, 50, 74–75. The cover text reads: "have not—a way to make globalization work for everybody else."

52. Roland Barthes, "The World as Object," in *Critical Essays* (Evanston: Northwestern University Press, 1972), 3–12, esp. 5.

53. Examples comparable to the continued subjugation of the Banda nutmeg trade may be readily multiplied throughout recent memory, in the widespread problems of migrant and sweatshop labor. For a report on contemporary exploitation and child labor in the cocoa-producing heartland of the Ivory Coast, see Norimitsu Onishi, "The Bondage of Poverty that Produces Chocolate," *New York Times*, 29 July 2001, sect. 1, 1.

54. See Manus van Brakel and Maria Buitenkamp, *Sustainable Netherlands: A Perspective for Changing Northern Lifestyles* (Amsterdam: Friends of the Earth, 1992).

55. See Alex Hittle, *The Dutch Challenge: A Look at How the United States' Consumption Must Change to Achieve Global Sustainability* (Washington, D.C.: Friends of the Earth, 1994).

56. While he implicates other developed nations too, Rees charges that those with the highest population density are "living even further beyond their own ecological means." See William E. Rees and Mathis Wackernagel, "Ecological Footprints and Appropriated Carrying Capacity: Measuring the Natural Capital Requirements of the Human Economy," in *Investing in Natural Capital: The Ecological Economics Approach to Sustainability*, ed. A-M. Jannson et al. (Washington, D.C.: Island Press, 1994), 374; also cited in David C. Korten, *When Corporations Rule the World* (Bloomfield, Conn.: Kumarian Press, and San Francisco: Berrett-Koehler Publishers, 2001), 40 and note 13.

Chapter 11. Of Nutmegs and Botanists

1. For example, Alfred W. Crosby Jr., *The Columbian Exchange: Biological and Cultural Consequences of 1492* (Westport, Conn.: Greenwood Publishing Company, 1972).

2. James Walvin, *Fruits of Empire: Exotic Produce and British Taste, 1660–1800* (Basingstoke and London: Macmillan, 1997); Henry Hobhouse, *Seeds of Change: Five Plants That Transformed Mankind* (London: Sidgwick & Jackson, 1985).

3. For example, Steven Shapin, "Pump and Circumstance: Robert Boyle's Literary Technology," *Social Studies of Science* 14 (1984): 481–520.

4. Geoffrey C. Bowker and Susan Leigh Star, *Sorting Things Out: Classification and Its Consequences* (Cambridge, Mass.: MIT Press, 1999); Frans A. Stafleu, *Linnaeus and the Linnaeans: The Spreading of Their Ideas in Systematic Botany* (Utrecht: Oosthoek, 1971), chap. 9.

5. On explanatory asymmetry, see Simon Schaffer and Steven Shapin, *Leviathan and the Airpump: Hobbes, Boyle and the Experimental Life* (Princeton: Princeton University Press, 1985), chaps. 1 and 8.

6. For the Dutch spice monopoly, see Julie Berger Hochstrasser's chapter in this volume.

7. See Pierre Poivre, *Oeuvres complettes* (Paris, 1797), 279–80; Bernard de Jussieu and Michel Adanson, "Rapport fait à l'académie des sciences, sur le transport des plants de cannelliers et de gérofliers, à l'Isle de France," *Histoire de l'Académie Royale des Sciences* 1772 (Paris, 1775), 56–61. My analysis draws on Sid-

ney W. Mintz, *Sweetness and Power: The Place of Sugar in Modern History* (New York: Viking, 1985).

8. Dupont's biography is in Poivre, *Oeuvres*, 3–71. Similarly, see Richard H. Grove, *Green Imperialism: Colonial Expansion, Tropical Island Edens and the Origins of Environmentalism, 1600–1860* (Cambridge: Cambridge University Press, 1995), chap. 5.

9. Jean-Baptiste-Christophe Fusée Aublet, *Histoire des plantes de la Guiane Françoise, rangées suivant la méthode sexuelle,* 4 vols. (Paris, 1775), vol. 1, vij; for Aublet's background, see the preface and Madeleine Ly-Tio-Fane, *The Odyssey of Pierre Poivre* (Port Louis, Mauritius: Esclapon Ltd., 1958), introduction.

10. Louise Audelin, "Les Jussieu, une dynastie de botanistes au XVIIIe siècle (1680–1789)" (diss., École des Chartes, Paris, 1987).

11. Aublet, *Histoire,* vol. 1, vij. On Aublet and Poivre's dispute, see Louis Malleret, *Pierre Poivre* (Paris: École Française d'Extrême-Orient, 1974), chap. 6.

12. Malleret, *Poivre,* 110, 199–210; Pierre Poivre, "Relation abrégée des voyages faits par le Sieur P*** pour le Service de la Compagnie des Indes," *Revue de l'histoire des colonies françaises* 1918, 1st trimester, 8–86.

13. For their political affiliations, see Michèle Duchet, *Anthropologie et histoire au siècle des lumières* (Paris: François Maspero, 1971), 118–34; Malleret, *Poivre,* 213.

14. Aublet, *Histoire,* vol. 4, 89, 91.

15. Ibid., 91, 85–87.

16. Keith Michael Baker, *Inventing the French Revolution: Essays on French Political Culture in the Eighteenth Century* (Cambridge: Cambridge University Press, 1990).

17. Maurice Linÿer de la Barbée, *Le Chevalier de Ternay,* 2 vols. (Grenoble: Editions des 4 Seigneurs, 1972), vol. 1, part 3.

18. Malleret, *Poivre,* 405–11, 445–54.

19. Ibid., 635–37; de Jussieu and Adanson, "Rapport fait à l'académie des sciences," 61.

20. For other attempts, see Ch. de La Roncière, "Le routier inédit d'un compagnon de Bougainville," *La Géographie* 35 (1921): 217–50, 243; Malleret, *Poivre,* 618.

21. Guillaume-Joseph-Hyacinthe-Jean-Baptiste Le Gentil de la Galaisière, *Voyage dans les Mers de l'Inde, fait par ordre du Roi,* 2 vols. (Paris, 1779), vol. 1, 689–90; Malleret, *Poivre,* 645–48.

22. Madeleine Ly-Tio-Fane, *The Triumph of Jean Nicolas Céré and His Isle Bourbon Collaborators* (Paris: Mouton & Co, 1970).

23. Aublet, *Histoire,* vol. 4, "Observations sur la Vanille," 93.

24. Madeleine Ly-Tio-Fane, *Pierre Sonnerat 1748–1814: An Account of His Life and Work* (Cassis, Mauritius: Imprimerie & Papeterie Commerciale, 1976), 55, 60, 95; Pierre Sonnerat, *Voyage à la Nouvelle Guinée* (Paris, 1776), 194–96.

25. Malleret, *Poivre,* 664–67; Henri-Alexandre Tessier, "Mémoire sur l'importation du Géroflier des Moluques aux Isles de France de Bourbon & de Sechelles, & de ces Isles à Cayenne," *Observations sur la Physique, sur l'Histoire Naturelle et sur les Arts* 14, no. 2 (1779): 47–54; idem, "Mémoire sur l'importation et les progrès des arbres à épicerie dans les colonies françaises," in *Mémoires de l'Académie Royale des Sciences* 1789 (Paris, 1794), 585–96; Malleret, *Poivre,* 669–71.

26. Bruno Latour, *Science in Action: How to Follow Scientists and Engineers through Society* (Milton Keynes: Open University Press, 1987).

27. Even Dupont (Poivre, *Oeuvres,* 42–43) denied Poivre's standing as a bota-

nist. For "empire and botany" accounts, see David Philip Miller and Peter Hanns Reill, eds., *Visions of Empire: Voyages, Botany, and Representations of Nature* (Cambridge: Cambridge University Press, 1996); Grove, *Green Imperialism*; Richard Drayton, *Nature's Government: Science, Imperial Britain and the "Improvement" of the World* (New Haven: Yale University Press, 2000).

Chapter 12. Out of Africa

I would like to thank Starr Douglas, at Royal Holloway, University of London, for providing the Smeathman quote from her research in the Swedish archives. Additional appreciation is extended to Bruce Mouser, Tony Tibbles, Marlène Elias, Philippe Pétout, Kåre Lauring, and the volume editors.

1. Henry Dethloff, *A History of the American Rice Industry, 1685–1985* (College Station: Texas A&M University Press, 1988).

2. James Clifton, "The Rice Industry in Colonial America," *Agricultural History* 55 (1981): 266–83; Peter Coclanis, *The Shadow of a Dream* (New York: Oxford University Press, 1989).

3. David Doar, *Rice and Rice Planting in the South Carolina Low Country* (1936; Charleston, S.C.: Charleston Museum, 1970).

4. Peter H. Wood, *Black Majority* (New York: Norton, 1974), 57–58.

5. W. Andriesse and L. O. Fresco, "A Characterization of Rice-Growing Environments in West Africa," *Agriculture, Ecosystems and Environment* 33 (1991): 377–95.

6. Gomes Eanes de Azurara, *The Chronicle of the Discovery and Conquest of Guinea*, 2 vols. (London: Hakluyt Society, 1899), vol. 2, 263–64.

7. G. Crone, *The Voyages of Cadamosto* (London: Hakluyt Society, 1937), 70; Walter Rodney, *A History of the Upper Guinea Coast, 1545 to 1800* (New York: Monthly Review Press, 1970), 21; A. Carreira, *Os portugueses nos rios de Guiné, 1500–1900* (Lisbon, self-published, 1984), 27–28.

8. O. Ribeiro, *Aspectos e problemas da Expansão Portuguésa* (Lisbon: Estudos de Ciencias Políticas e Sociais, 1962), 143–47; T. B. Duncan, *Atlantic Islands: Madeira, the Azores, and the Cape Verdes in Seventeenth Century Commerce and Navigation* (Chicago: University of Chicago Press, 1972), 168; Carreira, *Portuguêses nos rios*, 47–62; George Brooks, *Landlords and Strangers: Ecology, Society and Trade in Western Africa, 1000–1630* (Boulder, Colo.: Westview Press, 1993), 130–47, 279.

9. Brooks, *Landlords and Strangers*, 149.

10. John Atkins, *Voyage to Guinea, Brasil, and the West Indies* (London, 1735); Francis Moore, *Travels into the Inland Parts of Africa* (London: Edward Cave, 1738), 165–82; Theodore Canot, *Adventures of an African Slaver* (New York: Albert & Charles Boni, 1928); Rodney, *History of the Upper Guinea Coast*, 21; Philip Curtin, *Economic Change in Pre-Colonial Africa* (Madison: University of Wisconsin Press, 1975), 100–111.

11. Paul Richards, "Culture and Community Values in the Selection and Maintenance of African Rice," in *Valuing Local Knowledge: Indigenous People and Intellectual Property Rights*, ed. S. Brush and D. Stabinsky (Washington, D.C.: Island Press, 1996), 209–29, esp. 213–14.

12. Valentim Fernandes (c. 1506–10) in Th. Monod, A. Teixeira da Mota, and R. Mauny, eds., *Description de la Côte Occidentale d'Afrique* (Bissau: Centro de Estudos da Guiné Portuguêsa, Guinea Bissau, 1951), 47; Diogo Gomes (c. 1456) in Th. Monod, R. Mauny, and G. Duval, eds., *De la Première Découverte de la Guinée:*

Récit par Diogo Gomes (Fin XV siècle) (Bissau: Centro de Estudos da Guiné Portuguêsa, 1959), 42, 66.

13. Rodney, *History of the Upper Guinea Coast*, 20–21.

14. Daniel C. Littlefield, *Rice and Slaves* (Baton Rouge: Louisiana State University Press, 1981), 93–95.

15. Moore, *Travels*, 37.

16. Brooks, *Landlords and Strangers*, 318.

17. Richard Jobson, *The Golden Trade* (1623; Devonshire, U.K.: Speight and Walpole, 1904), 68.

18. P. E. H. Hair, Adam Jones, and Robin Law, eds., *Barbot on Guinea: The Writings of Jean Barbot on West Africa, 1678–1712*, 2 vols. (London: The Hakluyt Society, 1992), vol. 1, 186.

19. Moore, *Travels*, 127.

20. Littlefield, *Rice and Slaves*.

21. Ribeiro, *Aspectos e problemas*.

22. Ibid.

23. The sole archaeological excavation on *glaberrima* shows the cereal present between 300 B.C. and 300 A.D. in Mali. Older dates derive from research in botany and historical linguistics. See Susan K. McIntosh, "Paleobotanical and Human Osteological Remains," in *Excavations at Jenne-jeno, Hambarketolo and Kaniana in the Inland Niger Delta (Mali)*, ed. S. K. McIntosh (Berkeley: University of California Press, 1994), 348–53; Roland Portères, "African Cereals: Eleusine, Fonio, Black Fonio, Teff, Brachiaria, Paspalum, Pennisetum, and African Rice," in *Origins of African Plant Domestication*, ed. J. Harlan, J. De Wet, and A. Stemler (The Hague: Mouton, 1976), 409–52; Christopher Ehret, *The Civilizations of Africa: A History to 1800* (Charlottesville: University Press of Virginia, 2002).

24. Wood, *Black Majority*, 57–58.

25. Ibid., 25–26, 36, 62, 143–45.

26. A. S. Salley, "Introduction of Rice into South Carolina," in *Bulletin of the Historical Commission of South Carolina*, vol. 6 (Columbia, S.C.: The State Company, 1919), 11.

27. While *glaberrima* rice is always red in color, some *sativa* rice varieties also are.

28. Judith Carney, *Black Rice: The African Origins of Rice Cultivation in the Americas* (Cambridge: Harvard University Press, 2002), 147–52.

29. P. Collinson, "Of the Introduction of Rice and Tar in Our Colonies," *Gentleman's Magazine* 36 (June 1766): 278–80.

30. Hair et al., *Barbot on Guinea*, vol. 1, 282 n. 13, vol. 2, 681; G. Dow, *Slave Ships and Sailing* (Salem: Marine Research Society, 1927), 57, 73; Boubacar Barry, *Senegambia and the Atlantic Slave Trade* (Cambridge: Cambridge University Press, 1998), 79, 107–8, 117–18; Bruce Mouser, ed., *Slaving Voyage to Africa and Jamaica: The Log of the Sundown, 1793–1794* (Bloomington: Indiana University Press, 2002), 90 n. 295.

31. Bruce L. Mouser, "Who and Where Were the Baga? European Perceptions from 1793 to 1821," *History in Africa* 29 (2002): 337–64, esp. 357 n. 42.

32. Mouser, *Slaving Voyage*, 45 n. 170, 86 n. 282, 99 n. 317; Mouser, "Who and Where," esp. 357 n. 42.

33. Carney, *Black Rice*, 25–27, 48–52.

34. Elizabeth Donnan, *Documents Illustrative of the History of the Slave Trade to America*, 4 vols. (1930–35) (Washington, D.C.: Carnegie Institution, 1932), vol. 3, 121, 376.

35. Dow, *Slave Ships*, xxiii–xxiv.

36. African rice shatters with mechanical milling. See National Research Council, *Lost Crops of Africa* (Washington, D.C.: National Academy of Science, 1996).

37. Smeathman to Drury, Sierra Leone, 10 July 1773, MS D.26, Uppsala University, Sweden.

38. Mouser, "Baga," 8 n. 32; George Howe, "Last Slave Ship," *Scribner's Magazine* 8, no. 1 (July 1890): 113–29; Dow, *Slave Ships*, xxii.

39. Robert Harms, *The Diligent: A Voyage through the Worlds of the Slave Trade* (New Haven: Yale University Press, 2002), 311–12.

40. The cereal being milled may be millet. See Leif Svalesen, *The Slave Ship Fredensborg* (Boomington: Indiana University Press, 2000), 107.

41. John Drayton, *A View of South Carolina* (1802; Columbia: University of South Carolina Press, 1972), 125.

42. Theophilus Conneau, *A Slaver's Log Book or 20 Years' Residence in Africa* (Englewood Cliffs, N.J.: Prentice-Hall, 1976), 239.

43. Salley, "Introduction of Rice," 172.

44. Wood, *Black Majority*, 30–32, 105–14; Terry Jordan, *Trails to Texas: Southern Roots of Western Cattle Ranching* (Lincoln: University of Nebraska Press, 1981).

45. Wood, *Black Majority*, 143–45.

46. Sam B. Hilliard, "Antebellum Tidewater Rice Culture in South Carolina and Georgia," in *European Settlement and Development in North America: Essays on Geographical Change in Honour and Memory of Andrew Hill Clark*, ed. James Gibson (Toronto: University of Toronto Press, 1978); David Whitten, "American Rice Cultivation, 1680–1980: A Tercentenary Critique," *Southern Studies* 21, no. 1 (1982): 215–26.

47. *South Carolina Gazette*, Charleston, 19 January 1738.

48. Clifton, "Rice Industry," 276.

49. R. F. W. Allston, "Essay on Sea Coast Crops," *De Bow's Review* 16 (1854): 589–615; Whitten, "American Rice," 9–15.

50. R. F. W. Allston, "Memoir of the Introduction and Planting of Rice in South Carolina," *De Bow's Review* 1 (1846): 320–57.

51. Clifton, "Rice Industry," 273.

52. D. Richardson, "The British Slave Trade to Colonial South Carolina," *Slavery and Abolition* 12 (1991): 125–72.

53. Dale Rosengarten, "Social Origins of the African-American Lowcountry Basket" (Ph.D. diss., Harvard University, 1997), 273–311.

54. Karen Hess, *The Carolina Rice Kitchen: The African Connection* (Columbia: University of South Carolina Press, 1992).

55. José Almeida Pereira, *Cultura do arroz no Brasil* (Teresina, Brazil: EMBRAPA, 2002).

56. Manuel Vianna e Silva, *Elementos para história do arroz em Portugal* (Coimbra: Nova Casa Minerva, 1955).

57. Ribeiro, *Aspectos e problemas*, 153.

58. Frederick H. Hall, W. F. Harrison, and D. W. Welker, eds., *Dialogues of the Great Things of Brazil* (Albuquerque: University of New Mexico Press, 1987), 197.

59. Carney, *Black Rice*, 77, 151.

60. António Correira, *As Companhias Pombalinas de Grão-Para e Maranhão e Pernambuco e Paraíba* (Porto, Portugal: Editorial Presença, 1983).

61. Carney fieldwork interviews, August 2002.

62. "Relaciones de Yucatan," in R. C. West, N. P. Psuty, and B. G. Thom, *Las*

Tierras Bajas de Tabaso (Villahermosa: Gobierno del Estado de Tabasco, 1987), 316.

63. Richard Price and Sally Price, eds., *Stedman's Surinam* (Baltimore: Johns Hopkins University Press, 1992), 208–19.

Chapter 13. Collecting Naturalia in the Shadow of Early Modern Dutch Trade

Much of this essay was written while in residence at the Max Planck Institute for the History of Science, Berlin. I am especially grateful to Lorraine Daston, Anke te Heesen, Nick Hopwood, Paula Findlen, Claudia Stein, and Fernando Vidal for their valuable commentary.

1. For recent literature on the VOC, see Femme Gaastra, *De Geschiedenis van de VOC* (Zutphen: Walburg Pers, 2002); Leo Akveld and Els M. Jacobs, eds., *De Kleurrijke Wereld van de VOC* (Bussum: THOTH, 2002). For the rate of import of goods, see Els M. Jacobs, *In Pursuit of Pepper and Tea: The Story of the Dutch East India Company* (Zutphen: Walburg Pers, 2001); for pepper, see 83.

2. The literature on early modern Dutch collecting is not, as in the case of Italy, integrated with the literature on the history of science and medicine. See O. Impey and A. Macgregor, eds., *The Origins of Museums: The Cabinet of Curiosities in Sixteenth- and Seventeenth-Century Europe* (1988; London: House of Stratus, 2001); Ellinoor Bergvelt and Renée Kistemaker, eds., *De Wereld binnen Handbereik: Nederlandse Kunst- en Rariteitenverzamelingen 1585–1735*, 2 vols. (Zwolle: Waanders, 1992). Cf. David Freedberg, "Science, Commerce, and Art: Neglected Topics at the Junction of History and Art History," in *Art in History: History in Art*, ed. David Freedberg and Jan de Vries (Santa Monica, Calif.: Getty Research Institute, 1991), 377–428; Paula Findlen, *Possessing Nature: Museums, Collecting, and Scientific Culture in Early Modern Italy* (Berkeley: University of California Press, 1994); Roelof van Gelder, "Paradijsvogels in Enkhuizen: De Relatie tussen Van Linschoten en Bernardus Paludanus," in *Souffrir pour Parvenir: De Wereld van Jan Huygen van Linschoten*, ed. Roelof van Gelder, Jan Parmentier, and Vibeke Roeper (Haarlem: Uitgeverij Arcadia, 1998), 30–50.

3. Jan Huyghen van Linschoten, *Itinerario, Voyage ofte Schipvaert: Naer Oost ofte Portugaels Indien inhoudende een Corte Beschryvinghe der selver Landen Ende See-Custen* (Amsterdam: Cornelis Claesz, 1596). The volume includes the extremely informative "Beschryvinghe van de gantsche custe van Guinea" and "Reys-Gheschrift vande Navigatien der Portugaloysers in Orienten." See van Gelder et al., *Souffrir pour Parvenir*.

4. The phrase is taken from the title of the key study by Charles Ralph Boxer, *The Dutch Seaborne Empire 1600–1800* (London: Penguin Books, 1965).

5. As quoted in Benjamin Schmidt, *Innocence Abroad: The Dutch Imagination and the New World, 1570–1670* (Cambridge: Cambridge University Press, 2001), 155. Schmidt's account contains an excellent analysis of the reuse of sources in the production of such accounts as van Linschoten's.

6. Klaas van Berkel, "Citaten uit het Boek der Natuur: Zeventiende eeuwse Nederlandse Naturaliënkabinetten en de Ontwikkeling van de Natuurwetenschap," in Bergvelt and Kistemaker, *Wereld binnen Handbereik*, 169–91, 174; Schmidt, *Innocence Abroad*, 156–57; van Gelder, "Paradijsvogels in Enkhuizen," 42.

7. See Schmidt, *Innocence Abroad*, passim; in this volume, see Daniela Bleichmar's chapter.

8. In this volume, the chapter by Anke te Heesen in particular bears this out further.

9. The account was written in November 1593 by Duke Philip Ludwig II of Hanau-Münzenburg and is cited in van Gelder, "Paradijsvogels in Enkhuizen," 36–37.

10. Lorraine Daston and Katharine Park, *Wonders and the Order of Nature, 1150–1750* (New York: Zone Books, 1998), 149. See also Findlen, *Possessing Nature.*

11. Raphael Pelecius to Caspar Bauhin, 20 August 1594 (Basel UB, MS. Fr.Gr. II.1, p. 42, IV), as cited and translated in Brian Ogilvie, "Observation and Experience in Early Modern Natural History" (Ph.D. thesis, University of Chicago, 1997), 279.

12. Hugo Grotius, *Poemata* (Leiden: Hieronymus de Vogel, 1639), 276.

13. See, for an important corrective, Harold J. Cook, "Physicians and Natural History," in *Cultures of Natural History*, ed. N. Jardine, J. A. Secord, and E. C. Spary (Cambridge: Cambridge University Press, 1996), 91–105; and Harold J. Cook, "The Cutting Edge of a Revolution? Medicine and Natural History Near the Shores of the North Sea," in *Renaissance and Revolution: Humanists, Scholars, Craftsmen, and Natural Philosophers in Early Modern Europe*, ed. J. V. Field and Frank A. J. L. James (Cambridge: Cambridge University Press, 1993), 45–61.

14. Cook, "Physicians and Natural History," 91.

15. See, most recently, Florike Egmond, "Een mislukte benoeming: Paludanus en de Leidse Universiteit," in *Souffrir pour Parvenir*, 51–64.

16. Fabio Garbari, Lucia Tongiorgi Tomasi, and Alessandro Tosi, *Giardino dei Semplici* (Ospedaletto: Pacini, 1991); Margharita Azzi Visentini, *L'Orto botanico di Padova e il Giardino del Rinascimento* (Milan: Polifilo, 1984); Findlen, *Possessing Nature*, esp. "Living Museums," 256–61. On the Leiden garden, see H. Veendorp and L. G. M. Baas Becking, *Hortus Academicus Lugduno Batavus: The Development of the Gardens of Leyden University, 1587–1937* (Haarlem: Enschede, 1938); and L. Tjon Sie Fat and E. de Jong, eds., *The Authentic Garden: A Symposium on Gardens* (Leiden: Clusius Stichting, 1991), esp. E. de Jong, "Nature and Art: The Leiden Hortus as '*Musaeum*,'" 37–60.

17. Erik de Jong, *Natuur en Kunst: Nederlandse Tuin- en Landschapsarchitectuur 1650–1740* (Amsterdam: THOTH, 1993), *Inventaris van de Rariteyten opde Anatomie en inde Twee Gallerijen van des Universiteyts Kruythoff* (c. 1617), 232ff.

18. See Maurits Sabbe, *De Meesters van de Gulden Passer: Christoffel Plantin, Aartsdrukker van Philips II, en zijn Opvolgers, de Moretussen* (Amsterdam: P. N. Van Kampen & Zoon, 1937), 99.

19. Hondius must have come to know Porret's collection as a student in Leiden; he enrolled in 1596 and was among the students who benefited from the early years of activity in the Leiden garden and anatomical theater alike. Though he studied botany diligently while at Leiden, he followed in his father's footsteps in entering the ministry. See *Moufe-schans* (Leiden, 1621); and P. J. Meertens, *Letterkundig Leven in Zeeland* (Amsterdam: N. V. Noord-Hollandsche Uitgevers Maatschappij, 1943).

20. Frits Lugt, *Répértoire des Catalogues de Ventes Publiques Intéressant l'Art ou la Curiosité: Première Période vers 1600–1825* (The Hague: M. Nijhoff, 1938), no. 2.

21. Findlen, *Possessing Nature*, 153.

22. See Daston and Park, *Wonders*, 158.

23. The collection belonged to Francesco Calzolari; see ibid., 154–55; Findlen, *Possessing Nature*, 37–38.

24. Nicholas Thomas, *Entangled Objects: Exchange, Material Culture, and Colonialism in the Pacific* (Cambridge: Harvard University Press, 1991), 125.

25. Karl H. Dannenfeldt, *Leonard Rauwolf: Sixteenth-Century Physician, Botanist, and Traveller* (Cambridge: Harvard University Press, 1968), 228–30.

26. Ibid. See also H. Kellenbenz, "From Melchior Manlich to Ferdinand Cron: Germany's Levantine and Oriental Trade Relations (Second Half of XVIth and Beginning of XVIIth)," *Journal of European Economic History* 19 (1990): 611–622.

27. Translated from Latin in Ludovic Legré, *La botanique en Provence au XVIe siècle: Léonard Rauwolff, Jacques Renaudet* (Marseilles: H. Aubertin and G. Rolle, 1900), 103. The letter is dated 7 September 1583; the original is in the UB Leiden, Codex Vulcanius 101.

28. F. de Nave, ed., *Botany in the Low Countries (End of the 15th Century–ca. 1650)* (Antwerp: Plantin Moretus Museum), 109ff.

29. The standard reference remains F. W. T. Hunger, *Charles de l'Escluse (Carolus Clusius) Nederlandsch Kruidkundige 1526–1609*, 2 vols. (The Hague: M. Nijhoff, 1927; 1943).

30. See J. Heniger, "De Eerste Nederlandse Wetenschappeiljke Reis naar Oost-Indië, 1599–1601," *Jaarboekje voor Geschiedenis en Oudheidkunde van Leiden en Omstreken* 65 (1973): 27–49, esp. 36–37.

31. Ibid., Bijlage I and II.

32. Ibid., 44.

33. Ibid., 44–45.

34. K van Berkel, "Een onwillige Mecenas? De Rol van de VOC bij het Natuurwetenschapelijk Onderzoek in de Zeventiende Eeuw," *VOC en Cultuur: Wetenschappelijk en Culturele Relaties tussen Europa en Azië ten tijde van de Verenigde Oostindische Compagnie*, ed. J. Bethlehem and A.C. Meijer (Amsterdam: Schiphouwer en Brinkman, 1993), 43.

Chapter 14. Accounting for the Natural World

For discussion and suggestions I would like to thank Lorraine Daston, Valentin Groebner, Gloria Meynen, and Paul White, as well as the participants in the Potsdam conference and the editors of this volume. I would also like to thank Roy Vickery of the Department of Botany, Natural History Museum, London for providing me with Figure 4.5 and Pam Selwyn, who translated the text.

1. Daniel Defoe, *Robinson Crusoe* (London: Penguin Books, 1994), 68.

2. Ibid., 69. I would like to thank Laura Otis for drawing my attention to this table.

3. Ibid. On the interpretation of the mercantile aspects of the novel, see Maxmillian E. Novak, "Robinson Crusoe and Economic Utopia," *The Kenyon Review* 25 (1963): 474–90; and Gustav Hübener, "Der Kaufmann Robinson Crusoe," *Englische Studien* 54 (1920): 367–98.

4. This quotation, like all those that follow in this paragraph, can be found in Eduard Winter et al., eds., *D. G. Messerschmidts Forschungsreise durch Sibirien 1720–1727*, 5 vols. (Berlin: Akademie-Verlag, 1962–77), vol. 3, 214–20.

5. See here also Staffan Müller-Wille's chapter in this volume and Lisbet Koerner, *Linnaeus: Nature and Nation* (Cambridge: Harvard University Press, 1999), 33–55, where she draws parallels between Linnaeus's nomenclature, his list making, and his "cameralist" ideas.

6. In what follows, Messerschmidt's recording techniques will be described using the example of his botanical studies. Cf. Doris Posselt, "Die Erforschung der Flora Sibiriens in der ersten Hälfte des 18. Jahrhunderts unter besonderer Berücksichtigung der Aufzeichnungen von D. G. Messerschmidt," 3 vols. (Ph.D. diss., Friedrich-Schiller-Universität, Jena, 1969). Messerschmidt's working methods are described in more detail in Anke te Heesen, "Boxes in Nature," *Studies in the History and Philosophy of Science* 31, no. 3 (2000): 381–403.

7. Messerschmidt performed the work of a draftsman, geographer, secretary, natural scientist, and archivist almost without help. See Peter Simon Pallas, "Nachricht von D. Daniel Gottlieb Messerschmidts siebenjähriger Reise durch Sibirien," *Neue Nordische Beyträge* 3 (1782): 97–107, esp. 99. Such a "solo mission" was unusual for research purposes, and the fact that he was largely on his own can only be explained by the fact that his journey was one of the first designed by the Academy in Saint Petersburg. For details, see te Heesen, "Boxes," 387–89.

8. Winter, et al. *Messerschmidts Forschungsreise*, vol. 1, 241.

9. Ibid., vol. 2, 194.

10. *Index Botanicus Siberia*, F. 98, op. 1, Nr. 19, Archive of the Academy of Sciences, Saint Petersburg; Joseph Pitton de Tournefort, *Institutiones rei herbariae* (Paris, 1700).

11. *Sibiria Perlustrata*, F. 98, op. 1, Nr. 22, Archive of the Academy of Sciences, Saint Petersburg; see also Posselt, *Erforschung*, 75–88.

12. Winter et al., *Messerschmidts Forschungsreise*, vol. 2, 244; see also ibid., vol. 3, 238.

13. Ibid., vol. 3, 238.

14. Ibid., 216.

15. Ibid., vol. 1, 335.

16. Ibid., vol. 3, 70.

17. Ibid.

18. Ibid., 54.

19. Johann Heinrich Zedler, *Grosses vollständiges Universal Lexicon aller Wissenschaften und Künste*, 68 vols. (Halle and Leipzig, 1733), vol. 4, 1767.

20. On the household economy and the metaphorical field of the "house (hold)" (*Haus*), see Irmintraut Richarz, *Oikos, Haus und Haushalt: Ursprung und Geschichte der Haushaltsökonomik* (Göttingen: Vandenhoeck & Ruprecht, 1991).

21. Winter et al., *Messerschmidts Forschungsreise*, vol. 1, 335.

22. He goes beyond the recommendations of his teacher Friedrich Hoffmann in his work on the tasks and duties of the physician. See Friedrich Hoffmann, *Politischer Medicus, oder Klugheits-Regeln, nach welchen ein junger Medicus seine Studia und Lebensart einrichten soll* (Leipzig, 1752), 31–35. On guidelines for travelers who reported to the Royal Society, see, for example, John Woodward, *Brief Instructions for Making Observations in All Parts of the World* (London, 1696). For an overview of this sort of travel manual, see Justin Stagl, "Das Reisen als Kunst und Wissenschaft (16.–18. Jahrhundert)," *Zeitschrift für Ethnologie* 108, no. 1 (1983): 15–34.

23. Cf., Geoffrey T. Mills, "Early Accounting in Northern Italy: The Role of Commercial Development and the Printing Press in the Expansion of Double-Entry from Genoa, Florence and Venice," *Accounting Historians Journal* 21, no. 1 (1994): 81–96.

24. Cf., Richard Brown, ed., *A History of Accounting and Accountants* (1905; New York: A. M. Kelley, 1968), 143. For one of the first publications in German, see Wolffgang Schweicker's 1549 introduction "Zwifachen Buchhalten." On

bookkeeping more generally, see Anthony G. Hopwood, *Accounting and Human Behaviour* (London: Haymarket, 1974), and idem, "The Archeology of Accounting Systems," *Accounting, Organizations and Society* 12, no. 3 (1987): 207–34.

25. Johann Beckmann, "Italienisches Buchhalten," in *Beyträge zur Geschichte der Erfindungen* (Leipzig, 1783–1805), 177–85. On "cameralism," a German fiscal theory, see Keith Tribe, *Governing Economy: The Reformation of German Economic Discourse 1750–1840* (Cambridge: Cambridge University Press, 1988).

26. See Johann Wolfgang von Goethe, *Wilhelm Meister*, in *Goethes Werke* (Weimar: Weimarer Sophien-Ausgabe, 1898), vol. 21, 51. On the influence of the mercantile ideal on Goethe's scientific work, see Myles W. Jackson, "Natural and Artificial Budgets: Accounting for Goethe's Economy of Nature," *Science in Context* 7, no. 3 (1994): 409–31. On Goethe's note taking, see Ernst Robert Curtius, "Goethes Aktenführung," *Die Neue Rundschau* 62 (1951): 110–21. For an account of the connections between bookkeeping, diary-writing, and notebooks among early nineteenth-century scientists, see Hans-Otto Sibum, "Narrating by Numbers: Keeping an Account of Early 19th Century Laboratory Experiences," in *Preprint 173 of the Max-Planck-Institut für Wissenschaftsgeschichte* (Berlin, 2001); and idem, "The Bookkeeper of Nature: Benjamin Franklin's Electrical Research and the Development of Experimental Natural Philosophy in the Eighteenth Century," in *Reappraising Benjamin Franklin*, ed. J. A. Leo Lemay (Newark: University of Delaware Press, 1993), 221–42, esp. 225–27.

27. Johann Georg Krünitz, *Ökonomisch-technische Enzyklopädie oder allgemeines System der Land-, Haus- und Staats-Wirthschaft (Staats-, Stadt-, Haus- und Landwirtschaft und der Kunstgeschichte) in alphabetischer Ordnung* g, 242 vols. (Berlin, 1776), vol. 7, 187.

28. William Webster, *An Essay on Book-Keeping, According to the True Italian Method of Debitor and Creditor, by Double Entry* (London, 1719), 2–4.

29. Several publications have appeared on the production of demonstrability and objectivity in natural history. Cf. Theodore M. Porter, "Quantification and the Accounting Ideal in Science," *Social Studies of Science* 22 (1992): 633–52; and Lorraine Daston, "The Moral Economy of Science," *Osiris* 10 (1995): 3–24.

30. Georg Christoph Lichtenberg, *Schriften und Briefe*, ed. W. Promies, 2 vols., 2d ed. (Munich: Hanser, 1973), vol. 1, 352.

31. Ulrich Joost sees more a metaphorical connection than a procedural correspondence between Lichtenberg's notation practice and accounting; see Ulrich Joost, "Schmierbuchmethode bestens zu empfehlen," in Georg Christoph Lichtenberg, *Noctes: Ein Notizbuch*, ed. Ulrich Joost (Göttingen: Wallstein Verlag, 1992), 112–27, esp. 115–20.

32. Christoph Meinel, "Enzyklopädie der Welt und Verzettelung des Wissens: Aporien der Empirie bei Joachim Jungius," in *Enzyklopädien der frühen Neuzeit: Beiträge zu ihrer Erforschung*, ed. Franz M. Eybl et al. (Tübingen: Niemeyer, 1995), 162–87, esp. 170–73. See also the entry "Exzerpt" in *Historisches Wörterbuch der Rhetorik*, ed. Gert Ueding (Tübingen: Niemeyer, 1992), vol. 3, 183–85. Also related to the excerpt are the genres of the abridgment (*Brevier*), the florilegium, and the collectanea; on this, see also Joost, "Schmierbuchmethode."

33. On the instruction of "learned reading," see Helmut Zedelmaier, "De ratione excerpendi: Daniel Georg Morhof und das Exzerpieren," in *Mapping the World of Learning: The Polyhistor of Daniel Georg Morhof*, ed. Françoise Waquet (Wiesbaden: Harrassowitz, 2000), 75–92, esp. 78; the "Aufschreibesysteme" refer to Friedrich Kittler, *Aufschreibesysteme 1800–1900* (Munich: Fink Verlag, 1985). On the use of note cards and files from the perspective of media history, see Edward Tenner, "Wissen à la carte: Die Veränderung der geistigen Arbeit

durch neue Techniken der Informationsverwaltung," *Kultur & Technik* 1 (1992): 26–34; Stefan Rieger, *Speichern/Merken: Die künstliche Intelligenz des Barock* (Munich: Fink Verlag, 1997); and Markus Krajewski, *Zettelwirtschaft: Die Geburt der Kartei aus dem Geiste der Bibliothek* (Berlin: Kadmos Verlag, 2002).

34. Gerd Ueding, "Beredsamkeit aus Erfahrung—Georg Christoph Lichtenbergs Sudelbücher," *Photorin* 9 (1985): 1–18, esp. 3. Mary Poovey posits a relationship between accounting and rhetoric in her *A History of the Modern Fact: Problems of Knowledge in the Sciences of Wealth and Society* (Chicago: University of Chicago Press, 1998), chap. 2. See also Grahame Thompson, "Early Double-Entry Bookkeeping and the Rhetoric of Accounting Calculation," in *Accounting as Social and Institutional Practice*, ed. Anthony G. Hopwood and Peter Miller (Cambridge: Cambridge University Press, 1994), 40–66.

35. While today the word *commonplace* usually refers to a platitude, in the sixteenth and seventeenth centuries it was understood as a mnemonic tool, a quotation, a valuable observation that was worth noting. Cf., Francis Bacon, *Of the Advancement of Learning, or the Partitions of Science* (London, 1674), 162. See also Sister Joan Marie Lechner, *Renaissance Concepts of the Commonplaces* (New York: Pageant Press, 1962); Ann Moss, *Printed Commonplace-Books and the Structuring of Renaissance Thought* (Oxford: Clarendon Press, 1996); Richard Yeo, "Ephraim Chamber's Cyclopedia (1728) and the Tradition of Commonplaces," *Journal of the History of Ideas* 57 (1996): 157–75; and Ann Blair, "Humanist Methods in Natural Philosophy: Commonplace Book," *Journal of the Histories of Ideas* 53 (1992): 541–51, and idem, "Annotating and Indexing Natural Philosophy," in *Books and the Sciences in History*, ed. Marina Frasca-Spada and Nick Jardine (Cambridge: Cambridge University Press, 2000), 69–89. See also Lotte Mulligan, "Robert Hooke's 'Memoranda': Memory and Natural History," *Annals of Science* 49 (1992): 47–61. On the relationship between the commonplace and "brief facts," see Lorraine Daston, "Perché i fatti sono brevi?," *Quaderni storici* 108, no. 3 (2001): 745–70.

36. For an extensive discussion of Locke and the commonplace book, see Lucia Dacome, "Policing Bodies and Balancing Minds: Self and Representation in Eighteenth-Century Britain" (Ph.D. diss., University of Cambridge, 2000).

37. John Locke, *New Method of Making Common-Place-Books* (London, 1706), ii.

38. Ibid., v.

39. Ibid., 4.

40. See Dacome, "Policing Bodies," 70–72. On reading in the educational context before Locke, see Lechner, *Renaissance Concepts*, 154ff.

41. Krünitz, *Ökonomisch-technische Enzyklopädie*, vol. 135 (1824), 697.

42. Messerschmidt had become acquainted with this storage practice during his studies in Jena and in Halle under Friedrich Hoffmann. On this, see Johann Schröder, *Johann Schröders vollständige und nutzreiche Apotheke* (Leipzig and Frankfurt, 1718).

43. On Sloane's collection, see E. St. John Brooks, *Sir Hans Sloane: The Great Collector and His Circle* (London: Batchworth Press, 1954); John F. M. Canon, "Botanical Collections," in *Sir Hans Sloane: Collector, Scientist, Antiquary, Founding Father of the British Museum*, ed. Arthur MacGregor (London: British Museum, 1994), 136–49; J. E. Dandy, ed., *The Sloane Herbarium: An Annotated List of the Horti Sicci Composing It; with Biographical Accounts of the Principal Contributors* (London: British Museum, 1958); Roy Vickery, "Sloane," *Botanical Journal of Scotland* 46 (1994): 594–98; and Jessie M. Sweet, "Sir Hans Sloane: Life and Mineral Collection; Part III: Mineral Pharmaceutical Collection," *Natural History Magazine* 5, no. 36 (1935): 145–64.

Chapter 15. Surgeons, Fakirs, Merchants, and Craftspeople

With thanks to Bénédicte Bilodeau, Yves Cohen, Peter Dear, Jean-Marc Drouin, Norman Gritz, Maneesha Lal, Piyali Markovits, Jan Marontate, Father Gérard Moussay, Amina Okada Gyanendra Pandey, Dominique Pestre, Carmen Salazar-Soler, Simon Schaffer, Londa Schiebinger, and (not least) Sanjay Subrahmanyam.

1. See Henrika Kuklick and Robert E. Kohler, "Introduction," in *Science in the Field*, ed. idem, special issue of *Osiris* (2nd series) 11 (1996): 1–14.

2. Marie-Noëlle Bourguet, "La collecte du monde: Voyage et histoire naturelle (fin XVIIᵉᵐᵉ siècle–début XIXᵉᵐᵉ siècle)," in *Le Muséum au premier siècle de son histoire*, ed. Claude Blanckaert, Claudine Cohen, Pietro Corsi, and Jean-Louis Fischer (Paris: Muséum National d'Histoire Naturelle, 1997), 163–96; John Law, "On the Methods of Long-Distance Control: Vessels, Navigation and the Portuguese Route to India," in *Power, Action and Belief: A New Sociology of Knowledge?* ed. idem (London: Routledge & Kegan Paul, 1986), 234–63.

3. See, for example, Robert Boyle, *General Heads for the Natural History of a Country, Great or Small, Drawn out for the Use of Travellers and Navigators* (London, 1692), 9–12.

4. Benjamin Schmidt, "Inventing Exoticism: The Project of Dutch Geography and the Marketing of the World," in *Merchants and Marvels: Commerce, Science, and Art in Early Modern Europe*, ed. Pamela H. Smith and Paula Findlen (New York: Routledge, 2002), 347–69.

5. Jesús Bustamente García, "Francisco Hernández, Plinio del Nuevo Mundo: Tradición clásica, teoría nominal y sistema terminológico indígena en una obra renacentista," in *Entre dos mundos: Fronteras culturales y agentes mediadores*, ed. Berta Ares Queija and Serge Gruzinski (Seville: Escuela de Estudios Hispano-Americanos, 1997), 243–68; James H. Merrell, *Into the American Woods: Negotiators on the Pennsylvania Frontier* (New York: W. W. Norton, 1999); Nicholas Thomas, *Entangled Objects: Exchange, Material Culture, and Colonialism in the Pacific* (Cambridge: Harvard University Press, 1991).

6. Denys Lombard and Jean Aubin, eds., *Asian Merchants and Businessmen in the Indian Ocean and the China Sea* (Delhi: Oxford University Press, 2000); Ashin Das Gupta, *The World of the Indian Ocean Merchant 1500–1800* (Oxford: Oxford University Press, 2001).

7. Richard Grove, "Indigenous Knowledge and the Significance of South-West India for Portuguese and Dutch Constructions of Tropical Nature," *Modern Asian Studies* 30 (1996): 121–43.

8. See, for example, Ray Desmond, *The European Discovery of the Indian Flora* (Oxford: Oxford University Press, 1992).

9. Manuscripts collection, MSS 1915, 1916, 1916bis, 1916ter, and 1917 to 1926 (hereafter *Jardin de Lorixa*), Central Library, Muséum National d'Histoire Naturelle (hereafter BMNHN).

10. MS. 1915: "Preface," fol. IIIv, BMNHN. All translations are mine.

11. See Philippe Haudrère, *La Compagnie française des Indes au XVIIIᵉ siècle: 1719–1795*, 4 vols. (Paris: Librairie de l'Inde, 1989).

12. Colonies, Série C² 115, fol. 358, Centre des Archives d'Outre-mer, Aix-en-Provence (hereafter CAOM).

13. Claude Chaligne, "Chirurgiens de la Compagnie des Indes: Histoire du service de santé de la Compagnie, 1664–1793" (Thesis, Faculty of Medicine, University of Paris V, 1961), 42–46.

14. MS 1915: "Preface," fol. IIIr, BMNHN.

15. Chaligne, "Chirurgiens de la Compagnie," dedication and 85.

16. L'Empereur to Delavigne, 20 January 1699, V 959, fol. 153, Séminaire des Missions Étrangères de Paris (hereafter MEP).

17. L'Empereur to Delavigne, 6 January 1701, V 990, f. 533, MEP.

18. Johan Blussé and Ilonka Ooms, eds., *Kennis en Compagnie: De Verenigde Oost-Indische Compagnie en de moderne Wetenshap* (Amsterdam: Balans, 2002); Kapil Raj, "Eighteenth-Century Pacific Voyages of Discovery, 'Big Science,' and the Shaping of a European Scientific and Technological Culture," *History and Technology* 17 (2000): 79–98.

19. L'Empereur to Delavigne, 4 December 1702, V 958, fol. 207, MEP.

20. Augusto da Silva Carvalho, "Garcia d'Orta. Comemoração do quarto centenário da sua partida para a India em 12 de Março de 1534," *Revista da Universidade de Coimbra* 12 (1934): 61–246, esp. 103, 126.

21. MS 1915: "Avis au lecteur," fols. IVv–Vr, BMNHN.

22. L'Empereur to Abbé Raguet, 20 January 1727, F⁵ 19, CAOM.

23. L'Empereur to Delavigne, 6 January and 29 January 1701, V 990, fols. 533, 539, MEP; and Inde, Notariat de Chandernagore, O 2: Power of attorney, dated 7 September 1712, CAOM.

24. L'Empereur to Delavigne, 20 January 1699 and 6 January 1701, V 957, fol. 153 and V 990, fol. 533, MEP.

25. MS 1915: title page and index, BMNHN.

26. L'Empereur to Delavigne, 6 January 1701, V 990, fol. 533, MEP.

27. L'Empereur to de Jussieu, 25 December 1729, GGA/52766/1, Laboratoire de phanérogamie (LP), MNHN.

28. L'Empereur to Delavigne, 20 January 1699, V 957, fol. 153, MEP.

29. Monique Dussolin, "Etude d'un groupe social: les Européens à Chandernagor, 1ère moitié du XVIIIe siècle" (Mémoire de Maîtrise, Université de Paris VII, 1971), 60–90, esp. 67–68.

30. L'Empereur to de Jussieu, 25 December 1729, GGA/52766/1, LP, MNHN.

31. Jeremiah P. Losty, *Krishna: A Hindu Vision of God: Scenes from the Life of Krishna Illustrated in Orissan and Other Eastern Indian Manuscripts in the British Library* (London: The British Library, 1980); John Guy, *Palm-Leaf and Paper: Illustrated Manuscripts of India and Southeast Asia* (Melbourne: National Gallery of Victoria, 1982).

32. Hendrik Adriaan van Reede tot Drakestein, *Hortus Indicus Malabaricus*, 12 vols. (Amsterdam, 1678–93), vol. 3 (1682), viii.

33. *Jardin de Lorixa*, vol. 3 (MS 1917, BMNHN), pls. 18, 19, 20, 21, 22; *Hortus Malabaricus*, vol. 1, figs. 12, 13, 14, 15, and vol. 3, figs. 26, 27, 28.

34. *Jardin de Lorixa*, vol. 4 (MS 1918, BMNHN), pl. 9; *Hortus Malabaricus*, vol. 1, fig. 37.

35. M. K. Brett, "Indian Painted and Dyed Cottons for the European Market," in *Aspects of Indian Art*, ed. Pratapaditya Pal (Leiden: E. J. Brill, 1972), 167–71.

36. MS 1915: "Avis au lecteur," fol. IVv, BMNHN.

37. L'Empereur to Delavigne, 29 January 1701, V 990, fol. 539, MEP.

38. L'Empereur to Abbé Raguet, 20 January 1727, F⁵A 19, fols. 83r–84v, CAOM.

39. Charles de Montalambert, S.J. to the Directors of the Compagnie des Indes, 27 December 1726; Colonies C2 74, fols. 45r–50r, CAOM; and Pierre Christophe Lenoir to Abbé Raguet, 25 September 1728, Colonies F⁵A, fols. 135r–137r, CAOM.

40. Only a few of these letters have survived. See L'Empereur to de Jussieu, 25 December 1729 and 25 November 1733, GGA/52766/1 and 2, respectively, LP, MNHN; and Dossier Antoine de Jussieu, "Extraits de la correspondance d'Antoine de Jussieu," fol. 22, Académie des Sciences, Paris (AS).

41. L'Empereur to the Directors of the Compagnie des Indes, 25 January 1737, C² 285, fols. 11r–12r, CAOM; four letters from L'Empereur to Abbé Raguet and two from Raguet to L'Empereur, F⁵ 19, fols. 83r–84v, 116r–117v, 140r–142v, 169r–171r, 161r–162r, 179r–180r, CAOM.

42. Directors of the Compagnie des Indes to the Chandernagore Council, 21 January 1733, Inde, A102, pp. 150–52, CAOM; Raguet to L'Empereur, 20 November 1730, F⁵ 19, fols. 179r–180v, CAOM; Maurepas to Du Fay, 16 March 1739, Marine, B² 307, fol. 465r–v, Archives Nationales (AN).

43. Inde, Notariat de Chandernagor, O 17 (1742) N° 39/13ᵉ, unpag., Declaration, dated 13 February 1742, by Nicolas L'Empereur of his debts before his death, CAOM.

44. L'Empereur to the Directors of the Compagnie des Indes, 25 January 1737, C² 285, fols. 11r–12r, CAOM.

45. "Mémoire pour Messieurs de la Compagnie des Indes," undated, de Jussieu manuscripts, MS 284, BMNHN.

46. L'Empereur to de Jussieu, 25 December 1729, *Post scriptum*, GGA/52766/1, LP, MNHN.

47. L'Empereur to de Jussieu, 25 December 1729, GGA/52766/1, LP, MNHN.

48. De Jussieu manuscripts, MS 1116, undated, BMNHN.

49. Pierre Bourdieu, "The Specificity of the Scientific Field and the Social Conditions of the Progress of Reason," *Social Science Information* 14 (1975): 19–47; Lorraine Daston, "The Ideal and Reality of the Republic of Letters in the Enlightenment," *Science in Context* 4 (1991): 367–86.

50. This adds a material-economic dimension to Lorraine Daston's rather abstract list of moral-cultural values of science in "The Moral Economy of Science," *Constructing Knowledge in the History of Science*, ed. Arnold Thackray, *Osiris* (2nd series) 10 (1995): 2–26.

51. Michel Callon, "Some Elements of a Sociology of Translation: Domestication of the Scallops and the Fishermen of St. Brieuc Bay," in *Power, Action and Belief*, ed. John Law (London: Routledge & Kegan Paul, 1986), 196–233.

52. See William Roxburgh, *Plants of the Coast of Coromandel*, 3 vols. (London, 1795–1820); and Henry Noltie, *Indian Botanical Drawings 1793–1868 from the Royal Botanic Garden* (Edinburgh: Royal Botanic Garden, 1999).

53. For instance, Johann Koenig (1728–85), a student of Linnaeus, made his name as a botanist in the service of the Nawab of Arcot in southern India before being employed by the English East India Company.

Chapter 16. Measurable Difference

I am grateful to the many colleagues and friends whose contributions were precious at different stages of this research: to Patrice Bret, Jean-Marc Drouin, François Regourd, and Staffan Müller-Wille for providing me with details crucial to the development of my argument; to Mary Baine Campbell, Paula Findlen, and Mary Terrall for their insightful comments and help with the English language; to Londa Schiebinger and Claudia Swan for their indefatigable support. The Centre National de la Recherche Scientifique granted me the time and resources to revise and write the final version of this work.

1. MS 56, "Pièces relatives au voyage de M. Joseph Martin . . . ," fols. 1–33, Bibliothèque Centrale du Muséum National d'Histoire Naturelle, Paris (hereafter BMNHN). On the techniques of natural history, see Anke te Heesen's chapter in this volume.

2. On botany in relation to European political economy and colonial expansion, see Marie-Noëlle Bourguet and Christophe Bonneuil, eds., "De l'inventaire du monde à la mise en valeur du globe: Botanique et colonisation (fin XVIIᶜ siècle–début XXᶜ siècle)," *Revue française d'histoire d'outre-mer* 322–23 (1999): 9–170; Richard Drayton, *Nature's Government: Science, Imperial Britain, and the "Improvement" of the World* (New Haven: Yale University Press, 2000); Lisbet Koerner, *Linnaeus: Nature and Nation* (Cambridge: Harvard University Press, 1999); David Philip Miller and Peter Hanns Reill, eds., *Visions of Empire. Voyages, Botany, and Representations of Nature* (Cambridge: Cambridge University Press, 1996); E. C. Spary, *Utopia's Garden: French Natural History from Old Regime to Revolution* (Chicago: University of Chicago Press, 2000).

3. For examples, see Daniela Bleichmar on New World medicinal plants, this volume, and Londa Schiebinger on drugs and medical knowledge in the West Indies, this volume. On science as "action at a distance," see Bruno Latour, *Science in Action: How to Follow Scientists and Engineers through Society* (Cambridge: Harvard University Press, 1987), chap. 6.

4. R. P. Louis Feuillée, *Journal des observations physiques, mathématiques et botaniques, faites . . . sur les côtes orientales de l'Amérique méridionale, et dans les Indes occidentales*, 2 vols. (Paris, 1714).

5. John Woodward, *Brief Instructions for Making Observations in All Parts of the World* (London, 1696), 19; Marie-Noëlle Bourguet and Christian Licoppe, "Voyages, mesures et instruments: une nouvelle expérience du monde au siècle des Lumières," *Annales. Histoire, Sciences Sociales* 52 (1997): 1115–51, esp. 1117–23.

6. Feuillée, *Journal*, vol. 1, preface (2–3).

7. Ibid., vol. 2, 705–7.

8. "Climat," in *Encyclopédie, ou Dictionnaire raisonné des arts, des sciences et des métiers* (Paris, 1753), vol. 3, 532–33; Clarence J. Glacken, *Traces on the Rhodian Shore: Nature and Culture in Western Thought, from Ancient Times to the End of the Eighteenth Century* (Berkeley: University of California Press, 1967), part 4.

9. Carl Linnaeus, *Species Plantarum* (Stockholm, 1753), vol. 1, 1, 380. On Linnaeus and Kalm, see Staffan Müller-Wille's chapter in this volume.

10. AJ 15/511 (letter from Macao, 9 January 1787), Archives Nationales, Paris.

11. René-Antoine Ferchault de Réaumur, "Règles pour construire des thermomètres dont les degrés soient comparables," in *Mémoires de l'Académie royale des sciences* [hereafter *MARS*] *pour 1730* (Paris, 1732), 452–507; "Observations des thermomètres . . . pendant l'année 1732 et partie de l'année 1733," in *MARS 1733* (Paris, 1735), 417–38; "Observations . . . pendant l'année 1735," in *MARS 1735* (Paris, 1738), 546–75; "Observations . . . pendant l'année 1736," in *MARS 1736* (Paris, 1739), 469–502; "Observations . . . pendant l'année 1739," in *MARS 1739* (Paris, 1741), 457.

12. Réaumur, "Règles pour construire des thermomètres," 453.

13. Henri-Louis Duhamel du Monceau, "Observations botanico-météorologiques pour 1740," in *MARS 1741* (Paris, 1744), 149–51; R. P. Louis Cotte, *Traité de Météorologie* (Paris, 1774), xvii–xx.

14. AA 7, 6, 64/3, "Observations botanico-météorologiques faites au Canada par Gaultier de 1742 à 1748," Archives de l'Observatoire, Paris; Joseph-Nicolas Delisle, "Observations du thermomètre faites pendant les grands froids de la Sibérie," in *MARS 1749* (Paris, 1753), 1–14.

15. On the history of precision and measurement, see Norton Wise, ed., *The Values of Precision* (Princeton: Princeton University Press, 1996); Marie-Noëlle Bourguet, Christian Licoppe, and H. Otto Sibum, eds., *Instruments, Travel, and Science: Itineraries of Precision from the Seventeenth to the Twentieth Century* (London: Routledge, 2002).

16. On the correlation between imperial expansion and the shaping of nature into tabulated information, see the chapter by Antonio Lafuente and Nuria Valverde on Spanish America in this volume.

17. James McClellan III, *Colonialism and Science: Saint-Domingue in the Old Regime* (Baltimore: Johns Hopkins University Press, 1992), 163–80; François Regourd, "Sciences et colonisation sous l'Ancien Régime: Le cas de la Guyane et des Antilles françaises, XVIIe–XVIIIe siècles," 4 vols. (thesis, Université de Bordeaux III, 2000), esp. vol. 3, 457–69.

18. Médéric Louis Élie Moreau de Saint-Méry, *Description topographique, physique, civile, politique et historique de la partie française de l'Isle de Saint-Domingue,* 3 vols. (1797–98; Paris: Société Française d'Histoire d'Outre-Mer, 1984), vol. 1, 218, 249, 753. Regourd, "Sciences et colonisation," vol. 3, 461.

19. Quoted in Regourd, "Sciences et colonisation," vol. 3, 462, 477.

20. Jean-Baptiste Thibault de Chanvalon, *Voyage à la Martinique* (Paris, 1763). On his views on colonial management, see Chandra Mukerji's and Emma Spary's chapters in this volume.

21. On relations with the Société Royale de Médecine, see Regourd, "Sciences et colonisation," vol. 3, 461–62. Jean-Baptiste Cassan, "Mémoires sur le climat des Antilles et sur les maladies qui sont particulières à la zone torride," *Mémoires de la Société médicale d'émulation de Paris,* 5, year XI-1803, 25–180.

22. Thibault de Chanvalon, *Voyage à la Martinique,* 135/5–135/7.

23. Cassan, "Mémoires," 159.

24. *Affiches américaines,* 18 January 1780, no. 3, supp.; Nicolas-Joseph Thiery de Menonville, *Traité de la culture du nopal et de l'éducation de la cochenille dans les colonies françaises de l'Amérique,* 2 vols. (Paris, 1786), vol. 1, xxi–xxii.

25. *Affiches américaines,* 18 September 1784, no. 38. On Mozard, see McClellan, *Colonialism and Science,* 165–67.

26. Quoted in McClellan, *Colonialism and Science,* 166–68.

27. *Affiches américaines,* 21 January 1784, no. 3; 27 August 1785, no. 35, supp.; 14 April 1787, no. 30. On the role of local newspapers in the emergence of colonies' collective consciousness, see Benedict Anderson, *Imagined Communities: Reflections on the Origin and Spread of Nationalism,* 2d ed. (London: Verso, 1991).

28. Hypolite Nectoux, "Observations faites sur la culture du caféier," 1789, MS 308, BMNHN; MS n.a.f. 9551, fols. 19–26, Bibliothèque Nationale de France (hereafter BNF). Among Nectoux's papers are various drafts of the memoir: an annotated and undated copy, 1789, MS n.a.f. 9551, fols. 29–36, BNF; and a revised version, read to the Institut in 1803, MS n.a.f. 9545, fols 140–46, BNF. See also his correspondence with Richard and Thouin, MS n.a.f. 9545, fols. 16–17, 22–23, BNF. On Nectoux, see Patrice Bret, "Le réseau des jardins coloniaux: Hypolite Nectoux (1759–1836) et la botanique tropicale de la mer des Caraïbes aux bords du Nil," in *Les Naturalistes français en Amérique du Sud, XVIᵉ–XIXᵉ siècles,* Yves Laissus, ed. (Paris: CTHS, 1995), 185–216; idem, "Des 'Indes' en Méditerranée? L'utopie tropicale d'un jardinier des Lumières et la maîtrise agricole du territoire," *Revue française d'histoire d'outre-mer,* 322–23 (199): 65–90; idem, "La plantation idéale des Lumières: Nature, esthétique et équilibre dans la caféière du jardinier-botaniste Nectoux," in *Le sucre, de l'Antiquité à son destin antillais,* Danielle Begot and Jean-Claude Hocquet, eds. (Paris: CTHS, 2000), 215–42.

29. Nectoux, "Observations" (1803), fol. 145.

30. Ibid. Nectoux was probably using a Réaumur thermometer (the equivalent of his figures, in centigrade degrees, is 12.5 to 27.5 C).

31. Nectoux, "Observations" (c. 1789), fol. 36.

32. Spary, *Utopia's Garden*, 99–154.

33. F10/210, Plantes et arbres exotiques, dossier Basset, "Sur l'établissement de la culture des plantes coloniales en France," Archives Nationales, Paris (hereafter AN).

34. On the introduction of African rice in South Carolina, see Judith Carney's chapter in this volume.

35. F10/210, dossier Basset, "Rapport à la commission des subsistances par le Conseil d'agriculture," 3 Ventôse, year II, AN.

36. "Essai tenté pour la naturalisation de plusieurs végétaux," *Décade philosophique* 2, no. 15 (30 Fructidor, year II): 340–44.

37. *Décade philosophique*, no. 34 (10 Fructidor, year IX–1801), 387.

38. André Thouin, "Observations sur l'effet des gelées précoces qui ont eu lieu les 18, 19, et 20 vendémiaire an XIV," *Annales du Muséum d'histoire naturelle* 7 (1806): 85–114.

39. Ibid., 108–09.

40. On the early history of biogeography, see Gareth Nelson, "From Candolle to Croizat: Comments on the History of Biogeography," *Journal of the History of Biology* 11, no. 2 (1978): 269–305; and James E. Larson, *Interpreting Nature: The Science of Living Form from Linnaeus to Kant* (Baltimore: The Johns Hopkins University Press, 1994), chap. 4. On the French context, and the role of Augustin-Pyramus de Candolle, see Jean-Marc Drouin, "Bory de Saint-Vincent et la géographie botanique," in *L'Invention scientifique de la Méditerranée: Égypte, Morée, Algérie*, Marie-Noëlle Bourguet, Bernard Lepetit, Daniel Nordman, and Maroula Sinarellis, eds. (Paris: E.H.E.S.S., 1998), 139–57.

41. Alexandre de Humboldt, *Essai sur la géographie des plantes* (Paris, year XIII-1805). On Humboldt and Caldas, see Jorge Cañizares-Esguerra's chapter in this volume. On his work in relation with European naturalists, see Marie-Noëlle Bourguet, "Landscape with Numbers: Natural History, Travel and Instruments in the Late 18th and Early 19th Centuries," in *Instruments, Travel, and Science*, Bourguet, Licoppe, and Sibum, eds., 96–125.

42. Robert Brown, *General Remarks, Geographical and Systematical, on the Botany of Terra Australis* (London, 1814), 5–8.

43. Abbé Jean-Louis Giraud-Soulavie, *Histoire naturelle de la France Méridionale, II, 1: Les végétaux* (Nîmes and Paris, 1783), 35–36.

Contributors

Daniela Bleichmar received her Ph.D. from the History Department at Princeton University. She is a Mellon Post-Doctoral Fellow in Early Modern Visual Culture at the University of Southern California, where she is working on a book on the interactions among visual culture, natural history, and colonialism in early modern Europe and the Spanish Americas.

Marie-Noëlle Bourguet is Professor of History at the Université Paris 7-Denis Diderot and an associate member of the Centre Alexandre Koyré, Paris, in history of science. She is the author of *Déchiffrer la France: La statistique départementale à l'époque napoléonienne* (Paris, 1988) and has co-edited *L'invention scientifique de la Méditerranée: Égypte, Algérie, Morée* (Paris, 1998) and *Instruments, Travel and Science: Itineraries of Precision from the Seventeenth to the Twentieth Century* (London, 2002). She is currently writing a book on Alexander von Humboldt and the history of scientific travel.

Michael T. Bravo is University Lecturer in Geography at the Scott Polar Research Institute at Cambridge University, where he is also a member of the Department of the History and Philosophy of Science. He is the author of *The Accuracy of Ethnoscience* (Manchester University Press, 1996) and has coedited *Narrating the Arctic: A Cultural History of Nordic Scientific Practices* (2002). He is currently working on a book about artisans, knowledge, and imperial expansion in the North Atlantic.

Jorge Cañizares-Esguerra is a member of the History Department at SUNY-Buffalo and a University of Texas Harrington Faculty Fellow. His book *How to Write the History of the New World: Histories, Epistemologies, and Identities in the Eighteenth-Century Atlantic World* (Stanford University Press, 2001) won the 2001 AHA Atlantic History Prize and the John Edwin Fagg Prize on Spanish and Latin American History. His *Toward a Wider Atlantic: Nature Narratives and Identities 1550–1900* is forthcoming from Stanford University Press.

Judith Carney is Professor of Geography at the University of California, Los Angeles. She is the author of *Black Rice: The African Origins of Rice*

Cultivation in the Americas (Harvard University Press, 2001), which won the African Studies Association's Melville Herskovits Prize (2002) and the James D. Blaut Award of the Association of American Geographers (2003). Her work includes publications on gender, development, and environmental issues in West Africa and Brazil as well as research on the historical botany of the black Atlantic.

Harold J. Cook is Professor of the History of Medicine at University College London and Director of the Wellcome Trust Centre for the History of Medicine at UCL. He has authored *The Decline of the Old Medical Regime* (Cornell University Press, 1986) and the prizewinning *Trials of an Ordinary Doctor: Joannes Groenevelt in 17th-Century London* (Johns Hopkins University Press, 1994). He is currently working on a book about medicine and natural history in the Dutch Golden Age.

Anke te Heesen, Research Scholar at the Max Planck Institute for the History of Science, Berlin, is currently working on the notation systems of scientists. She is the author of the book *The World in a Box* (University of Chicago Press, 2002), coeditor with E. C. Spary of *Sammeln als Wissen* (Wallstein Verlag, 2001), and editor of *1900: Der Zeitungsausschnitt in den Wissenschaften* (special issue of the journal *Kaleidoskopien*, 2002). Her research concerns the history of scientific collections, museums, and the cultural history of science.

Julie Berger Hochstrasser is Associate Professor of the History of Early Modern Northern European Art at the University of Iowa. She has received fellowships from the Fulbright Foundation and the Center for Advanced Studies in the Visual Arts in Washington, D.C., and has published widely on Dutch still life painting relative to trade, food lore, material culture, and time. She is the author of *Still Life and Trade in the Dutch Golden Age* (Yale University Press, 2004).

Antonio Lafuente, Ph.D. in physics, works at the Instituto de Historia, Consejo Superior de Investigaciones Científicas (CSIC), Madrid. He has been a Visiting Scholar at the University of California, Berkeley. His most recent publications include a critical edition of Voltaire's *Elements de la philosophie de Newton* (Círculo de Lectores, 1998), *Guía del Madrid científico: Ciencia y Corte* (Doce Calles, 1998), *Georges-Louis Leclerc, Comte de Buffon* (1707–88) (CSIC, 1999), and *Los mundos de la ciencia en la Ilustración española* (Residencia de Estudiantes/fecy, 2003).

Andrew J. Lewis is Assistant Professor in the Department of History at American University in Washington, D.C. He earned his Ph.D. from Yale

University in 2001. He is currently completing a manuscript on natural history in early republic America, a project that examines zoology, botany, antiquities, popular science, and rival epistemologies between ordinary and elite Americans as they approached the natural world.

Chandra Mukerji is Professor of Science and Technology Studies and Cultural Studies at the University of California, Davis. She is the author of *Territorial Ambitions and the Gardens of Versailles* (Cambridge University Press, 1997), *From Graven Images: Patterns of Modern Materialism* (Columbia University Press, 1983), and *A Fragile Power: Science and the State* (Princeton University Press, 1989). She is currently writing a book on state power, material memory, and engineering of the Canal du Midi.

Staffan Müller-Wille is Research Scholar at the Max Planck Institute for the History of Science, Berlin. He is the author of the book *Botanik und weltweiter Handel* (1999) and is currently engaged in a research project on the cultural history of heredity. His current research interests include the history and epistemology of taxonomy, genetics, and anthropology.

Kapil Raj is Maître de Conférences at the École des Hautes Études en Sciences Sociales in Paris. He has published extensively on the reception and practice of modern science in India. His current research concerns the construction of field practices (mainly in natural history, social statistics, and geography) in the context of the European encounter with South Asia. His book on this theme, *Relocating Modern Science: Intercultural Encounter and Knowledge Construction, Europe and South Asia (17th–19th Centuries)*, is due to be published by Permanent Black in Delhi in 2005.

Londa Schiebinger is Professor of History of Science and the Barbara D. Finberg Director of the Institute for Research on Women and Gender at Stanford University. She is the author of *The Mind Has No Sex? Women in the Origins of Modern Science* (Harvard University Press, 1989), the prizewinning *Nature's Body: Gender in the Making of Modern Science* (Beacon, 1993), and *Has Feminism Changed Science?* (Harvard University Press, 1999). She is the editor of *Feminism and the Body* (Oxford University Press, 2000), coeditor of *Feminism in Twentieth-Century Science, Technology, and Medicine* (University of Chicago Press, 2001), and section editor of the *Oxford Companion to the Body* (2001). Her new book is *Plants and Empire: Colonial Bioprospecting in the Atlantic World* (Harvard University Press, 2004).

E. C. Spary is an affiliated research scholar at the Department of History and Philosophy of Science, University of Cambridge. Her major publications include *Cultures of Natural History,* coedited with N. Jardine and J. A. Secord (Cambridge University Press, 1996); *Utopia's Garden: French Natural History of the Old Regime to Revolution* (University of Chicago Press, 2000); and *Sammeln als Wissen,* coedited with Anke te Heesen (Wallstein Verlag, 2001). Her research interests cover the history of natural history, medicine, chemistry, and the arts in eighteenth-century Europe, especially France.

Claudia Swan is Associate Professor of Art History at Northwestern University. Her *Art, Science, and Witchcraft in Early Modern Holland: Jacques de Gheyn II (1565–1629)* is forthcoming from Cambridge University Press. She has also published *The Clutius Botanical Watercolors* (1998), a collection of late sixteenth-century watercolors used in the instruction of medicine at Leiden University. She has been a member of the Institute for Advanced Study in Princeton (1998–99) and a fellow at the Max Planck Institute for the History of Science, Berlin (2002). She is currently working on a book entitled *The Aesthetics of Possession: Art, Science, and Collecting in Early Modern Holland.*

Nuria Valverde holds a Ph.D. in the history of science from the Consejo Superior de Investigaciones Científicas (CSIC), Madrid. Her work focuses on the construction of scientific networks in eighteenth-century Spain and public understandings of them. She has collaborated on two exhibitions: "Madrid, Ciencia y Corte" (directed by A. Lafuente, 1999) and "Monstruos y seres imaginarios en la Biblioteca Nacional" (directed by A. Lafuente and J. Moscoso, 2000).

Index

Acknowledgments

The idea for this project emerged one snowy night over dinner with Matthias Kroß and Martin Schaad of the Einstein Forum, located in the cobblestoned market square of old Potsdam. That conversation and the many to follow launched an international workshop, "Botany in Colonial Context," in May 2001 in the forum's beautiful eighteenth-century buildings. Cosponsorship by the Deutsche Forschungsgemeinschaft allowed us to gather scholars in a variety of disciplines from all across Europe and the United States to discuss colonial botany in the early modern world. We wish to thank Martin, Matthias, Sigrid Weigl, Susan Neiman, Rüdiger Zill, and the forum staff for making the three-day conference an intellectually profitable and highly pleasurable experience.

The acclaim that the conference received in the European press emboldened us to undertake producing a volume. For some years we had been discussing the history of botany across our own disciplinary boundaries (one of us is a historian of science and the other a historian of art). Broadening that conversation to include such a range of authors and subjects as we gather here has been deeply rewarding; the subject of colonial botany is by its nature interdisciplinary and, as we hope this volume begins to demonstrate, responds well to various lines of questioning and historical pressures.

Working across these boundaries would not have been possible without the institutional support from which we have benefited over the past years. We are especially grateful to our energetic and agreeable editors at the University of Pennsylvania Press, Jerry Singerman and Erica Ginsburg, whose enthusiasm for the project and skill in shepherding us through the process have been evident throughout. They have helped make the task of organizing essays from scholars from numerous countries and academic traditions an enjoyable one. We are also indebted to the Pennsylvania State University Office of Research and Graduate Studies for kindly providing funds for translation. Coincidentally, Londa Schiebinger was in residence at the Max Planck Institute for the History of Science, Berlin, at the inception of this project, and Claudia Swan was in residence there as we prepared for publication; together, we are grateful to Lorraine Daston, director, for fostering the sort of interdisciplinary thinking that also informs this volume. Our greatest debt is, of course, to the authors of the essays, especially for carrying particular threads of conversation through time, space, and the pages of this book. We thank them for their time and effort and for making this process a pleasure.

Finally, we thank Pamela Selwyn and Jeremy Rogers for their care in providing accurate translations; our hardworking research assistants, Mary Faulkner, Sarah Goodfellow, and Katherine Maas, all Ph.D. candidates at Pennsylvania State University; Emily Long and Deborah Nelson in the Department of Art History at Northwestern University; Dr. Bobbie Arndt, who assisted in matters concerning Spanish language and literature, and Margaret Olszewski.

Milton Keynes UK
Ingram Content Group UK Ltd.
UKHW041012101024
449496UK00002B/61